Introduction to the Practice of Fishery Science

Revised Edition

William F. Royce
College of Ocean and Fishery Sciences
University of Washington
Seattle, Washington

Academic Press

San Diego   New York   Boston
London   Sydney   Tokyo   Toronto

Copyright © 1996, 1984 by ACADEMIC PRESS, INC.

Academic Press, Inc.
A Division of Harcourt Brace & Company
525 B Street, Suite 1900, San Diego, California 92101-4495

United Kingdom Edition published by
Academic Press Limited
24-28 Oval Road, London NW1 7DX

Library of Congress Cataloging-in-Publication Data

Royce, William F.
    Introduction to the practice of fishery science / William F.
Royce. — Rev. ed.
        p.   cm.
    Includes bibliographical references (p.   ) and index.
    ISBN 0-12-600952-X (case)
    1. Fisheries.   I. Title.
SH331.R68   1995
333.95'6—dc20   95-12388
                                                     CIP

PRINTED IN THE UNITED STATES OF AMERICA
95   96   97   98   99   00   QW   9   8   7   6   5   4   3   2   1

Contents

Preface   xi

Part I
The Challenges to Fishery Scientists

Summary   1

1
Expansion of Fishery Problems

1.1   Market Fisheries   4
1.2   Recreational Fisheries   6
1.3   Competition for Water   7
1.4   Common Property Resources   7
1.5   Aquaculture   8
1.6   Changing Goals and Laws   9
        References   11

2
Work of Fishery Scientists

2.1   Scope of Fishery Science Institutional Activities   14
2.2   Scope of Fishery Research   20
       References   24

3
Professional Careers

3.1   Knowers and Doers   26
3.2   Application of the Sciences   27
3.3   Historical Trends   27
3.4   Serving Society as Strategic Thinkers   29
3.5   Ethical Responsibilities   30
3.6   Functions of Professional Fishery Scientists   32
3.7   Role of Professional and Scientific Societies   35
3.8   Career Planning Strategy   36
       References   38

Part II
The Traditional Sciences

Summary   39

4
The Aquatic Environment

4.1   Hydrology and the Water Cycle   41
4.2   Streams and Rivers   43
4.3   Bottoms and Basins   47
4.4   Physical Properties of Water   51
4.5   Dissolved Materials   65
4.6   Photosynthesis   72
4.7   Circulation   74
4.8   Summary: The Ocean Environment of Fish and Fishing   90
       References   91

5
Food Chain and Resource Organisms

5.1 Identification and Nomenclature 92
5.2 Classification 96
5.3 Major Groups 97
References 133

6
Biology of Aquatic Resource Organisms

6.1 Senses 136
6.2 Respiration 145
6.3 Osmoregulation 146
6.4 Schooling 147
6.5 Food and Feeding 148
6.6 Reproduction and Early Development 154
6.7 Age and Growth 166
References 184

7
Ecological Concepts

7.1 Single-Species Populations 186
7.2 Communities and Ecosystems 198
7.3 Trophic Relationships 207
7.4 Evolutionary Adaptation 211
7.5 Ecology and Fishery Science 213
References 214

8
Analysis of Exploited Populations

8.1 Application of Ecological Concepts: A Chapter Summary 216
8.2 An Example of the Catch Trend in a New Fishery 218
8.3 Stocks and Migrations 220
8.4 Collection of Basic Data 234
8.5 Availability and Gear Selectivity 239
8.6 Recruitment 241
8.7 Stock Size 247

8.8   Survival and Mortality   252
8.9   Growth   258
8.10 Yield Models   259
      References   265

9
Aquacultural Sciences

9.1   Control of Environment   268
9.2   Amenable Species   272
9.3   Selective Breeding   274
9.4   Nutrition   277
9.5   Pathology and Medicine   281
      References   286

Part III
Application of Sciences

Summary   287

10
The Capture Fisheries

10.1 Fishing Gear   290
10.2 Statistical Perspectives on the Capture Fisheries   295
10.3 Trends in Production and Use   303
10.4 Fishing Systems   307
10.5 Role of Fishery Scientists   311
      References   313

11
The Culture Fisheries

11.1 Perspectives on the Culture Fisheries   314
11.2 Major Aquacultural Systems   318
11.3 Seed Production   327
11.4 Health and Quality Maintenance   330
11.5 Some Public Policy Issues   332
11.6 Role of Fishery Scientists   333
      References   334

12
Food Fishery Products

12.1   Characteristics of Fish Flesh   336
12.2   Methods of Preservation   341
12.3   Quality   344
12.4   Marketing Strategy   349
12.5   Role of Fishery Scientists   352
       References   353

13
Regulation of Fishing

13.1   Origins of Public Policy   355
13.2   Principles of Public Action   356
13.3   Current Regulatory Objectives   358
13.4   Regulatory Decisions   361
13.5   Establishment of a New Fishery Regulatory Regime   364
13.6   Implementation of the New Law of the Sea   368
13.7   Role of Fishery Scientists   368
       References   370

14
Aquatic Environmental Management

14.1   Trends in Water Use Policy   372
14.2   Environmental Impact Assessment   373
14.3   The Scope of Aquatic Environmental Change   374
14.4   Strategy for Protection   380
14.5   Environmental Improvement   389
14.6   Role of Fishery Scientists   395
14.7   An Outstanding Success   396
       References   397

15
Fishery Development and Restoration

15.1   Role of Government   399
15.2   Perspectives on Fishery Development   400
15.3   National Assistance   408
15.4   International Assistance   409

15.5   Project Planning and Execution   411
15.6   Institutional Development   412
15.7   Role of Fishery Scientists   415
       References   417

Appendix A
A Code of Professional Ethics

Code of Practices   418

Appendix B
Excerpts Pertaining to Fisheries from the Law of the Sea

Text   421

Bibliography   425
Index   443

Preface

The 1984 edition of this book grew out of my conviction that at the heart of any applied science is a determination to bring to the solution of significant problems both a mastery of relevant technical information and a sensitivity to social values. The 1996 edition continues in that direction but gives extra emphasis to our fishery management failures and the accelerating professional challenges.

The continuing goal of this revised text is to describe—for the student and prospective student, among others—the fishery scientist's role in environmental and resource issues as we approach the twenty-first century. It is an attempt to set forth the traditional expectations of our profession, and also the obligations and challenges of helping people who are increasingly concerned about the use of precious resources in our beleaguered environment. The total production from our ocean fisheries topped out at 100.3 million metric tons in 1989 and has since been declining despite the expanded investments in gear and vessels. The continuing augmentation of wild stocks as well as the expanding markets must be satisfied by increased aquacultural production. For this reason, a new chapter on aquacultural sciences has been added.

The practicing fishery scientist's horizons have been extended beyond biological research, fishery management, or teaching, valuable as these activities are. An accelerating professional challenge is to work toward the solution of complex environmental problems with businesses, government agencies, and the many groups of people, because much of what we do to our atmosphere and to our land has an impact on our waters. Because these issues require the application of the latest evidence, the references cited have been updated.

A pervasive problem is allocation of scarce fishery resources. It requires scientific information about the fish stocks or the fish farming practices, but is environmental in scope and is touched at every point by the need for policy decisions at state, national, and international levels. Fishery scientists must define problems in terms of the biological, social, and economic alternatives that will shape the decisions to be made. This broad view must include appreciation of human values as well as knowledge of fish biology and population dynamics.

The profession is rapidly expanding and changing. Earlier work in the field earned public trust and gained worldwide acceptance in national and international laws requiring the use of fishery science. Fishery scientists must develop a strategic vision that includes consideration of the sociopolitical evolution of the issues, and must work with specialists from many other social and environmental fields. A characteristic approach is to form a multi-disciplinary team.

Conforming to the organization established in the 1984 edition, the book is divided into three major sections. Part I is about professional careers and the expanding challenges to the profession, with a summary of the work of organizations that employ fishery scientists. Part II is a brief introduction to the traditional sciences that apply to the aquatic environment and its organisms, their biology, ecology, populations, and culture. Part III is an attempt to present an overall qualitative concept of the activity of fishery scientists in this last decade of the century. It provides a perspective on fishery problems in six major areas, and the ways in which the many kinds of scientists are attacking them.

Global patterns of fisheries use and over-use confirm that the approach adopted here must be, as far as possible, worldwide rather than just North American. All countries continue to face a complex mixture of social, political, and technical problems in managing their fisheries, their aquatic environment, and their aquaculture. The emphasis is on principles and issues rather than on details of the techniques or sciences. These may be found in the major references listed for each chapter, which have been carefully selected and updated to reflect findings since 1984.

Just as before, I am heavily indebted to dozens of colleagues in the scientific fields of universities and government, as well as to fishery managers in about 40 foreign countries where I have worked, for their perspectives on the issues.

William F. Royce

Part I

The Challenges to Fishery Scientists

Summary

Fisheries are human activities associated with fishing and aquaculture.

Fishery science is a body of scientific knowledge and also a profession practiced by fishery scientists.

The practice of fishery science has traditionally been associated with basic research, which it still is, but it has become predominantly a professional service to clients.

The service to clients—

includes definition of the full scope of problems, evaluation of alternatives, and implementation of timely solutions.

involves team approaches using many diverse sciences, law, and engineering.

The fishery scientist—

needs knowledge of the sciences, expert judgment, and ability to communicate.

adheres to professional standards.

organizes optimal use of living aquatic resources.

seeks problem solutions consistent with democratic principles.

finds solutions within financial, social, and time limits.

balances long-term and short-term solutions.

leads teams with many different scientific and professional skills.

has a specialty in at least one science and a synoptic view of whole problems.

The profession, although still small by comparison with medicine or engineering, has been growing rapidly, and has been broadly accepted by the public.

Fishery problems are increasing as worldwide catches approach or exceed maximum sustainable levels, as waters are used for many purposes, and as more aquaculture becomes feasible.

Employment, formerly predominantly in government and universities, is expanding rapidly in the private sector.

1 | Expansion of Fishery Problems

The use of aquatic resources has been a challenge to humans throughout our existence on earth. The first people who lived on the shore of a sea, river, or lake found food in the animals and plants, fibers in the plants, and ornaments in the shells. They gradually learned to catch fish in traps, to make hooks from bone and metal, and to twist fibers into lines and nets. They learned to use reeds and wood to build rafts that would carry them along the shore to find better fishing and carry their catch home. As people began to farm, they discovered that they could raise finfish, shellfish, and aquatic plants in ponds or keep them in enclosures along the shore, and thus have food available when needed. They learned to make bigger boats, to navigate, and to make maps, and thus they were able to fish in waters that were beyond the sight of land. They learned to preserve fish by drying them in the sun, and by smoking or salting them, although the catch had to be brought ashore within a few hours. They also had fun spearing or angling, and became known for their prowess. However, people have never completely mastered the resources of the sea. Some of its inhabitants are beyond our reach or power. The legends continue that more and bigger fish remain in the sea than have ever been caught.

Today people can reach the deepest part of the ocean with a line; we can even visit it personally in a vehicle. We can fish easily in every ocean of the world without ice on its surface, and bring the catch home in good condition. We can subdue the largest animals on earth, the great whales. We have discovered that practically all the waters that are not frozen, or are not more than 50°C, contain an intricate web of life in enormous variety from bacteria to whales that is sustained as an organic production system by energy from the sun. We know that because water covers nearly three-fourths of the earth, it receives much more energy than the land, and it is therefore a larger reservoir of energy and organic material.

Now that the living resources are known, our society's challenges are to understand, protect, and use this great production system for the continuing benefit of humans. Solving the problems is a continuous and ever more challenging process of scientific work and social decision in which fishery scientists play a major role.

1.1 MARKET FISHERIES

When the weary people of the world began to rebuild their countries and economies after World War II, they looked first to their supply of food. Those who had been fishermen went back to sea if they could find a ship. By 1948 they were operating enough old or makeshift vessels to bring the world production of fish back to the prewar level of about 20 million metric tons (MMT) per year. This total included about 17.1 MMT from the sea and about 2.5 MMT from fresh water.

The old ways of fishing were not followed for long because the wartime navies had brought many changes in our knowledge of the ocean, and in the design and equipment of our vessels. The oceans had been studied as never before so that submarines could be found and caught, a problem fundamentally the same as finding and catching fish. The current systems and temperature structure of the oceans were much better known. Ingenious acoustical equipment had been devised to probe beyond the limits of fishing. Accurate new navigational systems and instruments came into common use. New synthetic fibers replaced hemp in ropes, lines, and nets. Moreover, many men had gone to sea during the war, learned to live with it, and learned about the new equipment.

Soon people responded to the need for food with the new vessels that were designed to fish farther from port, navigate with new equipment, find fish with sonar instruments, catch fish with synthetic nets, dress fish with new machines, and freeze fish on board. All of these things did not happen immediately; mistakes were made, some ideas were discarded, and some gear was modified.

But ingenious and persistent people prevailed, and a revolution in ocean fishing developed. The new stern trawlers were as much of an advance over the trawlers of the 1930s as the reaper was over hand-threshing equipment. New fibers for nets that did not rot and a new power block for retrieving purse seines were as dramatic an improvement over the earlier purse seiners as a moldboard plow over a wooden stick. New sonar instruments and gear that could catch schools of fish midway between surface and bottom increased the productivity of herring and cod fishermen just as the productivity of new varieties of corn surpassed the productivity of the old. New filleting, packaging, and freezing equipment that could be taken to sea facilitated landing the catch in good condition ready for market, and reduced the worry about spoilage, just as similar processing on land had done for agricultural products.

The new fleets manned by skillful and daring people ranged all of the seas. Vessels were able to fish the North Pacific one month and the South Atlantic two months later. When catches were disappointing, they could move to new grounds—perhaps to another ocean. All of the known fish stocks in the world could now be fished.

The results of this new technology have become apparent. The ocean catches that had reached about 20 MMT annually in the 1940s doubled to about 40 MMT by 1960, and doubled again to about 80 MMT by the early 1980s. This led to hopes that the wild stocks in the oceans could supply a larger proportion of the world's protein food. However, the total ocean production has topped out at about 100 MMT during the late 1980s, and is probably beginning to decline (Table 1.1).

It has also become clear that about 90% of the major commercial fishery stocks in the world's oceans are being fished destructively, that most catchable species formerly discarded have become marketable, and that many stocks are now supplying substantially less than their maximum sustainable levels under wisely controlled fishing (FAO, 1992; Miles, 1989; Royce, 1993). Furthermore, many of these stocks occur across national boundaries, and unprecedented

TABLE 1.1   Commercial Fish Catches (Millions of Metric Tons)[a]

Year	United States	World
1950	2.7	25.9
1960	2.8	40.2
1970	2.8	65.6
1980	3.6	72.0
1985	4.9	86.4
1986	5.2	92.8
1987	6.0	94.4
1988	5.9	99.1
1989	5.8	100.3
1990	5.9	97.6
1991	5.5	97.1
1992	5.6	98.1
1993	4.9	98.0
1994 (est.)	5.0	97.0

[a]Live weight including molluscan shells. Sources: U.S. Department of Commerce, the United Nations, and the Food and Agriculture Organization.

international regulatory cooperation for reaching maximum sustainable levels of production is required.

Such action will be a major challenge to fishery scientists, who will need to influence both national and international action on the stocks that typically cross the national boundary lines in the oceans.

1.2 RECREATIONAL FISHERIES

As more and more people live in increasingly crowded suburbs and cities, many retain a love of nature, a need for the peace and quiet of the forest, and an urge to meet the challenges of the wild. They may stroll through a city park, swim from an ocean beach, camp in a wilderness, or drive through scenic areas. The scale of the activity can be modified to suit the elderly or the most active. To all, nature offers a revitalization and escape from the irritations and pressures of the cities.

A surprisingly large proportion of the people go fishing, either as a primary objective or as a supplement to boating, hiking, or camping. In the United States about one-fourth of all adults go fishing each year. They may seek solitude, caring little about the catch, or seek catch limits and trophies to satisfy the urge to compete. They may travel by city bus to a pier, or by private plane to a remote lake. The demand for recreational fishing is widespread through all parts of the country, among people of all income levels and all kinds of employment, and is increasing as people can travel farther more easily. The total recreational catch by anglers is larger than the commercial catch in most freshwater areas and in some marine areas near centers of population.

Outside the United States the demand for recreational fishing is also increasing. Some anglers seek trout that have been stocked in mountain streams around the world. Others are discovering the huge freshwater fish of the Nile or the Amazon, or the marlins and tunas off the coasts of tropical countries. These anglers are an important fraction of tourists, whose money is sought by all countries.

Anglers in the United States are numerous enough to be a potent political force. Long ago they requested artificially reared fish to augment the natural stocks, which resulted in major programs of hatchery operation. Sport fishers so outnumber commercial fishers (about 220 to 1) that any conflict between the groups is often resolved in favor of the sport fishers. Recreational use of the fish stocks is declared to be the primary use in many official policy statements.

Because anglers seek to escape from metropolitan areas, and because their fishing is frequently combined with other activities such as boating and hiking, in which enjoyment of the environment plays a large part, they are especially sensitive to the condition of the environment. Their trip satisfaction may include a number of factors, such as getting outdoors, finding clean waters, and socializing with family or friends. Their concern has been a primary force in cleaning up the environment (Spencer, 1993) and fishery scientists must work closely with them.

The major technical problems with respect to maintaining the recreational stocks are similar to those of the commercial fisheries, but the problems of allocation among users are much more complex. Recreational fishers place a qualitative value on their catch and their experience. Their expenditures for travel, lodging, and equipment frequently exceed by far the food value of their catch. The basic issue for the fishery manager is how to allocate the experiences among those who want to participate.

1.3 COMPETITION FOR WATER

The demand for water is at least as insistent as the demand for food and recreation. Bodies of water serve many human needs, a large proportion of which conflict with the needs of natural living resources. Water may be diverted to supply cities, create electricity, provide transportation, supply industry, or irrigate lands, leaving behind streams that are much smaller than before and dams that may block migratory fish. Smaller streams may be eliminated completely by diversion through pipes; marshes or estuaries may be filled to create new land. The water in a major river system is now fought over by the offstream users who want to consume it and the instream users who want to use it for transportation, power generation, waste dilution, and food or recreation. The modern battleground is usually in the thickets of water law: the native peoples' rights, the private riparian rights, or the paramount public rights. Further, the old laws may exacerbate inequities as water requirements approach the level of supply, and consequently the laws are being changed.

Fishery scientists face complex issues and play key roles in aquatic environmental management. The fishery agencies commonly become advocates for the living resources and the quality of the aquatic environment. What are the effects of pollution or changed water flows on the fish as individuals, or on the populations of fish? How can natural waters be improved by removing barriers to migration, enhancing populations, controlling unwanted animals, or adding fish shelters? What is the expected impact on fish and fisheries of proposed water impoundments, diversions, dredging, and filling? How can unavoidable impacts be mitigated?

1.4 COMMON PROPERTY RESOURCES

One of the older legal principles in most parts of the world applies to mobile resources that cross private or political boundary lines, and therefore are regarded as common property. Fish and water are included in this category of resources, along with air, wildlife other than fish, underground aquifers and oil pools, and scenic beauty. No single user has exclusive rights to such resources, and furthermore, fish and water are renewable resources that can be available indefinitely if they are used properly.

From these circumstances arise the special problems of managing the resources for the public. Users of a private resource can plan its use for their benefit for whatever time they wish, whereas the users of a common property resource are in direct competition with everyone else who wants to use it. If users of a common resource were to restrain their use to prevent damage to the resource or to prolong use, other users could also benefit as much or for as long a time.

In addition, there is a strong public feeling that everyone has a right to use such resources. When people exercise their right to go commercial fishing, the result in many fisheries in all parts of the world is, first, a reduction of the level of sustainable yield below the maximum and, next, a reduction of long-term profits for the fishers. Restoration of the level of yield can be achieved by reducing the efficiency of the fishermen, but such measures cannot restore the profits. That can be done only by allocating the right to fish to a smaller number of fishers.

Further complications ensue when a fish stock is sought by both recreational and commercial fishers. The former cherish their right to fish and are not concerned with a profit, whereas the latter have a similar right plus the need to make a profit if they are to continue fishing. Still more complications arise when the water needed by the fish is needed for other human uses.

These public perceptions of rights and principles are being exercised by ever-increasing populations on limited, if not already reduced, resources. The failures of governments to maintain such resources has been called the "tragedy of the commons" (Hardin, 1968).

1.5 AQUACULTURE

After the rate of increase in the total production from the world's fisheries began to decline in the 1970s, many scientists and entrepreneurs looked to aquaculture because both the demand for fish and its price relative to other protein foods were increasing. Some began to forecast a shift from "hunting"-type fisheries toward "culture"-type fisheries that would be analogous to the gradual shift from hunting to agriculture that occurred over several millenia.

The interest persists, and aquaculture began to grow rapidly during the 1980s, but some of the optimism has cooled as the constraints on aquacultural growth have been recognized. Most importantly, aquaculture had never received the amount of scientific attention that had been devoted to agriculture, and few aquatic animals and plants had been domesticated. Thousands of aquatic organisms can be maintained in captivity, a fact that has led some scientists to be overly optimistic. Only a handful of organisms, however, have been genetically adapted to intense cultivation, and these, plus another handful of wild species that are amenable to cultivation, are used to produce food by aquaculture. Some of the amenable species are also used for nonfood aquaculture, including bait fish, hundreds of species of ornamental fish, pearl oysters, and fish for stocking public waters.

If the optimists are correct in their prediction of a 10- to 20-fold increase in aquacultural production by the year 2000, the total may begin to exceed the production from the sea (see Chapter 11). This will mean a great challenge to, and a reorientation of, the profession of fishery science. Indeed, a shortage of scientists may be one of the constraints upon the expansion of aquaculture.

1.6 CHANGING GOALS AND LAWS

After different civilizations conquered their environment and partially domesticated it, they grew to admire nature. North Americans, especially, have had a love affair with the natural wonders of their environment, and their admiration coalesced first in the conservation movement during the early years of the twentieth century. Conservation then was the concept of sustained yield, multiple use, and the greatest good for the greatest number for the longest time. Its application to the living resources of the sea emerged in 1958 from the International Conference on the Law of the Sea, held in Geneva, Switzerland, where conservation was defined as follows: "the aggregate of the measures rendering possible the optimal sustainable yield from those resources so as to secure a maximum supply of food and other marine products. Conservation programmes should be formulated with a view to securing in the first place a supply of food for human consumption" (see Chapter 13).

This prevailing view of conservation was modified during the 1960s and 1970s as people began to challenge the implicit goal of economic growth, in part because of the deteriorating environment. The conservation concept of wise use was regarded as unwise by some. All resources, especially those regarded as common property, were seen to be in eventual danger with the continuing exponential growth of the human population, and its wanton use of the resources (Meffe, 1986).

Such fears have led to laws that give governments increased responsibility for all aspects of the environment. In particular, earlier laws that were to sustain harvest of commercially valuable species have been succeeded: first, by laws to reserve them for recreational use and, next, by other laws to preserve them for their aesthetic, ecological, educational, historical, and scientific values. In addition, laws have been directed at control of the adverse impact of industry on all aspects of the environment and the economy. Details of some of these environmental laws will be elaborated in later chapters, but the most important deserve mention here.

One major step in environmental management was taken by the United States in the *National Environmental Policy Act* of 1969 (NEPA). Its principal requirement was to establish policies that "utilize a systematic, interdisciplinary approach which will ensure the integrated use of the natural and social sciences and the environmental design arts in planning and decision making which may have an impact on man's environment." Another requirement was inclusion of a detailed environmental impact statement (EIS) in every proposal for legislation or major federal action. This requirement has been widely adopted in the United States, and similar policies have been elaborated in law by many countries and international agencies [United Nations Environmental Programme (UNEP), 1980].

Another major United States law is the *Endangered Species Act* of 1973, which expanded the authority of earlier acts of the same title. This established worldwide lists of "endangered" and "threatened" species and stringent requirements for the protection of all such animals and plants as well as the habitats of each. International components of this act have included implementation of the Convention on Nature Protection and Wildlife Preservation in the Western Hemisphere, prohibition of the importation of any listed species, and the requirement that federal action not jeopardize any such species in foreign countries.

Reconsideration of these issues in 1994 and 1995 has generated major controversies about government management versus property rights, scientific disagreements, the government and private costs of enforcement, and the incidental effects of many major human activities on endangered species. The evolution of the policy process is moving toward increased emphasis on regulation and less support for consumptive uses (Mangun, 1992).

Still another United States law is the *Marine Mammal Protection Act* of 1972, which prohibited, with few exceptions, taking any marine mammals. This act includes some new concepts and standards, such as optimum sustainable population, maximum productivity of the population or species, optimal carrying capacity of the habitat, and health of the ecosystem—concepts that have so far defied widespread agreement on their scientific bases. But the populations of many species of whales have recovered to levels in the 1990s that may well justify sustained harvest of many populations, especially those that may consume large quantities of fish (Schmidt, 1994).

Such laws, which are directed primarily at "preservation" of the environment in as natural a state as possible, are a substantial departure from the older concepts of "conservation," meaning wise use. The aspects that apply to fish and their environment have become a major challenge to fishery scientists, especially in the broad ecological research that is required. They have also brought fishery scientists into contact with a much larger public constituency than that of the recreational and commercial fishermen with whom they have been working.

At the same time that concern for preservation of the environment has been increasing, concern for improved conservation of marine fisheries has resulted in a radical change in the Law of the Sea (see Chapter 13). The new law has already become the basis for national legislation in coastal countries around the world. It provides first and foremost that each coastal state may establish an Extended Economic Zone (EEZ) up to 200 miles out from its coast. This area worldwide contains the fish stocks that now provide about 99% of the world's marine fish catch. The era of "freedom of the seas" for fishing is over, and the distant-water fishing fleets that once fished within a few miles of the coasts of other countries can now be controlled.

Many aspects of the new laws require scientific support in their implementation (Appendix B). For example, "The coastal state, taking into account the best scientific evidence available to it, shall ensure through proper conservation and management measures that the maintenance of the living resources is not endangered by overexploitation." Other provisions require the coastal state to promote the objective of optimal yield, to determine its capacity to harvest the living resources in the EEZ, and to give foreigners access to any surplus. The coastal state may adopt any of the normal fishery regulations and systems of licensing and inspection to ensure compliance. It is obligated to seek agreement with neighboring states on the use of shared stocks.

By the early 1980s, most coastal countries had claimed a 200-mile EEZ and were involved with new policies, plans, and laws. The United States had passed the *Fishery Conservation and Management Act* of 1976, which was consistent with the progress in negotiations to that date and contained a provision that it would be amended to agree with a comprehensive treaty if such a compact were ratified.

Implementation of laws based on these concepts has major policy implications for fishery development in the less developed countries. Many of them will have greater use of important protein food supplies for their people as well as increased opportunities to export high-value fishery products. In order to do these things they need effective institutions with the ability to assess, allocate, and manage the resources that they control, all of which are activities that can be improved by use of fishery science.

REFERENCES

Food and Agriculture Organization (FAO), Marine Resources Service (1992). "Review of the State of World Fishery Resources," Fish. Circ. No. 710, Rev. 8. FAO, Rome.

Hardin, G. (1968). The tragedy of the commons. *Science* **162**, 1243–1248.

Mangun, W. R. (1992). Fish and wildlife policy issues. *In* "American Fish and Wildlife Policy: The Human Dimension" (W. R. Mangun, ed.), pp. 3–31. Southern Illinois Univ. Press, Carbondale.

Meffe, G. K. (1986). Conservation genetics and the management of endangered species. *Fisheries* **11**(1), 14–23.

Miles, E. L. (1989). "Management of World Fisheries: Implications of Extended Coastal State Jurisdiction." Univ. of Washington Press, Seattle.

Royce, W. F. (1993). Fisheries. *In* "International Year Book, 1993," pp. 246–247. P. F. Collier, New York.

Schmidt, K. (1994). Scientists count a rising tide of whales in the sea. *Science* **263**, 25–26.

Spencer, P. D. (1993). Factors influencing satisfaction of anglers on Lake Miltona, Minnesota. *N. Amer. J. Fish. Mgmt.* **13**(2), 201–209.

United Nations Environmental Programme (UNEP) (1980). "Environmental Impact Assessment: A Tool for Sound Development," Ind. and Environ. Guidel. Ser., Spec. Iss. No. 1, UNEP, Paris.

2 | Work of Fishery Scientists

Students of fishery science, whose studies largely involve biology, may well think of their departure from the university as a move from one living system to another. Any group or organization that they join will have structure, process, subsystems, relationships among components, and system processes that can usefully be compared with those of the cells, organs, organisms, or ecosystems that they have been studying (Miller, 1978).

One can begin the comparison by noting that the student role in the university is similar to the role of a cell in a colonial organism. At either the undergraduate or the graduate level, the student usually works independently and competes with other students for grades or evaluation by the faculty. But in almost every job he or she will be a member of a group in an organization. The employee will interact with other members much as organs interact within an organism, and will be a member of a group that has goals and objectives, which will probably be part of a larger organization with broader goals and objectives.

Students should learn all they can about the organization that they propose to join. What are the general features of its structure, process, subsystems, relationships among components, and system processes? What are its goals and objectives? What is its history? Is it meeting its new challenges effectively or showing signs of senility by failing to grow?

Many organizations employ fishery scientists, and the student needs to know in general about the work they do. What are their overall activities, and why do they employ fishery scientists? This chapter will discuss the scope of fishery research and the activities of employers of fishery scientists.

2.1 SCOPE OF FISHERY SCIENCE INSTITUTIONAL ACTIVITIES

A few decades ago most fishery scientists were employed by fishery resource agencies to perform research and by universities to teach fishery research. Now a major proportion of them are involved with other kinds of institutions and with professional application of fishery science to diverse environmental and business problems. More experienced fishery scientists are frequently in supervisory and managerial positions that require broad knowledge of how governments and businesses function.

2.1.1 Common Activities

All institutions, even those consisting of only one person, have service and managerial functions, and a prospective employee should have a concept of what these functions are.

The normal service functions of fiscal management, property management, personnel and payroll procedures, procurement, or secretarial work need no further discussion except to note that they must be provided and managed. Another service function of special importance to fishery institutions is library service, which frequently is difficult to arrange in remote locations.

But management of the institution is especially important because use of common property resources involves so many government agencies and so many diverse laws, at least in most of the developed countries. (One promising private aquaculture venture in the United States was sold because the owner did not have the managerial resources to satisfy the more than 30 government agencies that were involved.) The manager must see to policy development and planning, which are necessary for setting goals, objectives, schedules, priorities, and standards. The manager must deal with all constituencies such as commissions or legislatures in public institutions; owners, customers, and regulatory agencies in private institutions; and the public in both. The manager must also ensure that the purpose of the institution is fulfilled and that it functions effectively.

2.1.2 Fishery Resource Agencies

Fishery resource agencies have been established in almost all countries and in major provinces or states of larger nations. In the United States, the federal government has a National Marine Fisheries Service and a Fish and Wildlife Service. Most states have departments of fisheries, or of fish and game. Usually they are organized and staffed to do the following:

1. Collect and evaluate facts concerning the resources. This includes an ongoing collection of statistics about catch, fishing effort, participants, and

equipment, plus any necessary research on the problems identified by the public with respect to its use of the resources.

2. Provide public information. The basic statistics and results of investigations are commonly disseminated through regular publications, press releases, and special presentations. Some agencies conduct organized educational programs. Some publish fishery research and management journals with worldwide distribution.

3. Protect and improve the aquatic environment. Many fishery agencies have authority to review plans for construction that might adversely affect the environment, to establish water quality standards, and to control pollution.

4. Establish regulations as authorized by legislative bodies. These may be designed to prevent waste, allocate the catches among users, obtain optimal yields, or protect the health of consumers.

5. Enforce laws and agency regulations. This requires professional enforcement officers and is frequently a major part of agency activity, especially if high seas patrol is required.

6. Propagate, distribute, and salvage fish. In North America these functions support recreational fisheries. In many countries fishery resource agencies supply fish eggs or fish seed (the name given to easily transportable postlarval fish) to commercial growers.

7. Provide public facilities for use of the resources. These may include port facilities, market centers, public piers for recreational fishing, and access roads to fishing areas.

8. Provide social services and financial assistance to accelerate fishery development. Such assistance may include low-cost loans, special medical services, subsidies to improve equipment, help in organizing cooperatives, and special information services.

9. Advocate the cause of aquatic resources. Many fishery resource agencies, as the compilers of information about the resources, are expected to defend them from damage, a function that frequently subjects them to pressure from political groups.

2.1.3 Other Government Agencies

Almost all agencies that have responsibility for use of natural waters also have some responsibility for the fish in them. Their fishery activities, which vary greatly, always are secondary to their primary activities, so it is possible here merely to cite some examples:

1. Agencies with water impoundments. Urban water departments, irrigation districts, and public hydropower agencies are almost always involved with the fisheries in the impoundments and with managing their water consumption, water intakes, or water releases below the impoundments in ways that favor fish production. Some in the United States have been required to operate fish hatcheries to mitigate damage to the fisheries.

2. Public land agencies whose land includes streams and lakes. Such organi-

zations frequently have departments with activities similar to those of fishery resource agencies. And if their primary function causes changes in water quality, they may study, monitor, and mitigate the impacts.

3. Agricultural departments. Part of private aquaculture for food production is closely integrated with animal production or field crops and is an alternative use of agricultural water and land. The fish farmers may be supplied with varied services comparable to those supplied to other farmers.

4. Sewage departments. Disposal of treated sewage in natural waters usually requires monitoring of its impact on the water quality. And in some countries, sewage that is free of poisons is used to fertilize ponds for fish farming.

2.1.4 Intergovernmental Fishery Commissions and Councils

When governments share a common property resource, they frequently establish a commission and agree on specific limited responsibility for it. Such commissions usually have members from the supporting governments and a paid staff to fulfill their functions. Examples are:

1. International fishery commissions. These include the Northeast Atlantic Fisheries Commission, the International Commission for the Northwest Atlantic Fisheries, the International Pacific Halibut Commission, The International Pacific Salmon Fisheries Commission, The International Whaling Commission, and many others. Usually such commissions have a staff to collect and publish information such as catch statistics needed for periodic negotiations. A few, such as the Pacific Halibut and Pacific Salmon commissions, also have responsibility to conduct applied research.

2. Interstate fishery commissions in the United States. Three of these have been organized by Atlantic, Gulf, and Pacific coastal states to promote better utilization of the fisheries in each of the three areas.

3. River basin commissions. Almost all of the larger rivers of the world cross government boundaries and many commissions have been formed to secure optimal use of the water. Most of them are concerned with water quality and fish habitat as well as all other uses of the water.

4. Regional fishery management councils in the United States. Eight of these were created by the Fishery Conservation and Management Act of 1976. Each has members from the constituent states and the federal agencies that are involved. Their principal function is to prepare fishery management plans and each employs a staff of several people to conduct its business.

2.1.5 Private Organizations

Private employment of fishery scientists has grown rapidly since the 1970s and 1980s as fishery and environmental problems have increased. Most such employment is of the following kinds:

1. Fishing companies. Many large companies employ fish technologists for product development and quality control. Others employ fishery resource spe-

cialists for advice on fish resources, regulatory impacts, and fishing strategy. An alternative for smaller companies is to contract for such services from a consulting organization.

2. Environmental research or consulting and engineering design firms. Many proposals for construction of industrial facilities, highways, water impoundments, dredging, land reclamation, use of toxic materials, and water transportation involve large-scale use of water or impacts on natural waters. Many countries require environmental impact assessments and engineering design to minimize impacts. Such assessments and design are usually prepared by specialized private firms.

3. Industries with mitigation or monitoring requirements. When impacts cannot be avoided and construction has been allowed, a company may be required to monitor its activities and/or operate mitigation facilities such as fish hatcheries.

4. Fish health services. Aquaculturists, occasionally facing outbreaks of disease, may need advice from specialists; frequently these are veterinarians who have had part of their training in fish health problems.

5. Promotional and lobbying organizations. Hundreds of local, national, and international groups have been formed to promote various kinds of environmental issues. Many are nonprofit and devoted to advancing public understanding. International examples of importance to fisheries are the World Conservation Union (formerly the International Union for Conservation of Nature and Natural Resources, IUCN) and the International Game Fish Association. North American examples are the American Cetacean Society, the International Association of Fish and Wildlife Agencies, the Izaak Walton League of America, the National Fisheries Institute, and the Tuna Research Foundation, Inc. Most of these organizations employ a staff and distribute publications.

2.1.6 International Development Agencies

Most developed countries and several of the United Nations family of agencies assist developing countries. The overall goal of such assistance is socioeconomic development, especially of poorer peoples. The approach is to help those peoples gain the technology, skills, and organizations that will increase their productivity (see Chapter 15).

Although only a small percentage of funded projects are related to fisheries, there are several hundred fishery development projects in operation at any one time that include many possible kinds of assistance. Each is based on an estimate of priority needs, such as to help the government assess the abundance of resources or manage the fishing, making better gear and boats available to fishing villages, improving aquacultural methods, or improving processing and marketing methods. Others include providing better port facilities, loans to governments for facilities or to development banks for reloaning to fish businesses, creation and management of extension services, building institutions, and environmental management.

The essential elements, if the assistance is to be successful, are money and enhancement of the problem-solving capability within the country. Once financing is obtained, the work is planned using the advisory services of experts of many kinds, by collaborative work on projects within the country or among the countries of the region, by education and training, and by providing access to world experience with the problems through conferences, publications, and library services.

Much of the financing is provided by the World Bank and the regional development banks in Asia, Africa, and Latin America. Together they have supplied well over $1 billion for fishery projects since 1965. Most of the loans go to governments for support of production facilities, and a major proportion of these are channeled through national development banks to provide credit to private individuals, cooperatives, or government corporations. Recently, increased emphasis has been given to helping poorer peoples, and loans are made to governments to strengthen their abilities to assess resources, train fishermen and fish farmers, and provide extension services. Characteristically, the projects of these development banks include meticulous planning and recurring post-evaluation.

The preeminent technical fishery development organization in the world is the Department of Fisheries of the Food and Agriculture Organization of the United Nations (FAO Fisheries). FAO is an independent specialized agency of the United Nations that was founded in 1945. In the early 1990s, it was reporting on the fisheries of about 60 countries and more than 200 fishing areas. The reports included about 70 groups of species, some with more than 100 individual species. Funding for FAO fishery development comes predominantly from the United Nations Development Programme (UNDP) in New York and secondarily from a few national donors. FAO also has a Regular Programme that is financed by its member countries and provides advice to the development banks on many of their projects.

FAO Fisheries also maintains a major mechanism for collaboration in fisheries through regional commissions. The regions include the Indo-Pacific, Indian Ocean, Mediterranean, East Central Atlantic, and Southeast Pacific. It is a primary source of information about world fisheries through publications that include the Yearbooks of Fishery Statistics, *Fishery Technical Papers, Marine Science Contents Tables, Aquaculture Journal, Fisheries Abstracts,* and others. It also organizes many technical conferences that bring together the world's leading experts on major fishery issues.

Other United Nations agencies also have important programs for fisheries. The United Nations Educational, Scientific, and Cultural Organization (UNESCO) encourages development of marine science by establishing centers of research and education and by providing facilities such as laboratories and oceanographic ships. Much of its program is supported through the Intergovernmental Oceanographic Commission (IOC), which promotes scientific investigation of the nature and resources of the oceans.

The United Nations Environmental Programme (UNEP) organizes workshops and conferences and publishes extensively about environmental management and conservation. Recent examples are a workshop on pollution of the Mediterranean, a conference on international trade in endangered species, and a series of publications on coastal zone management.

Most developed nations also assist the developing countries through bilateral programs and a substantial number of these have included fishery projects (U.S. National Academy of Sciences, 1981); for example:

1. The Canadian International Development Agency (CIDA) and Canada's International Development Research Centre (IDRC) have provided funds and mobilized expertise from Canadian governments and universities. Recently about half of their projects have had a high fisheries content.

2. The Norwegian Agency for International Development (NORAD) has devoted 5 to 9% of total bilateral assistance to fisheries projects. Such projects include fishing and processing facilities, vessel surveys of resources, and establishment of training centers.

3. The Swedish International Development Authority (SIDA) has supported long-term development programs through bilateral agreements with developing countries. In 1980, 3 out of 20 such programs were fisheries oriented. A major program was to develop small-scale fisheries and improve life in fishing communities in five countries around the Bay of Bengal.

4. The United States Agency for International Development (AID) is the principal U.S. agency for carrying out nonmilitary assistance to developing countries. In the 1980s it supported about 30 fisheries and aquaculture projects that were intended to assist the rural poor. Several other agencies also support scientific and development activities related to fisheries or aquaculture. The National Oceanic and Atmospheric Administration (NOAA) supports cooperative education and training projects through its Sea Grant International Program, which in the early 1990s had projects in about 20 developing countries. NOAA has also been involved in cooperation with the People's Republic of China, with projects including the study of marine and freshwater aquaculture. The U.S. Department of the Navy has assisted with training and loan of equipment to countries for surveying and charting harbors and coastal areas. The Peace Corps has supported about 300 volunteers on fishery projects, most of whom worked on freshwater aquaculture. The Smithsonian Institution has supported a marine biology center in Tunisia.

2.1.7 Universities, Museums, and Research Stations

Most universities in all parts of the world have departments giving instruction and doing research in sciences related to fisheries. These sciences include biology, marine biology, zoology, limnology, oceanography, chemistry, and mathematics. Only a few of these give special emphasis to fishery science, but when the university programs are combined with those of museums and research stations the number is substantial. An older listing of hydrobiological and fisheries institutions in the world provided details on about 700 institutions (Hiatt, 1963). Of these about 35 were in Canada and 143 in the United States. A more recent listing of university curricula in the marine sciences and related fields listed 285 institutions in the United States and Canada (U.S. Dept. Commerce, 1979). Only about a dozen were identified as giving degrees in fishery sciences, but this figure is probably too low because degree programs for some institutions were not included.

In the United States, fisheries and other aquatic sciences in universities are partially supported by two programs of the federal government. The Cooperative Fishery Research Unit Program that was authorized in 1960 is administered by the Fish and Wildlife Service of the Department of Interior. Twenty six fishery units and three combined fishery and wildlife units have been established at state universities to perform research and train graduate students. The Fish and Wildlife Service also operates a Fisheries Academy for continuing education at Leetown, West Virginia.

The Sea Grant College Program in 1983 (Fritz, 1983) involved 170 individual colleges, universities, and research institutions, of which 16 universities with more substantial activities were designated as Sea Grant Colleges. This program, which was started in 1966, has been administered by the National Oceanic and Atmospheric Administration of the U.S. Department of Commerce. A large proportion of the activities have been in fisheries, and include research, education, and marine advisory services.

2.2 SCOPE OF FISHERY RESEARCH

Virtually all fishery institutions, public or private, now use fishery research. They may do it themselves, contract for it to be done, or depend on the published work of others.

2.2.1 Basic Research

Research is the glamorous part of fishery science. It involves a broad spectrum of inquiry from the most basic fundamentals of aquatic environment and life to an examination of the potential impact of a pollutant or a fishing regulation. If the research is directed toward a general understanding of a phenomenon it is usually called *basic*. If it is directed toward a specific issue and required for a decision, it is usually called *applied*.

Fishery scientists become involved in a large array of basic sciences if we judge by the topics included in major fishery journals. Most of these fall into five groups.

1. Descriptive Biology
   a. Ichthyology

   b. Physiology
   c. Anatomy, gross and micro-
   d. Behavior
2. Aquaculture
   a. Pathology and microbiology
   b. Genetics
   c. Nutrition
3. Ecology
   a. Limnology
   b. Oceanography
   c. Hydrology
   d. Population dynamics
   e. Analytical chemistry
4. Social Sciences
   a. Economics of natural resources
   b. Cultural anthropology
   c. Evolution of fishery law
5. Fish Preservation
   a. Chemistry of fish flesh
   b. Microbiology of fish flesh

Such a list is far from complete, and it is obvious that the sciences listed have been chosen because of their potential application to problems related to the living aquatic resources or their environment. But all basic research activity in fisheries has some common characteristics.

First, most basic research is in general fields that include activities other than those concerned with the fisheries or the aquatic environment. For example, oceanography includes many physical, chemical, geological, and even biological studies that rarely have direct application to fishery problems, even though certain studies of circulation and productivity may be critically important. For example, ichthyology, the branch of zoology dealing with fishes, includes study of many fishes of little importance as resources.

Second, basic research in support of fishery sciences is most commonly performed in universities, occasionally in special government laboratories, rarely in private laboratories that lack government support. This practice tends to determine the type of graduate education in fishery science that is usually oriented toward basic research. On the other hand, employment in basic fishery research outside universities is rare—probably less than 5% of the total.

Third, such research work is frequently an individual activity, including data collection, analysis, laboratory work, and preparation of the report. It is directed toward a general goal of increasing understanding rather than toward a specific social problem. If it is a student effort, its schedule is related to the schedule of the educational program. If it is a faculty effort, its schedule depends on availability of students.

Fourth, much of the work has a broad usefulness not confined to a geographic region, a particular body of water, or even to aquatic organisms. It results in reports that are distributed internationally and it engenders international or national symposia. It is judged, not on its immediate usefulness, but on its originality, the extent to which it advances the general understanding of the problem, and its potential for simplifying and enhancing the subsequent action on fishery or environmental problems.

Fifth, such general usefulness carries little weight in the usual government process of justifying appropriations. Agencies are always beset by demands for immediate results and frequently cut basic research, the "nice to know" projects, before cutting investigation of a pressing problem. Consequently a common technique for budgeting basic research is allocation of a small proportion of research funds to special individuals or groups.

Finally, a general knowledge of basic research areas and some experience in performing research are essential for all fishery scientists, whether they are employed to do research or not. Above all, the research experience should emphasize the need for integrity in the science.

2.2.2 Applied Research

Applied research is usually the first activity after a societal problem has been identified. Applied researchers draw on the understanding developed by the basic research studies listed earlier and then inquire further into fishery problems that may be grouped as follows:

1. Assessment of Fish Stocks
   a. Migratory and reproductive behavior
   b. Rate of growth
   c. Abundance
   d. Effects of fishing
   e. Effects of environmental changes
2. Control of Fishing
   a. Effects of regulations
      i. Fish stocks
      ii. Catches
      iii. Economic yield to fishermen
      iv. Social structure of fishing communities
   b. Enforcement strategy
3. Aquaculture
   a. Control of reproduction
   b. Selective breeding
   c. Seed production
   d. Diet formulation
   e. Feeding methods
   f. Disease diagnosis and control
   g. Fish pond or enclosure management

4. Environmental Protection
   a. Baseline studies
   b. Pathology of environmental stress
   c. Bioassay
   d. Monitoring
   e. Prediction of effects of environmental change
   f. Environmental quality evaluation
   g. Risk evaluation
   h. Environmental toxicology
5. Fish Capture
   a. Fish behavior and physiology
      i. Migratory routes
      ii. Aggregations and schooling
      iii. Sensory systems
      iv. Habitat preferences
      v. Reproductive behavior
   b. Hydroacoustics
   c. Fishing gear behavior
6. Fishing Strategy
   a. Location, abundance, and composition of stocks
   b. Optimal characteristics of fishing vessels
   c. Laws and regulations
   d. Forecasts of catches
7. Fish Preservation and Marketing
   a. Preservation and packaging systems
   b. Quality improvement and control
   c. Market analysis
   d. Product development
   e. Forecasts of demand
   f. Inventory policy

This listing is no more inclusive than the listing of basic sciences given earlier. But these items also have many things in common, some of which contrast sharply with the basic sciences.

First, all of these applied research activities are planned on the basis of the general understanding that has been developed from basic research. The applied researcher has a mandate to discover the best available scientific basis for action. The work should not be planned as an end in itself, but as the first step toward action on a social problem.

Second, such research is usually performed in government laboratories, some of which are in nonfishery agencies that are involved with the aquatic environment. Sometimes, universities perform it on a contract basis rather than on a grant basis. Many industries perform such research or retain consulting groups when they have an obligation to guard against damage to the aquatic environment, or to mitigate the consequences of previous environmental damage. Some fishing companies support such research as an aid to development of operational policies or an improvement in fishing practices.

Third, the researcher will commonly be a member of a team that includes people collecting data, people collating and summarizing large amounts of data, and researchers working on other kinds of research related to the same social problem. He or she will probably have definite objectives and a schedule for producing a report.

Fourth, the research will apply to a particular situation (although it may be an example of how to approach a general problem). The report will be prepared to communicate results to a diverse group of scientists and lay people. Many of these people will judge it, not on its originality or potential usefulness, but on its immediate value to them.

Fifth, useful results will generate company or government support for continuing, possibly expanding, work.

Finally, since almost every action of a resource agency is based on applied research, or starts with additional applied research, it is important for all echelons in the agency to be familiar with how the research is done, how it must be interpreted, and how credible it is.

REFERENCES

Fritz, E. S. (1983). Fisheries science in the National Sea Grant College Program. *Fisheries* 8(2), 11–14.

Hiatt, R. W. (1963). "World Directory of Hydrobiological and Fisheries Institutions." Amer. Inst. of Biol. Sci., Washington, D.C.

Miller, J. G. (1978). "Living Systems." McGraw–Hill, New York.

U.S. Department of Commerce (1979). "University Curricula in the Marine Sciences and Related Fields: Academic Years 1979–1980, 1980–1981." Office of Sea Grant, Washington, D.C.

U.S. National Academy of Sciences (NAS) (1981). "International Cooperation in Marine Technology, Science, and Fisheries: The Future U.S. Role in Development." National Academy Press, Washington, D.C.

3 | Professional Careers

Fishery scientists began to be identified more than a century ago as scientists became involved with improving our use of living aquatic resources, protecting the waters in which the resources live, or improving the practice of aquaculture. These scientists, who were usually trained in biology or limnology, began career-long activities in fisheries. The activities have proliferated as society, more and more concerned about living aquatic resources, has welcomed the scientific contributions. But fishery science has not acquired a public identity as clearly understood as those of other learned professions, so some definitions are in order.

Fishery science has two different meanings. First, it is a body of scientific knowledge pertaining to the fisheries and their environment. It is also called *fisheries science, fisheries biology,* or just *fisheries*. Second, it is a profession that expands and uses the body of scientific fishery knowledge to obtain optimal benefits for society from the living resources of the waters. In this sense it includes research of many kinds in addition to biological in order to understand the full scope of the socioeconomic and political issues related to the problems of the fisheries and aquatic environment.

In addition, it is important to recognize that people fish for things other than fish, such as sponges, crustaceans, mollusks, marine mammals, and even aquatic plants. And we sometimes use the word *fish* collectively to refer to any of the things that we catch.

These definitions do not have official sanction, nor are they based on any recorded consensus. They reflect my understanding of common usage at the beginning of the 1990s, a usage that has been rapidly evolving during the past four decades.

3.1 KNOWERS AND DOERS

The term *fishery scientist* denotes a knower of connected facts, but most fishery scientists go beyond just knowing to doing something about a fishery. This makes them comparable to engineers, physicians, or other professionals who use their sciences to solve human problems. So what do we mean by *fishery scientist?*

The fishery scientists' designation emerges not from arbitrary definition but from interpretation and description of what they do. They are more concerned with fisheries than with fish, with sociopolitical decisions than with sets of facts, with programs than with systems, and with institutions than with ideas. Their doing is directed toward the immediate and practical fishery issues, but will include information from statisticians, who predict the changes in populations; the resource economists and cultural anthropologists, who predict the impacts of the changes on people; and the political scientists, who advice on resource policies.

Fishery scientists have collected the fishery knowledge from much of the world—population dynamics from Russia and England, carp culture from China and India, trout and molluscan culture from Europe, salmon and catfish culture from the United States, fish preservation old and new from many countries, aquatic environmental protection from the United States, commercial fishery organization from Japan, recreational fishing and fishery management from North America, and fishery development from the Food and Agriculture Organization of the United Nations.

They have steadily gained public confidence, through trusted leadership, and the public has responded by asking more of them. Now they are asked to shepherd all of the living aquatic resources, including the threatened and endangered, and to help maintain the quality of water as an index of our use of watersheds and oceans.

Their work takes them worldwide, for every country uses fish. They are helping all coastal countries make optimal use of the marine resources in the extended economic zones and are guiding the great transition from capture to culture fisheries.

Their leaders have become statesmen who rewrite the ancient ocean laws and negotiate international agreements; or executives who manage government organizations, fishing companies, fishery publications, and consulting firms; or publicists who inform people about the fisheries and aquatic environment; or curators of museums and great public aquaria; or educators who supply the fresh new generations of fishery scientists.

Few fishery scientists remain in ivory towers. They understand and are equally at ease with fishermen and diplomats, researchers and executives, partisans and polemicists, conservationists and environmentalists.

They have a passion for science, each with a scientific specialty plus an understanding of the need for scientific teamwork and the use of many sciences. But above all they are professionals who use the sciences, who weigh, combine, and doubt, and who then provide advice on decisions without cunning or deceit.

3.2 APPLICATION OF THE SCIENCES

A few fishery scientists may still adhere to the old ivory tower notion that they merely present the facts and stay aloof from decisions and application. But research results in any field are never enough for solution of a complex problem and the researcher invariably must bear a share of responsibility because a solution based on the research will carry the prestige of science.

The reasons for this are inherent in the process. The action to be taken involves a forecast of something to be expected. It will be based on a judgment about the adequacy of the research and whether the researcher used a model appropriate for the situation to be expected. The decision maker will use the research, of course, but will consider his or her own opinions and experience, even perhaps with some wishful thinking.

Fishery decisions usually require several kinds of sociopolitical and ecological research that may indicate different courses of action. Such diverse views can be minimized if the researchers are organized into teams and required to reach a consensus.

Fishery decisions are seldom permanent because there is no way of ensuring the acceptance or adequacy of the research results nor the applicability of the research to changing environmental conditions. They are usually tentative for a trial period and subject to review. Many regulatory decisions are routinely reviewed every year.

In many respects the role of the fishery scientist has become professional, similar to the role of an architect or engineer, who creates a design or a system and stays involved with the implementation. The scientist will know the limitations of the research and will receive reports and reactions that will improve subsequent research. In almost every case, he or she will be part of an organization, and will be only as effective as the organization.

3.3 HISTORICAL TRENDS

Fishing and aquaculture are ancient human activities that received little scientific attention until early in the nineteenth century. Then the genesis of one major activity in fishery science followed the discovery by fish culturists in Europe (Fry, 1854) that the reproductive process of salmonids (trout and salmon) was easy to control. The eggs and sperm could be removed manually from the fish without harming them, the fertilized eggs could be hatched in incubators, and the young fish could be fed easily in captivity. Furthermore, the eggs or young fish could be transported to other fish farms or stocked in natural waters. Since each female could produce a few thousand eggs, most of which could be hatched successfully, people began to think that this practice offered a solution to the declines in fish stocks that had suffered from overfishing or pollution.

Later, in the middle of the nineteenth century, the introduction of steam propulsion and otter trawling into the North Sea fisheries led to a challenge of the prevailing belief that the sea fisheries were inexhaustible. Many scientists were convinced that overfishing was taking place. A North Sea Convention, concluded in 1882 among most of the countries bordering the North Sea, provided for more orderly fishing, but there was still no agreement about how to deal with too much fishing or destructive fishing practices.

The governmental and private actions that followed the growth of public concern about the fisheries (Benson, 1970) led to careers in fisheries by scientists and therefore the birth of the profession. Salmonid culture increased rapidly in Europe and was brought to North America in the 1850s (Fry, 1854). By 1870 some 200 private persons were practicing fish culture in the United States, and the states themselves were organizing departments of fisheries. In 1870, a group of fish culturists organized the American Fish Culturist's Association (Sullivan, 1981). The name was changed to the American Fisheries Society (AFS) in 1884, and it has become the major professional fisheries society in the world with about 11,000 members in the early 1990s.

In 1871, the U.S. Congress authorized creation of a Commission on Fish and Fisheries that soon started its scientific work in Woods Hole, Massachusetts. Similar government organizations started at about the same time in many other countries.

In addition to the national and private organizations, a unique international organization emerged early in the twentieth century (Went, 1972). European scientists from several countries recommended to their governments that international cooperation must be organized to provide a rational basis for exploitation of the sea and that some specific investigations should be started immediately to promote and improve the international fisheries. Their efforts resulted in the inaugural meeting of the International Council for the Exploration of the Sea (ICES) in 1902.

ICES grew and flourished, in spite of two world wars, and is now a major international scientific oceanic organization in the world. Eighteen countries were members in the late 1980s. It provides for free scientific exchange among government, academic, and private scientists on issues frequently in dispute among the member nations. A similar organization was established in the late 1980s for the Pacific area, named the Pacific International Council for the Exploration of the Sea (PICES).

The other major step in the international development of fishery science came after the formation of the Food and Agriculture Organization of the

United Nations (FAO) in 1945. It included a Department of Fisheries that, in the 1980s, employed about 390 specialists, and was by far the biggest fishery consultancy in the world (see Section 3.3.6).

Out of these nuclei of the profession have grown the fishery organizations of the world. Almost every national government and provincial or state government in the larger countries has a department of fisheries. Employment of professionals is in the tens of thousands and growing. The scientific literature has become immense (Bibliography), the United Nations covering about 120 journals in its monthly periodical that reproduces the tables of contents of core journals in marine science and technology.

3.4 SERVING SOCIETY AS STRATEGIC THINKERS

Fishery science was formally recognized in North America during the 1960s as one of the group of natural resource or conservation professions that included soil conservation, range management, wildlife management, forestry, watershed management, and park and recreational development. With the passage of environmental legislation, these groups have tended to coalesce into environmental management, and many of the aquatic aspects of the latter have become inseparable from fishery science.

The profession of fishery science grew as society recognized its usefulness, and its obligations to society grew with the recognition. Its recognition and obligations are similar in many respects to those of other learned professions. Professionals are characterized by:

Special education, knowledge, and skills in the chosen field
Continuing efforts to remain professionally competent
Loyalty to employers and clients
Dignity and pride in the profession
Awareness of obligations to serve society
Ethical conduct

Some learned professions are licensed in most countries but the environmental or conservation professions are not—at least not in the early 1990s. In fishery science in North America, peer recognition is achieved through certification as a Professional Fishery Biologist by the American Fisheries Society.

Such recognition is a privilege extended to those who go beyond technical training and beyond matters of intellectual difficulty. As a distinguished engineer has said (Wickenden, 1981).

Professional status is therefore an implied contract to serve society, over and beyond all specific duty to client or employer. To possess and to practice a special skill, even of a high order, does not in itself make an individual a professional man. The difference between technical training and professional education is no simple matter of length. It is rather a matter of spirit and scope.

Most fishery scientists have been employed in public service. At first they usually served only in strictly scientific roles, but gradually they have entered senior administrative positions in agency directorates. Now it is customary to find professionals at all levels, from junior scientists to directors in international fishery agencies and the fishery agencies of most countries.

Total professional employment of fishery scientists in the United States and Canada has approximately doubled each decade since 1940. The rate of increase has varied with changes in government programs and with shifts of proportion in various kinds of employment. No complete data are available, but data from the American Fisheries Society and the Sport Fishing Institute have been presented from time to time and are probably indicative of professional involvement. Membership by individuals in AFS has increased at a compound annual rate of about 7% since 1940, and totals about 11,000 in 1994. Of these, about 630 are Canadians and about 300 are from other countries.

The distribution of employment of AFS members has recently been about half in government, about 30% in educational institutions, about 18% in private employment, and about 2% in retirement.

The recent growth in public concern about environmental issues and the ensuing legislation have led to accelerating employment of fishery scientists for their aquatic environmental skills. A significant proportion of industries that use large quantities of water or discharge effluents now employ fishery scientists. In addition, many environmental consulting and engineering firms employ fishery scientists to assist with design of water quality management programs or facilities.

Compensation of fishery scientists includes a job that is usually challenging and desirable, and government jobs are usually secure, with relatively good retirement benefits. Starting salaries and overall average salaries for fishery scientists have been close to the overall average salary for engineers and chemists in the United States as a whole.

3.5 ETHICAL RESPONSIBILITIES

Of all the characteristics of professions, ethics is probably the most important insofar as credibility of the individual and profession is concerned. Fishery scientists share the difficulty of maintaining ethical standards because, in work that requires good judgment as well as good science, they frequently face situations in which the various interests of clients or employers, of political entities, or of society at large are in conflict.

Judgement is required because of the limitations of the science and the demands of laws or agencies for advice that goes beyond the capability of the accepted science to produce conclusive recommendations. Frequently the demand is for a prediction of the consequences of a proposed environmental or fishery action. Or the demand may be for an action concerning an animal about which there is great public interest but no reliable scientific information. Any such situation poses a dilemma for the scientists. The following is an extreme example of a conflict in which many scientists were involved and, in the final decision, the scientific evidence was inconclusive.

The issue was the impact of electric power generation on Hudson River fish populations (Christensen *et al.*, 1981). The power was needed and could be generated much more cheaply if water from the river could be used for cooling and if a substantial quantity of the river could be pumped temporarily into high-storage reservoirs to be returned through turbines in peak electric demand periods. The river was significant as a spawning and nursery area for alewife, blueback herring, striped bass, and American shad. The conflict arose mostly because pumping the water killed eggs and young fish when they were sucked through the pumps. Consolidated Edison, the power company, started studies in 1965 and its basic methods were fully developed by the end of 1972. But the National Environmental Policy Act of 1969 (NEPA) had become law and courts decided that it applied to the proposed power plants.

Studies continued along with negotiations and hearings from 1972 through 1980. They involved eight federal agencies, seven consulting engineering or environmental firms, scientists from six colleges or universities, scientists from three state fishery agencies, and a scientist from the Fisheries Research Board of Canada. After the expenditure of all the effort and money for nearly 17 years, only some of the scientific issues were resolved, and some of the scientific methods could not satisfy the legal challenges. The contestants, the Environmental Protection Agency and Consolidated Edison, compromised with an out-of-court settlement that was reported as a "Peace Treaty for the Hudson."

Nothing in the record indicates that any scientists were unethical nor is this implied. But the lessons are that some issues are extremely important to society, that scientific methods or answers may not be obtainable, and that pressure may be intense for advocacy on the part of scientists toward one contestant's viewpoint. In retrospect, it might have been far better to have had lawyers seek a compromise early in the negotiations, rather than ask the scientists for something they could not deliver.

Such controversies are increasing because some new U.S. laws include concepts about which there is little scientific agreement, and require scientific evidence that is lacking or very expensive to obtain. This is true occasionally of NEPA and other U.S. acts such as the Endangered Species Conservation Act of 1969 and the Marine Mammal Protection Act of 1972.

Another kind of dilemma for the scientists arises when their scientific data are adequate for conclusions that are contrary to their employer's policies or to the political pressures on the agency. They may be under intense pressure to conceal the controversial parts of their findings, with their job or the job of their agency director perhaps at stake. If their conclusions are contrary to a position of their country in international negotiations, they may be considered unpatriotic if they use them. Such confrontations can sometimes be avoided if agreement can be obtained in advance on standards for the scientific data that will be used in the decision process.

Conflict situations are increasingly common for fishery scientists as they use their science to help solve society's problems. It is vital for them to hold to fundamental canons of behavior for learned professions, which have been stated well by the National Society of Professional Engineers as follows:

1. Hold paramount the safety, health, and welfare of the public in the performance of their professional duties.
2. Perform services only in areas of their competence.
3. Issue public statements only in an objective and truthful manner.
4. Act in professional matters for each employer or client as faithful agents or trustees.
5. Avoid improper solicitation of professional employment.

More detailed standards of professional conduct have been adopted by the American Fisheries Society (Appendix A).

3.6 FUNCTIONS OF PROFESSIONAL FISHERY SCIENTISTS

Research is the traditional and still a vital function of fishery science, but fishery scientists have been asked increasingly to contribute to all the activities of their employers. Presented herewith, to illustrate the kinds and levels of involvement, is an outline of the principal functions of fishery scientists, each discussed in general terms, and then with respect to two specific examples— Example 1, a business of raising trout in a commercial hatchery, and Example 2, government regulation of a large oceanic fishery.

3.6.1 Basic Research

The objective of basic research is to establish new understanding for general application. The problem is defined on the basis of existing knowledge in a specific field of science, which may be one of more than a dozen branches of the natural or social sciences (see Section 3.1.1).

Example 1. A critical aspect of the business of rearing trout for market is a correct diet because such trout receive only artificial food and feeding them is costly. Basic nutritional research is one of the several kinds of research that has made trout rearing practical. It has determined their requirements in terms of proteins, fats, carbohydrates, vitamins, and minerals for each species of trout at different sizes and under varied environmental conditions. The research is a lengthy laboratory process that usually requires feeding a diet, complete except for an item in question, to groups of trout of each size and under each environmental condition.

Example 2. Regulation of fishing on an ocean fish stock requires basic knowledge of the identity and life history of each species involved, in addition to basic knowledge of the environment and the economic performance of the fishermen. The facts needed about the fish include the name of the species, how it is identified, where it is found, what kind of environment it needs at all stages in its life, how it spawns, what it eats, what eats it, when and where it migrates, and how fast it grows. Such information is gathered from direct observation of the fish in fishing or research vessel catches and from recaptures of tagged fish. The facts are accumulated slowly and evaluated periodically by comparing the current field observations with published accounts of earlier observations.

3.6.2 Applied Research

Applied research is the predominant research activity. It is usually an investigation of the technical, social, and economic aspects of a fishery problem and it is frequently a part of a cycle of research, decision, monitoring, evaluation, and further action (see Section 3.1.2).

Example 1. After the nutritional requirements of trout have been established by basic research, the applied researchers can formulate diets from the available feedstuffs, which may include meals made from meat, fish or grains, scrap meat or fish, and by-products from human food processing. They can decide which mixtures are likely to be the most cost-effective and how best to deliver the food to the fish. Then they can conduct feeding trials under specific hatchery conditions and recommend practical diets.

Example 2. After fundamental information on each species of fish in a fishery has been obtained, the applied researcher will determine the impact of various amounts of fishing on the stock of fish. They will try to estimate the amount of fishing that will produce the maximum sustainable yield from the stock. This will be based on estimates of natural and fishing mortality rates, the rate of growth of individual fish, and the size of the stock. Such estimates are usually refined annually as more information accumulates, and are used for regulatory decisions.

3.6.3 Planning

The planner will determine the scope of the problem, recognizing that fisheries are human activities, and then, having analyzed alternative solutions, will recommend a final solution that probably can be achieved within the constraints of time, financing, social impacts, legal requirements, and public understanding.

Example 1. If the trout business is at a single location, the planner will probably be the implementer and supervisor as well. The planning phase will involve a review of the production cycle with estimates of the weight of fish on hand each week, decisions about the best diet at each age, estimates of the quantity of each component needed, determination of sources of each component, decisions about how the diet is to be mixed or otherwise prepared, a proposed schedule for delivery, and plans for repair or replacement of equipment. The plans should include alternatives in case the quantities have not been estimated correctly or some components are unavailable. The plans will include estimates of costs for each appropriate period in the production schedule.

Example 2. Regulation of fishing can be effective only if the people affected understand and accept (even if grudgingly) the regulations. It requires careful planning that includes justification of the need for action, determination of objectives to be accomplished, and evaluation of alternative ways of meeting the objectives. The evaluation will include full information about the condition of the stocks, the recent history of regulation, and how each alternative is likely to affect the stocks and the success of fishing on them. All pertinent information will be collected in a proposal that can be distributed to interested parties for comment. The proposals may be modified if necessary and submitted to the appropriate authority for action.

3.6.4 Implementation

Plans are implemented by organizations. Many people are involved with operations that include finances, production, information, services, logistics, communications, and enforcement, each activity suitable to the kind of organization involved.

Example 1. The operation of the trout hatchery according to the plans will probably include negotiations for financing production, hiring personnel and training them, repairing equipment, buying food and supplies, and the day-to-day production functions. It will also include processing, packaging, and selling the product.

Example 2. After a reasonable consensus on the proposed regulations has been reached, the plan will be formally approved and a regulation promulgated as a law. It will be published to inform the fishers and appropriate instructions will be developed for enforcement or monitoring operations. Then will come the day-to-day operations of surveillance and data collection that may involve many people using ships or aircraft or automobiles throughout the area of fishing or in the ports of landing. There will also be recurring evaluation of the progress of the fishery through periodic reports of catches and operations to supervisors and to the public.

3.6.5 Supervision

Supervision is the responsibility for all or part of a project. It is also helping employees with information, training, standard-setting, and guidance. It involves helping managers with planning, scheduling, unanticipated problems, and project completion.

Example 1. The trout hatchery supervisor will be responsible for quality and cost control, for growth and health of the fish, and for meeting production and sales schedules. He or she will be particularly watchful for events that can cause loss of the fish, such as failure of the water supply or appearance of an epidemic disease.

Example 2. Supervision of a project to regulate an ocean fishery will involve many diverse activities, each of which will have a supervisor. The activities will usually include enforcement, vessel operation, aircraft operation, statistical data collection, and applied fishery research. Each will require special expertise and the supervisors will normally report to a general supervisor.

3.6.6 Leadership of Organizations

Managers maintain organizations by being responsive to their customers or constituencies and by ensuring that their product is delivered. They constantly strive to improve performance and, in particular, prepare organizations for change by stopping obsolete programs and inaugurating new ones. Major parts of their work are usually management of money and people. Their special challenges are to know the historical progress of the organization and to communicate the internal changes necessary to keep it performing effectively (Wojcieszak, 1994).

Example 1. The manager of a commercial hatchery will regularly evaluate its product quality and its economic performance relative to its competitors. He or she will keep informed about the industry's activities and current economic trends, will change hatchery operations to introduce any proven efficiencies, and will try to ensure adequate financing and staffing by people who can perform in the organization.

Example 2. The manager of a fishery regulatory agency has multiple constituencies, from the fishers and their legislators to the press and the general public. Much time will be required to explain the purposes and activities of the organization. It will also be necessary to manage the budgets and employees in ways that satisfy the constituencies while, in addition, trying to anticipate their future demands.

3.7 ROLE OF PROFESSIONAL AND SCIENTIFIC SOCIETIES

The inheritance and progress of fishery science are reflected in its professional and scientific societies. Dozens of these provide for communication and joint action by fishery scientists.

The distinction between professional and scientific societies is only a matter of degree. Both are dedicated to exchange of scientific information, but the professional societies give more emphasis to relationships with clients, and the scientific societies to relationships among the member scientists.

The outstanding professional and scientific organization in fishery science is the American Fisheries Society. It publishes four scientific journals, its *Transactions, North American Journal of Fisheries Management,* the *Progressive Fish Culturist,* and the *Journal of Aquatic Animal,* and a bimonthly magazine, *Fisheries.* It circulates lists of major fishery books and markets them at a discount of 5%. It draws most of its 19,000 members (1993) from North America, but about 70 countries are represented. It held its 125th Annual Meeting in

1995. It has 4 regional divisions, 20 scientific discipline sections, and 53 chapters. Most of the divisions, sections, and chapters have newsletters and hold meetings. It has a program of professional certification, and takes positions with government bodies on many conservation and environmental issues.

Many societies important to fishery scientists are primarily scientific. North American examples include the American Society of Ichthyologists and Herpetologists, the Ecological Society of America, the American Society of Limnology and Oceanography, and the World Mariculture Society. Each of these publishes a major journal and holds scientific meetings.

Students who have yet to choose a career should become acquainted with the professional organizations in the fields they are considering. These will usually supply information on career options and employment opportunities. Those who have chosen a field of fishery science should become acquainted with the scientific publications in their field as well as the profession. Student membership in the American Fisheries Society and the more specialized societies is an essential part of beginning a professional career. Such memberships are usually available at reduced rates.

3.8 CAREER PLANNING STRATEGY

Choice of a career may be an awesome task for the student. Who can predict what circumstances will exist 5 to 10 years hence, when the student will be starting chosen work, let alone what the circumstances will be 40 to 50 years hence near the end of his or her working life? We can do little more than describe the recent trends in the present opportunities, the kinds of preparation needed for the various kinds of work, and the possibilities for subsequent career development.

3.8.1 Recent Trends

Along with the expansion of fishery problems (see Chapter 2), a striking trend in fishery science is the rapid increase in diversity in the sciences involved and in the kinds of employment offered. As for the sciences, (described more fully in Chapter 3) those regularly applied to aquaculture, to management of wild fish stocks, and to management of the aquatic environment now number in the dozens. Employment, which four decades ago was almost entirely public—in government or university—now is increasingly private.

The increase in diversity of employment includes a shift in function away from the traditional research toward supervisory and managerial positions, especially in business. There are now proportionately few people in basic research (except in universities), a large proportion in applied research, and a rapidly growing proportion in planning, implementation, supervision, and management. The shift in this direction has taken some fishery scientists to senior positions in governmental and intergovernmental fishery agencies, in fishing companies, in environmental consulting firms, and in academia.

Such changes have occurred worldwide as all coastal countries try to benefit from their nearby fishery resources, protect their environment, and grow more fish. And an increasing proportion of fishery scientists from the developed countries are assisting those in the lesser developed.

3.8.2 Preparation

Within such a broad range of career opportunities a student has no single optimum choice of courses, but may be guided by information available from a survey of fishery scientists in 1971 (Royce, 1972). The survey drew responses from 324 fishery scientists on the west coast of North America and asked them what undergraduate courses they found most useful and least useful. English and scientific writing were at the top of the most useful list, followed closely by biostatistics, zoology, fisheries, and mathematics. Foreign language was at the bottom of the list, approached closely by humanities, social sciences (except economics), and advanced courses in biological, physical, and chemical sciences.

An explicit comment about desirable coursework was made by an employer of a major number of U.S. fishery scientists [Hester, 1979 (Associate Director of the U.S. Fish and Wildlife Service)]:

> In 13 years as a faculty member, myself, I never had an employer comment that one of my former students didn't know enough about fisheries or wildlife biology. I have, however, on several occasions had comments about their inabilities to deal effectively with the public. My experience has led me to believe that even after graduation a biologist has a tremendous opportunity to learn biological information through books, meetings, workshops, as well as other forms of on-the-job training, and he or she is eager and hungry for this knowledge. But a biologist who has never been exposed to public administration, environmental law, public speaking, human psychology, legislative process, planning, economics, technical writing, policy formulation, or political science is unprepared in these areas and is unlikely to pursue any of these after graduation.

More recent analysts have reinforced the view that a broad training in the biological, ecological, sociological, and political aspects of fisheries is important, and that a lengthy focus on a narrow biological aspect may lead to neglect of the conceptual, organizational, and managerial aspects that are critical to managing people and their organizations. The students should be broadly trained in the whole management process even though each teacher may have an area of specialization. Their graduate programs would be a mixture of broad course work and field training (internship). They would learn about the professional dimensions of the private and government work, and how to be the coordinator and the leader of teams of specialists (Cole, 1992).

An important aspect of career preparation is to keep options open for a time, at least through early undergraduate years. If a professional career such as fisheries is the choice, these years can probably best be used to obtain communication skills and good general courses in the several basic sciences and mathematics. Then a decision can be made to pursue a major field that may or may not involve graduate work. The more specialized courses can be planned accordingly. If the decision is made in favor of basic research, then a doctorate program should be planned. But the student should be aware that the intense specialization required may reduce the options for employment outside research and teaching.

Obviously, a broad understanding is desirable, and it needs to include the structure of aquatic systems and their interrelationships with air and land. Also needed is understanding of the linkages and the functioning of organisms, populations, people, and governments.

The profession of fishery science, including the environmental aspects, is clearly headed in the direction of interdisciplinary approaches, and the supervisors and managers will be those who can grasp the whole problem and lead the teams. Furthermore, it is changing ever more rapidly, especially in the aspects of working with people (Ditton, 1995; Murphy *et al.*, 1995). The best strategy for the student who intends to become a professional is probably to acquire a scientific specialty, know how that specialty relates to the other fishery sciences, and add the synoptic view of how professionals work and what they do.

REFERENCES

Benson, N. G., ed. (1970). "A Century of Fisheries in North America," Spec. Publ. No. 7. Amer. Fish. Soc., Washington, D.C.

Christensen, S. W., Van Winkle, W., Barnthouse, L. W., and Vaughn, D. F. (1981). Science and the law: Confluence and conflict on the Hudson River. *Environ. Impact Assess. Rev.* 2, 63–68.

Cole, C. F. (1992). New dimensions to the education of fish and wildlife professionals. *The Environmental Professional* 14, 325–332.

Ditton, R. B. (1995). Fisheries professionals: Preparing for demographic change. *Fisheries* 20(1), 40.

Fry, W. H., ed. (1854). "A Complete Treatise on Artificial Fish Breeding." D. Appleton, New York.

Hester, F. E. (1979). Fisheries education from the federal perspective. *Fisheries* 4(2), 22–24.

Murphy, W. F., Cross, G. H., and Helfrich, L. A. (1995). Lifelong learning for agency fisheries professionals: What are the continuing education needs? *Fisheries* 20(7), 10–16.

Royce, W. F. (1972). Undergraduate education of fishery scientists. *Natl. Oceanic Atmos. Adm., Fish. Bull.* 70, 681–691.

Sullivan, C. R. (1981). The history, structure, and function of the American Fisheries Society. *Fisheries* 6(5), 25–29.

Went, A.E.J. (1972). Seventy years agrowing: A history of the International Council for the Exploration of the Sea 1902–1972. *Rapp. P.-v. Reun., Cons. Int. Explor. Mer.* 165, 1–252.

Wickenden, W. E. (1981). The second mile. *Eng. Educ.*, April 1981, pp. 687–689 (reprint).

Wojcieszak, D. B. (1994). Improving communication is the biologists' challenge of the 1990s. *Bioscience* 44(3), 122.

Part II

The Traditional Sciences

Summary

Fishery science includes both long-term inquiry into better knowledge of the living aquatic resources and short-term inquiry into immediate problems resulting from use of the resources.

The traditional sciences that have been used routinely in fishery inquiries include studies of the aquatic environment, identification and biology of the organisms, and studies of the ecological role of the resources.

Other sciences are needed increasingly because fishery problems are related to accelerating global environmental and social problems.

The aquatic environment occupied by the living resources includes the streams, lakes, and oceans to depths of about 500 m.

The biological productivity is sustained by energy from the sun through the organic production cycles in the waters.

The resource organisms are only a very small percentage of the species and biomass of living organisms in the waters. All depend on other organisms.

The communities of organisms that include the resource organisms may be affected in their entirety by fishing or aquaculture.

The populations of resource organisms change in numbers and total weight as they are affected naturally by changing environment, or by humans through fishing or aquacultural controls.

Measuring and predicting the changes are major activities of fishery scientists who are concerned with wild fish stocks and their management.

Measuring and predicting the impact of alternative environmental, reproductive, population, fish food, and fish health controls are major activities of fishery scientists who are concerned with the culture fisheries.

4 | The Aquatic Environment

Water is an alien environment for us, very different from the land and surrounding air with which we are familiar. Aquatic animals and plants respire, feed, grow, escape enemies, and reproduce under conditions that we cannot experience and only vaguely understand. Yet almost every part of the aquatic environment is inhabited by species that perform all of these functions. They exist in wondrous varieties, some in enormous numbers, and others so rare that they are known only from single specimens. Just a few are useful to humans and these occur in abundance in only a small fraction of the waters. These few plants and animals comprise the organic production systems that are of concern to fishery scientists. They must understand the functioning of the systems and the factors that restrict or enhance their production.

The production systems are complicated and incompletely understood. They are part of the concern of the marine and freshwater sciences that are collectively called *oceanography* and *limnology*, respectively. These are relatively new sciences, and they are rapidly adding to our knowledge of conditions and life in the waters. They include many matters of immediate importance to the fishery scientist, such as the productivity of the sea or the population dynamics of fish, as well as matters of remote interest, such as the geology of the seabed or the fascinating biology of the coral reefs.

Oceanographers and limnologists, as well as hydrologists, geologists, geographers, engineers, and other applied scientists, are concerned also with water itself as a major resource. All of the basic physical and chemical studies of the natural waters help applied scientists to understand and plan for the proper use of water for navigation, power, cooling, waste dilution, irrigation, and domestic supply. Similarly, the basic studies help with the solution of problems created when water causes damage through floods, droughts, or beach erosion.

Raising fish is just one of the many uses of water. These uses frequently conflict, but seldom is any one favored entirely, because multiple uses are generally possible and desirable. When fishing is involved, fishery scientists can help by seeing that the water is most effectively used for fish. They can do this best if they have an understanding of water as a resource as well as an environment producing fish.

It is beyond the scope of this volume to include a review of oceanography and limnology, but certain physical, chemical, and biological features of the waters that profoundly affect the production systems will be discussed. These are the features corresponding to the characteristics of the land that determine agricultural production, such as light, oxygen, climate, soil, plant nutrients, land topography, pests, and competitors. Special mention will be made of those aquatic features that are limiting or enhancing.

The summary to follow is necessarily descriptive rather than theoretical or analytical. Moreover, all of the topics include difficult questions of physics, chemistry, and geology that have received intensive research. The treatises listed in the Bibliography at the end of the volume will provide an introduction to the extensive literature in each field.

4.1 HYDROLOGY AND THE WATER CYCLE

Hydrology is broadly defined as the science dealing with the occurrence, circulation, distribution, and properties of the waters of the earth and its atmosphere. It embraces many sciences, including oceanography, limnology, and meteorology, but its usage is becoming restricted to the occurrence and distribution of water over the land. Hydrologists are commonly concerned with the development and use of water supplies on and under the surface of the land.

Water covers about 71% of the earth's surface. More than 98% of this accessible water is in the oceans, and most of the balance is in the accumulation of ice on the land masses. Only about 0.02% occurs as fresh water in the form of atmospheric water vapor, inland waters, and circulating groundwater. This fresh water is estimated to be somewhat less than the amount that evaporates annually from the oceans and the land (Table 4.1).

Despite the relatively small proportion of the earth's water that is transferred each year from the oceans to the atmosphere and the land, this part is extremely important. It forms the hydrologic or water cycle (Fig. 4.1). About 1 m of water evaporates annually from the surface of the oceans (Table 4.2), of which about 91% returns to the ocean in the form of rain and the remaining 9% returns to the land. Of the latter, about one-fifth soon runs off, another one-fifth is tempo-

TABLE 4.1   Area and Volume of Water on Earth[a]

Area of earth's surface	510	$\times$ 10^6 km^2
Area of all land	149	$\times$ 10^6 km^2
Area of all oceans and seas	361	$\times$ 10^6 km^2
Area of Atlantic Ocean	82	$\times$ 10^6 km^2
Area of Pacific Ocean	165	$\times$ 10^6 km^2
Area of Indian Ocean	73	$\times$ 10^6 km^2
Volume of all oceans and seas	1370	$\times$ 10^6 km^3
Volume of polar caps and other land ice	16.7	$\times$ 10^6 km^3
Volume of inland waters	0.025	$\times$ 10^6 km^3
Volume of circulating groundwater	0.25	$\times$ 10^6 km^3
Volume of atmosphere water vapor	0.013	$\times$ 10^6 km^3

[a]From Sverdrup *et al.* (1942) and Hutchinson (1957).

rarily stored in the earth as groundwater, and about three-fifths evaporates from freewater surfaces and from plants.

The average precipitation on land of 67 g/cm^2/year is distributed very unevenly. More than the average falls from the westerly winds of the temperate zones, especially near the mountains of the western sides of the continents. Much less falls from the easterly trade winds of the subtropical areas, although in the doldrums near the equator more than the average falls. Secondary monsoon systems develop in some subtropical parts of the world, where almost all of the precipitation falls in only a part of the year. Even in the wetter temperate

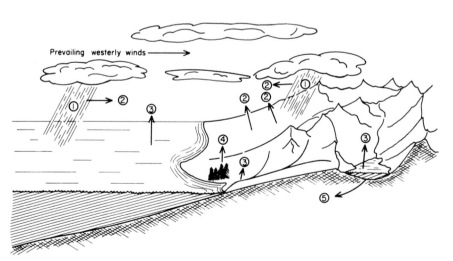

FIGURE 4.1   Diagram of the water cycle. (1) Precipitation as rain or snow. (2) Evaporation from falling particles and from the earth. (3) Evaporation from bodies of water. (4) Transpiration from plants and animals. (5) Drainage underground.

TABLE 4.2 Approximate Water Balance of Oceans and Continents[a]

	g/cm²/year	km³/year × 10³
Evaporation from ocean surfaces	106	383
Precipitation on ocean surfaces	96	347
Evaporation from land surfaces	42	63
Precipitation on land surfaces	67	99

[a]From Hutchinson (1957). These values are not known accurately and other figures will be found.

zones, precipitation is usually greater during the winter, evaporation is greater during the summer, and some precipitation is stored temporarily as snow or ice. Such seasonal variability results in cycles of change in the flow of rivers or the level of lakes; sometimes the changes are great enough to cause seasonal floods or drying of the flow.

The fishery scientist should always recognize that the fish must endure the extreme conditions that occur in rivers and estuaries. The low-water extreme is usually the most troublesome for the fisheries, because all users may require the water at this time.

4.2 STREAMS AND RIVERS

The water running off from the surface forms streams, where it is joined by the groundwater from seeps and springs. The smallest streams seldom flow far before they join another stream, frequently of about the same size, and together they continue to join other streams in a network. The smallest streams are designated as being of *order* one and, when they meet, form a stream of order two and so on to order four or five, which are substantial rivers. Surprisingly, the degree of branching of the several stream orders in the networks tends to be relatively constant among river basins in similar terrain.

A stream network occupies a *catchment area,* which is bounded by the highest land between it and adjoining catchment areas. Usually this area comprises the *drainage area,* although, in a few instances, the boundaries of the subsurface drainage area differ from those of the surface.

4.2.1 Flow of Water

Most of our planned use of a river and its channel depends on knowledge of its *hydrograph,* data on its discharge or water level over time. These data, collected over as many years as possible, provide a basis for estimating the maxima and minima as well as all of the seasonal changes in runoffs and levels. The hydrograph supplies essential information for all water users and designers of all structures in the channel. Most governments publish an annual update of the hydrographs of all streams being measured. One example is the Water Data Reports by the U.S. Geological Survey.

The discharge is usually measured by the velocity–area method. This requires measurement of the velocity of flow at numerous points in a cross section of the river and determination of the mean velocity in the area of the cross section, with adjustment for turbulence caused by rough bottom. This is done at various water levels, which are recorded each time, usually by a continuously recording gauge. After several measurements have been made, a graph of the discharge–level relationship is drawn to define the *rating curve* for that cross section of the river. Thereafter, the discharge is estimated whenever desired from the level and the rating curve.

Most users of discharge data need forecasts of mean flow at various seasons and the probability of high or low flows at any time. Especially important to water users, including fisheries, is the probability of low flows when several kinds of uses will be in competition. Especially important to engineers and communities along the banks is the probability of floods, which structures must withstand. Estimates of maxima and minima to be expected in 20-, 50-, or 100-year periods are commonly made from the statistical distributions of discharge.

The accuracy of such estimates depends on the adequacy of the hydrographs for the various parts of each stream. But long-term hydrographs may not be available or adequate for a particular location, so estimates will be made from precipitation (for which longer-term records are frequently available) and from the catchment area. Satisfactory estimates have proved, however, to be extraordinarily difficult because of the many variables in the relationship of precipitation and runoff. These include variations in rainfall due to local showers, variations in rate of surface runoff due to terrain, transpiration by vegetation (which in forested areas may be two-thirds or more of the precipitation), and variable temporary storage of the water as snow, ice, or liquid in different soils and subsoils. Much of the work of hydrologists is devoted to improving our understanding of the relationships among these many variables.

4.2.2 Flow of Sediments

As water goes through a drainage basin, it dissolves and suspends materials and pushes along heavier materials. There is continuous movement of weathered material from the catchment basin toward and through the stream channels. Many human activities accelerate the movement, and our influence is so widespread that the natural movement is usually obscured.

The dissolved materials, which are estimated to be more than half of the material carried by the world's rivers, include many salts and are especially prevalent in water draining from limestone areas. The finest suspended materials (Table 4.3), which are the size of clay particles, settle out very slowly and are usually called the *wash load*. The larger suspended particles such as silts

TABLE 4.3   Atterberg's Size Classification
of Soil Particles[a]

Grade limits (mm)	Name
2000–200	Blocks
200–20	Cobbles
20–2	Pebbles
2–0.2	Coarse sand
0.2–0.02	Fine sand
0.02–0.002	Silt
<0.002	Clay

[a]Adopted by the International Commission on
Soil Science.

and fine sands are called the *suspended load*. They are common in more rapidly flowing waters such as "muddy" flood waters. The larger particles, the size of coarse sands or greater, that comprise the *bed load,* are rolled along the bottom either as individual particles or as slowly moving dunes. Much larger particles, even the size of boulders, can be rolled down precipitous mountain streams during floods.

The transport of the different materials varies with the flow. The dissolved materials and the wash load are transported in similar concentrations at most flow levels, whereas the suspended and bed loads may increase greatly at high flows. Most of the latter are derived from stream banks and are moved mostly during higher flows. Almost all streams tend to meander, and thus their banks are in an intermittent process of relocation due to cutting by the faster water on the outside of curves in the channel and deposition from the loads of the slower water on the inside.

4.2.3 Examples of River Discharge

Any comparison of river discharge must start with the Amazon, which drains the largest catchment area in the world (Table 4.4). Its average water discharge of about 100,000 m³/s is about 20% of the total discharge of the world's rivers. Yet its total sediment discharge is less than half that of the Hwang Ho (Yellow) River of China, which annually carries nearly 2 mm of the basin's soil from the catchment area to the sea. The latter value, called the *denudation speed,* is only about 0.09 mm/year for the Amazon basin.

The denudation speed is driven by rainfall, but vegetative and soil factors are important. It is greatest from nonforested lands with fine erodible soils. Some of these are the semiarid lands used for agriculture, such as many of those in the Missouri basin. It may be much larger in local areas abused by humans, such as strip-mining areas in the United States, where denudation rates of 7 mm/year, or 1000 times the rate in nearby unmined areas, have been reported.

TABLE 4.4 Water and Sediment Discharges of Selected Rivers[a]

River	Catchment area (10^6 km^2)	Discharge Water 10^3 m^3/s	Discharge Water mm/year[b]	Discharge Sediment 10^6 tons/year	Discharge Sediment 10^{-3} mm/year[b]	Sediment as ppm of discharge
Amazon	7.0	100	450	900	90	290
Mississippi	3.9	18	150	300	55	530
Congo	3.7	44	370	70	15	50
Missouri	1.4	2	50	200	100	3,200
St. Lawrence	1.3	14	340	3	2	7
Hwang Ho (Yellow)	0.77	4	160	1,900	1,750	15,000

[a]From Jansen et al. (1979). By permission of Pitman Books Ltd., London.
[b]Data are for discharge from the entire catchment area. All data are rough approximations for the mouths of the rivers.

At the opposite extreme in sediment discharge are rivers that flow through lakes or reservoirs that stop all of the bed load and most of the suspended sediment load. The St. Lawrence River, which drains the Great Lakes of North America, is an example of minimal sediment transport.

4.3 BOTTOMS AND BASINS

Anyone who has watched a muddy river flood a valley or noticed the changes in a beach during a storm must appreciate the role of water as a transporter of sediments. The waves pound the beach to push some particles higher and carry others back out, and soon a bar appears. In addition, the wind blows dusts, and ice carries rocks and other debris far out over the water. Thus, the bottom underneath the water is covered with combinations of clay, silt, sand, gravel, stones, and boulders. To these are added the skeletons and wastes from plants and animals, and particles blown out of volcanos or formed by direct precipitation of chemicals from the water. The result is an infinite variety of bottom sediments, from the finest clays consisting of particles averaging less than 0.001 mm in diameter to blocks (Table 4.3).

Sediments cover almost all of the bottom, from the shores to the most distant ocean deeps. Bedrock is a rarity found seldom except in the cliffs of mountains or canyons that may mark the shore or be found far beneath the surface. The sediments in lakes and near the ocean shores, called *terrigeneous,* are generally coarse and mostly sand. The inorganic deposits, which make up only a fraction of the sediments of the ocean depths far from shore, are generally fine. Such *pelagic* deposits may be largely of inorganic origin, in which case they are known as red or brown clay; if they contain significant amounts of skeletal materials they are called *oozes.*

The study of the transportation of sedimentary particles and the formation of sediments is largely in the field of geology. Geologists are discovering many clues to the history of the earth beneath the water, partially because most of the rock now exposed at the surface of the earth was once a sedimentary deposit at the bottom of the sea. Deposits containing large amounts of organic matter may have produced petroleum.

The geologist's findings about the existing sediments are of interest to fishery scientists because a large fraction of the useful aquatic plants and animals live close to the bottom. Here they may attach, burrow, rest, hide, place their eggs, or feed on bottom-living plants and animals. Each type of bottom will favor different plants and animals according to their particular adaptation. A large part of the world's fisheries depend on species collectively called the *bottom fishes*—fishes that are caught within about 5 m of the bottom.

4.3.1 Lakes and Reservoirs

Water accumulates in a great variety of basins that have been formed by tectonic, volcanic, glacial, fluviatile, and other natural processes. In addition, people construct reservoirs to store water where and when they need it. Most lakes are parts of river systems, but some are in closed basins from which water escapes only by underground flow or by evaporation. The largest body of water in a closed basin, the Caspian Sea, has an area of 436,000 km². The largest lake in an open basin, Lake Superior, is 83,000 km². The deepest known, Lake Baikal in Siberia, is 1741 m.

The form of the basins is shown on *bathymetric* charts, which have contours (isobaths) of equal depths. From such charts are obtained the essential descriptive measurements or data for computation of length (l), breadth (b_x), maximum depth (z_m), area (A), volume (V), mean depth ($\bar{z}$ or V/A), and the length of shoreline (L).

From these data are calculated two measurements that are useful in a comparison of different lakes with respect to their productivity. One of these is the index of shoreline development (D_L), a ratio of the length of shoreline to the circumference of a circle with an area equal to that of the lake [$D_L = L/(2\sqrt{\pi A})$]. A D_L of 1 is a perfect circle; a larger index indicates the departure from a circle. The other is the index to the shape of the basin, a ratio of the mean depth to maximum depth ($\bar{z}/z_m$). This index has a value of 1.0 in a cylinder and 0.33 in a perfect cone.

Lakes have been classified according to the form of their basins in too many ways to repeat here. One classification in wide use makes a division into two types on the basis of the relative amount of life, a division closely related to the kind of basin. *Oligotrophic* lakes have scarce populations of plants and animals. They are usually deep, cold, and clear. Sparse sediments occur on the bottom, and oxygen is present at all times in deep water. They tend to have a low D_L and a large $\bar{z}/z_m$. *Eutrophic* lakes have abundant populations of plants and animals and organically rich bottom materials. Usually they are relatively shallow, warm, and turbid, with low oxygen at times in the deeper water layers. They tend to have a high index of shoreline development and a small ratio of mean depth to maximum depth.

4.3.2 Estuaries and Lagoons

Near the shores of the oceans are many bodies of water either continuously, intermittently, or not connected with the ocean. They have been classified into two major groups, not on the basis of the type of basin, which may be highly diverse, but on the basis of the circulation of their water. *Estuaries* are where fresh water and salt water meet and mix. The most generally accepted definition is a semienclosed body of water that has a free connection with the open sea and within which seawater is measurably diluted with fresh water derived from land drainage. Excepted from this definition are the large semienclosed seas, such as the Baltic. *Lagoons* are coastal bodies of either fresh water or salt water that may have an intermittent connection with the ocean but that usually have a stable salinity and little or no tidal exchange. They may be much saltier than the ocean if water flows occasionally in from the sea and evaporates. They may be fresh if elevated slightly above the ocean. Lagoons are usually small and shallow and can be considered another variety of lake.

Estuaries vary as environments according to the mixing process, which is determined largely by the tides, by the inflow of fresh water, and by the shape of the basin. The extremes are the shallow estuary that has an equal salinity from top to bottom and the deep estuary into which both river and sea flow and out of which flows a mixture of fresh water and salt water on the surface (see Section 4.7.5). The classification of estuaries has therefore been developed with consideration of both geologic origin and the mixing processes, which occur only partly as a consequence of the shape of the basin. No standard classification exists, but the following are the major types of estuaries in approximate order from the shallowest to the deepest.

1. Deltaic river mouths, such as the Mississippi, Amazon, and Niger. In these the river flows almost directly into the sea and there is almost no mixing zone of fresh water and salt water. Salt water occurs only as a wedge along the bottom.
2. Bar-built estuaries, such as the inland seas of the Netherlands. In these large, shallow seas the tidal turbulence and winds mix the water almost uniformly from top to bottom in most parts.
3. Drowned river mouths, such as Chesapeake Bay. In these the fresh water and salt water are mixed considerably, but not completely, from top to bottom.
4. Tectonic estuaries, caused by faulting and submergence of the land, such as San Francisco Bay. These are usually deep. Mixing of fresh water and salt water occurs in the surface layer only.
5. Fjords caused primarily by glacial action such as those in Norway, Chile, and Alaska. These are usually very deep. Mixing of the water occurs only in the surface layer.

4.3.3 Oceans

The immensity of the ocean basins is not entirely indicated by the proportion of the earth's surface in the oceans (71%), because much of the oceans is very deep. The average depth is 3800 m, about 4.5 times the average elevation of the land (840 m) above sea level. Great depths occur in the *trenches*, several of which are more than 10,000 m deep. The deepest point now known is in the Mariana Trench—11,033 m. Such depths are substantially greater than the altitude of the highest mountain—Mt. Everest is 8848 m high.

Charting the ocean depths has been undertaken for many centuries along the coasts where hazards to navigation exist. Most maritime nations have published charts of their coastal waters. Usually these charts show *isobaths*, lines of equal depth, in areas where the soundings are numerous, but beyond a short distance from the coast the soundings themselves are shown frequently because few reliable soundings have been obtained in the depths of the oceans.

Fortunately, much more detailed charts have become available since the development of echo sounders about four decades ago. The deep-sea charts now show *ridges, rises, trenches, seamounts, canyons, basins,* and other features. Much more remains, however, to be discovered and charted.

The geology of the deep-sea floor is a vital part of the knowledge of the earth's crust, so it is surprising that most of the knowledge of the sea bottom, its sediments, and its topography has come during the last four decades—indeed, largely since the end of World War II. It remains one of the least-known features of the earth, and its exploration is considered as challenging to many people as the exploration of outer space.

Fortunately for fisheries, knowledge of the area above a depth of 500 m, which is about the maximum that can be reached by ordinary commercial fishing gear, is much more precise. Most of it has been thoroughly charted, not only with respect to the depths and hazards to navigation, but also with respect to the fishing banks, bottom types, and hazards to fishing gear.

The predominant feature of the coastal zone, the *continental shelf,* is of utmost importance to the fisheries. It is the place where most of the world fishing occurs, either with gear that catches the bottom fishes or with gear that catches the pelagic, schooling fishes in the water above the shelf.

The continental shelf is indeed a shelf—a nearly flat plain extending out to a margin beyond which the bottom slopes steeply and is furrowed by canyons (Fig. 4.2). This zone is called the *continental slope.* The shelf margin occurs at an average depth of about 130 m, but most charts show an isobath at 100 fathoms (fm), or 200 m, which for practical purposes marks the margin of the continental shelves of the world. The depth of 200 m has been designated in some international treaties as the margin of the continental shelf.

The exact depth at the margin is not especially important to fisheries, because seaward from the actual margin the continental slope is usually very steep, rough, and difficult to fish. Minor *terraces* of flat bottom occur in a few places on the continental slope, but these are seldom important to the fisheries. The proportion of the ocean floor less than 200 m deep, most of which is continental shelf, is estimated at 7.6%. The proportion between 200 m and 1000 m is only 4.3%. When the latter figure is interpolated, it may be estimated that the area between 200 and 500 m is only 1.6% of the ocean floor. The other 90.8% of the ocean floor is beyond the reach of most commercial fishing gear.

The continental shelves of the world are widest off the coastal plains of the land and narrowest off the land with mountain ranges. Their widths range from near 0 to 1300 km, with an average of about 50 km. The shelf slopes an average of only about 0.2%, a slope so small that the shelf would appear level to the human eye. On the other hand, the average slope of the continental slope varies from about 3.5% off the flat coastal plains to about 6% off the mountainous coasts.

The major continental shelves are off northwest Europe, off Siberia in the

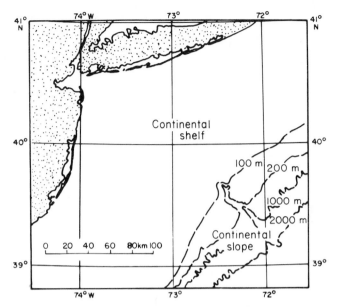

FIGURE 4.2 Bottom topography of part of the continental shelf and slope off the eastern United States.

Arctic Ocean, off Alaska in the eastern Bering Sea, in the several seas of eastern Asia, off South America east of Argentina, and off the North American coast between Cape Hatteras and Newfoundland (Fig. 4.3). All except the Siberian shelf are major fishing grounds.

On the continental shelves occur *banks*, which are elevations frequently important to fishermen but which are not hazards to navigation, *shoals* and *bars*, which are nonrocky hazards to navigation, and *reefs*, which are coral or rocky hazards to navigation.

4.4 PHYSICAL PROPERTIES OF WATER

The fishery scientist needs to understand some of the physical properties of water that affect its function as an environment for plants and animals. Some of the properties may influence the behavior of the organism, for example, temperature and light transmission; others may be useful as a means for fishermen to find animals, for example, sound transmission; still others may be useful in understanding the circulation of the water.

4.4.1 Temperature

The temperature of the water influences profoundly the lives of all aquatic plants and animals. Plants and most of the animals assume the same or almost

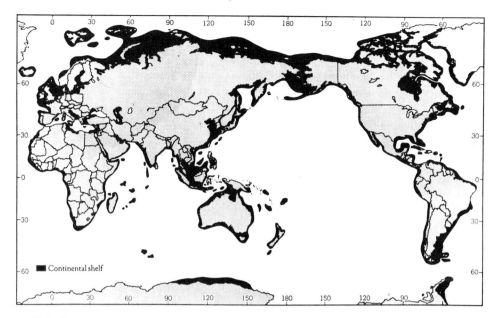

FIGURE 4.3 The continental shelves of the world. (From FAO Doc. 57/7/4725. "A Contribution to the UN Scientific Committee on the Effects of Atomic Radiations on the Specific Questions Concerned with the Oceanography and Marine Biology in Respect to the Disposal of Radioactive Wastes." FAO, Rome.)

the same temperature as the water. Such animals are called *cold-blooded,* or *poikilothermous.* Each plant or animal is adapted to a normal seasonal temperature regime and is commonly affected adversely by unusual temperatures. Many animals reproduce, feed, or migrate only within certain temperature limits, which may be narrow or broad according to the species (Table 4.5 and Fig. 4.4).

TABLE 4.5   Approximate Range of Temperatures Tolerated by Some Resource Animals

Yellowfin tuna	18°–32°C
Cod	1°–12°C
Pompano	15°–35°C
Rainbow trout	0°–25°C
American oyster	1°–36°C
(on exposed tide flats)	−49°C
Pacific salmon	
Fresh water	0°–27°C
Ocean	0°–20°C

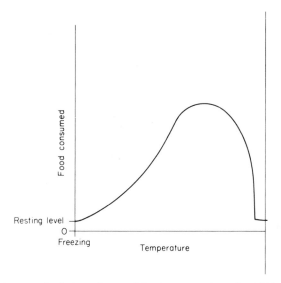

FIGURE 4.4 Diagram of relation of rate of food consumption of cold-blooded animals to temperature. [After N. S. Baldwin (1957). Food consumption and growth of brook trout at different temperatures. *Trans. Am. Fish. Soc.* **86**, 323–328.]

Temperature also controls density of water in ways that determine the entire temperature structure of all waters. It changes the solubility and physiological effects of solids and gases so that their effect on animals must be considered together with temperature. For a fishery scientist temperature is commonly the most important item of physical information about a body of water.

The earth and its atmosphere as a whole are warmed by radiation from the sun and cooled by an equal amount of radiation back to space; thus, on the average the temperature remains the same. The warming and cooling by radiation is not uniform in different parts of the earth's surface; the low latitudes are warmed more than they are cooled and the higher latitudes are cooled more than they are warmed. Both ocean and land transfer heat to the atmosphere. Wind and ocean currents move the heat around, generally from the tropics toward the poles. The balance of the heat received, radiated back, and transported latitudinally also remains about the same; thus, the mean annual temperatures of both air and water at each point on the earth remain about the same.

The waters are heated almost entirely by radiation from the sun. The radiation is almost all absorbed, and heat is lost continuously by back radiation, evaporation, and conduction. The waters are warmed by day and cooled at night, gradually warming more during the spring and summer and cooling during the autumn and winter. In temperate regions the waters are usually warmest in late summer, coldest in late winter.

The radiation from the sun warms the waters only in a thin surface layer.

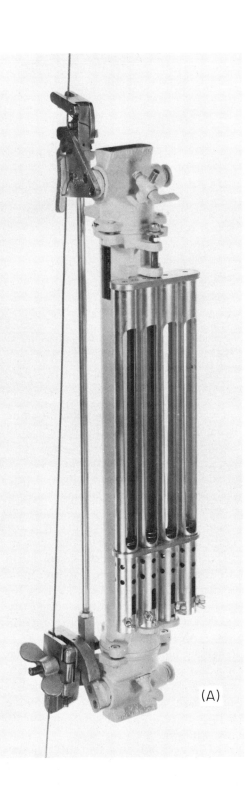

(A)

(B)

More than 90% of the heat is absorbed in the upper 20 m of the clearest ocean waters and in the upper 4 m of the average coastal waters. All waters below are warmed by mixing with the surface layers.

Temperatures are measured with a variety of instruments, which are chosen according to the depth, accuracy, and frequency of measurements desired. An occasional surface temperature is measured easily by a conventional mercury thermometer, but if a continuous record is desired, an electrical resistance thermometer may be installed with a recorder. At depths below the surface that can be reached with instruments on a wire, special reversing thermometers are used (Fig. 4.5). These are designed to indicate the temperature in a conventional manner when in one position but retain the same reading indefinitely after being inverted. They are lowered in the recording position and inverted by a "messenger" dropped down the wire. These thermometers are made to measure temperatures within ±0.02°C, an accuracy needed for subsequent computation of density (see Section 4.4.3). In shallow water (<100 m), electrical resistance thermometers are also used when accuracy need be no closer than ±0.5°C.

One of the most convenient instruments for repetitive measurements of temperature in the upper 500 m of water is the *bathythermograph* (Fig. 4.6). This is a rugged instrument that may be lowered to depth from a moving vessel. It contains a pressure sensor and a temperature sensor, which record a graph of temperature against depth on a glass slide.

The maximum surface temperatures in fresh water and salt water (excluding hot springs) occur in shallow sheltered waters, where 40°C may be attained. Temperatures at the surface of the open sea and of large lakes rarely exceed 30°C. The minima are the freezing temperatures of 0°C in fresh water and about −1.9°C in seawater. The seasonal variation is related to the size of the waters and is greater close to land. There, the water may reach 40°C in the heat of a summer day and may freeze in winter. Open ocean water commonly shows a change in temperature of less than 4°C in the tropics and the polar regions and less than 10°C in the intervening temperate zones (Fig. 4.7).

The surface temperatures prevail only in the upper layer, which is mixed by the wind. At the bottom of the mixed layer is a *thermocline*, a layer in which temperature changes rapidly with depth (see Section 4.4.3). Below the thermocline (except in tropical lakes) temperatures range from about 4°C in lakes down to about 1°C in the great depths of the oceans. Thus, ocean and lake waters below the thermocline are almost constantly and nearly uniformly cold.

FIGURE 4.5 A Nansen Bottle (A) and a reversing thermometer (B). The bottle is lowered with the valves open. When it is at the desired depth, a weight is dropped down the wire to strike the lever at the top and allow the bottle to topple over, closing the valves and inverting the thermometers. Each thermometer contains two mercury columns, one of which breaks when inverted to record the temperature and the other of which is used to correct the reading for the ambient temperature when read at the surface. (Courtesy of G. M. Mfg. and Instrument Corp., New York.)

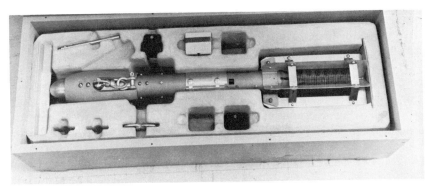

FIGURE 4.6 A Spilhaus bathythermograph in its storage case. The instrument contains a pressure element and a temperature element. It produces a graph of temperature versus depth to a maximum depth of 300 m. (Courtesy of G. M. Mfg. and Instrument Corp., New York.)

4.4.2 Pressure and Compressibility

The pressure at any depth in the water is equal to the weight of the atmosphere and the weight of the water column. For rough calculations the latter may be estimated as the equivalent of 1 atm for each 10-m increase in depth. One atmosphere is equal to the weight of 760 mm Hg, about 14.7 lb/in. or 1.013 kg/cm^2.

Water is only slightly compressible; for biological purposes it may be considered as not changeable in volume at all. Much more important biologically is the differential compressibility of water and gases. The latter compress proportionately to pressure. Any gas in the lungs or swim bladder of an animal swimming from the surface to a depth of only 10 m will compress to half of its volume at the surface. It will compress to only one-tenth of the volume at the surface when the animal goes to 90 m. When gases do not occur in animals and plants they are little affected by changes in pressure of 10 or 20 atm but may be seriously affected by changes of 200 atm. Such effects probably arise from the slightly different compressibility of water and skeletal parts.

4.4.3 Density

The density of water is controlled by temperature, dissolved solids, and pressure. It is usually defined in relation to the density of pure water at its temperature of maximum density, which weighs 1 g/cm^3. The relative density of water at different temperatures, salinities, and pressures is shown in Tables 4.6 and 4.7.

The density of seawater is generally between 1.01500 and 1.03000 g/cm^3, but oceanographers rarely report it in this way. Instead they use a convention of subtracting 1.00000 and multiplying by 1000 and designate the value as σ. Density can be measured directly by simple hydrometers when high accuracy is not required, as in estuaries, where σ may range from 0 to 25. Such determina-

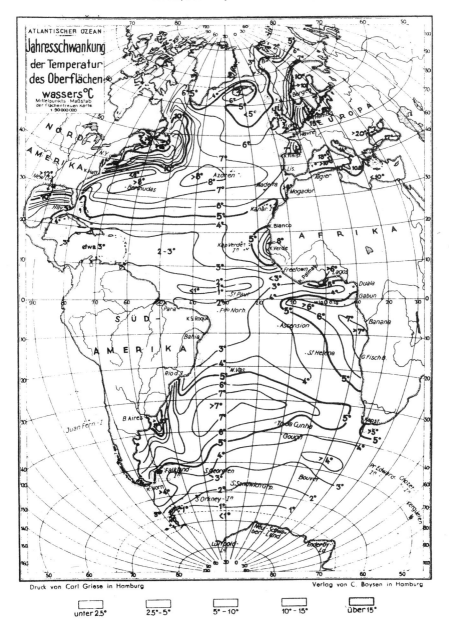

FIGURE 4.7 Annual variations of the sea temperature in the Atlantic Ocean. [From G. Schott (1942). "Geographie des Atlantischen Ozeans." Boysen, Hamburg.]

TABLE 4.6   Density of Water as a Function of Temperature and Salinity[a]

Salinity (‰)	Temperature (°C)				
	0	5	10	20	30
0	0.9999[b]	1.0000	0.9997	0.9992	0.9957
15	1.0120	1.0119	1.0114	1.0096	1.0069
25	1.0201	1.0198	1.0192	1.0172	1.0143
35	1.0281	1.0277	1.0270	1.0248	1.0218

[a]From M. Knudsen, ed. (1901). "Hydrographical Tables." G.E.C. Gad, Copenhagen.
[b]Density of ice at 0°C and 0‰ salinity is 0.9168.

tions can be reliable to about $\pm 0.2\sigma$, but this accuracy is not enough for work in lakes or oceans. In such cases density is usually calculated from temperature and salinity, both of which can be measured very precisely (Fig. 4.8).

Changes in density are so slight that they have little if any direct biological significance that can be separated from the individual effects of temperature, salinity, and pressure, but they are of great significance for the indirect estimation of water currents (see Section 4.7.3).

The vertical changes in temperature and the associated changes in density are, however, of great biological significance. Water near the surface is most dense at a temperature of about 4°C when fresh or between 4 and -2°C when salt. The water of lowest density is found at the surface. When density increases with depth are large, the water column is resistant to mixing by wind and current; when density differences are very small, the water column can be mixed to great depths. In consequence the seasonal changes in surface temperatures are accompanied by changes in the temperature structure of the water column.

Consider a lake in a temperate region in early spring when the entire water column is near the temperature of maximum density (4°C). A condition of

**TABLE 4.7   Density of Pure Water
as a Function of Depth at 0°C**[a]

Depth (m)	Density (g/cm³)
0	0.9999
250	1.0009
500	1.0020
1000	1.0042
2000	1.0084

[a]For more complete tables, see M. Knudsen, ed. (1901). "Hydrographical Tables." G.E.C. Gad, Copenhagen.

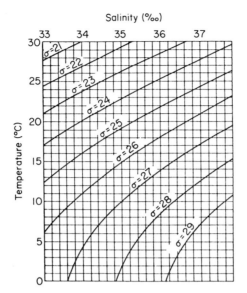

FIGURE 4.8   Temperature–salinity–density relationship.

instability prevails when the entire lake can be stirred by the wind (Fig. 4.9). As the surface water warms with the season, the mixing cannot distribute the warmed water evenly, and because the warmer water is less dense than the cooler water it remains on the surface. This layer will be mixed by wave action to a depth of only a few meters. Below this mixed layer will be a layer with a large temperature gradient, or the thermocline. Below the thermocline the water will remain at the temperature it had when last overturned, near 4°C, the temperature of maximum density in deep temperate-zone lakes. In the ocean the process and temperature structures are similar (Fig. 4.10), but the density is modified also by the salt content, the mixing processes are more complicated, and the temperature at maximum density is below 0°C in typical seawater.

The thermocline disappears seasonally in most freshwater lakes. When the surface waters cool in autumn, the density increases are again unstable. If the annual cycle of temperature change goes below and above the temperature of greatest density, two periods of mixing occur. In arctic lakes, where the temperature increases merely to 4°C or less, only one period of mixing will occur. Likewise, in tropical lakes only one period of mixing will occur, at the time of greatest cooling: the temperature below the thermocline may be much above 4°C. In the case of the oceans the thermocline is usually permanent in tropical seas, seasonal in temperate waters, and absent or poorly developed in arctic waters. In a few lakes and isolated ocean basins of extreme salinity unusual temperature conditions prevail. The discussion of these is beyond the scope of this volume.

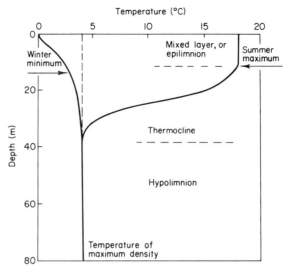

FIGURE 4.9 Diagram of water temperature in a lake at different seasons. As the surface temperature warms from the winter low, the water column becomes unstable at 4°C and then stable again as it warms further. In autumn, the reverse occurs.

The thermocline is a barrier, a zone of stable water through which mixing of the waters does not occur. It is a zone of large temperature change, which must be accommodated by animals swimming through it. In the ocean it may be at a depth of 1000 m or as shallow as 10 m; in lakes it is usually found between 5 and 50 m. In places sheltered from the wind its depths may be constant; in

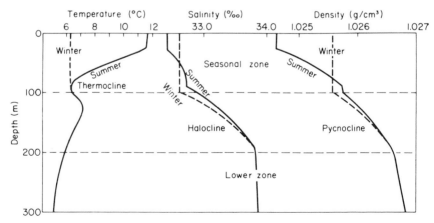

FIGURE 4.10 Diagram of temperature, salinity, and density structure of the subarctic Pacific Ocean off British Columbia. [From J. P. Tully (1965). Time series in oceanography. *Trans. Roy. Soc. Can.* 3(4), 366.]

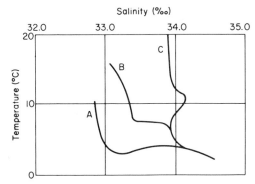

FIGURE 4.11 Temperature–salinity relationship of water masses of northeast Pacific Ocean. (A) Subarctic water in gyre of central Gulf of Alaska. (B) West wind drift near 45°N lat. (C) Subtropical water near 40°N lat. [Adapted from S. Tabata (1965). Variability of oceanographic conditions at Ocean Station "P" in the northeast Pacific Ocean. *Trans. Roy. Soc. Can.* 3(4), 367–418.]

other places its depth may change rapidly as winds and currents pile up or disperse the mixed surface layer.

Because density depends on both salinity and temperature, different combinations of salinity and temperature can produce the same density. Consequently the salinity–temperature–depth relationship is used directly to identify water masses. When temperature and salinity of a column of water are plotted, the shape of the curve is frequently characteristic of water from a particular region (Fig. 4.11).

4.4.4 Sound

Poets who have written emotionally about the quiet waters have been unaware that there are numerous sounds in water. The natural sounds of waves, cracking ice, and animals are transmitted in water much more efficiently than sounds in air. They do not pass efficiently through the air–water surface, so we seldom hear the sounds from below.

Sounds are produced by a large proportion of the aquatic mammals, fish, and shrimp. They produce sounds either by forcing air from lungs or air bladders or over vibrating membranes, by grinding teeth, by rubbing fins, or by snapping together parts of their skeletons. They also produce mechanical sounds incidentally when swimming, burrowing, or feeding, the last especially when feeding on crustaceans or mollusks. Their sounds have been described as thumps, grunts, growls, groans, knocks, thuds, clucks, boops, barks, whines, grates, scrapes, clicks, chirps, and snaps. The sounds may occur in a great chorus, as from a school of snapping shrimp. Most of the sounds produced by animals are of low frequency varying from about 10 to 1000 cycles/s (Hz), but some produced by mammals are of frequencies ranging up to 80,000 Hz, far above the limit of human hearing (about 10,000 Hz).

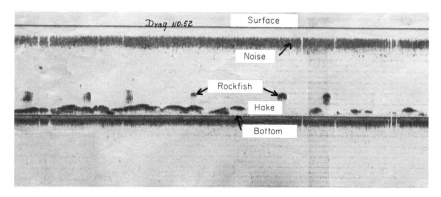

FIGURE 4.12   An echo-sounder record of schools of hake and rockfish in 150 m of water in Puget Sound. (Record by W. Pereyra, U.S. Natl. Marine Fisheries Service.)

Sound is important to the fisherman who uses either echo-sounding or echo-ranging equipment for either navigation or fish location. (*Echo sounding* is used here to designate the practice of directing a sound beam vertically down to the bottom and receiving an echo; *echo ranging* refers to the practice of directing a beam horizontally or at a substantial angle from the vertical.) Echo-sounding equipment is considered essential for all high-seas vessels as a navigational aid and as a help in locating fish near the bottom (Fig. 4.12). Echo-ranging equipment is used extensively by vessels for the location of fish in the depths, midway between surface and bottom. Some wholly new fishing techniques have developed around the use of these sonic devices.

Sound in water travels between 1400 and 1550 m/s. It travels slowest at the surface of fresh water at 0°C and increases in speed with increasing temperature, salinity, and pressure. The speed near the surface is about 4.0–4.6 times as fast as the speed in air.

Sound is reflected when it strikes the surface of material of different properties. It is reflected strongly by an air–water surface and by a rocky bottom, less strongly by a mud bottom and by the skeletons of animals, and least by the water-filled, fleshy bodies of plants or animals. It is reflected strongly by the air bladders of fish. Modern instruments will produce echo records that make it possible to distinguish between different types of bottom and various kinds of fish.

Sound not only is reflected by the surface and the bottom but also is refracted (bent) by waters transmitting sound at different speeds, especially in the region of the thermocline. Consequently, echo ranging at angles other than vertical frequently requires interpretation of multiple echoes that may be caused by reflection, refraction, or both. Such difficulties severely restrict the usefulness of near-horizontal echo ranging for fish location to 1000 m or less in many waters.

4.4.5 Electricity

Water conducts electricity with far less resistance than air. *Specific resistance* is that offered by a 1-cm cube with the current perpendicular to two parallel faces. The specific resistance, expressed in ohms, is the reciprocal of the *specific conductance,* a measure that is convenient to use at times.

The specific resistance of water varies according to salinity and temperature. The least saline natural waters have a specific resistance of about 50,000 ohms; average fresh water, about 6000 ohms; and hard waters, about 3000 ohms. Average seawater of 35‰ (per mille) salinity has a resistance of only about 18 ohms at 25°C and 33 ohms at 0°C.

A conductor such as seawater moving through the earth's magnetic field produces a gradient of electric potential that is strong enough to measure with field instruments. The electric potential is roughly proportional to the velocity of the water current, being altered somewhat by the conductance of motionless water layers or the bottom and by the imposition of earth currents. The natural electrical potential near the surface of the ocean varies from less than 0.05 μV/cm to about 0.5 μV/cm. The lesser value is close to the minimum practical limit of field instrumentation, so the (presumably) smaller potentials in fresh water have not been measured similarly (see Section 4.7.3).

4.4.6 Light

In addition to warming the surface of the waters, the light of the sun provides the energy for photosynthesis of organic material by plants and enables all of the animals that have eyes to see. It varies in daily and seasonal cycles, and these cycles regulate the metabolism and behavior of most animals and plants. The direction of the sun may be used by some animals to guide their migrations. The sun has a vital influence on all life; indeed it sustains life itself.

Natural waters reflect, scatter, and absorb light so greatly that only a thin surface layer is ever illuminated. In the clearest ocean water, only 1% of the light remains at 150 m; in average coastal water, at 10 m; in really turbid water, at only 1 m (Fig. 4.13). Researchers who descend in vehicles can barely see light from the sun below about 700 m in the clearest ocean water and below 80 m in average coastal water. Some fish may have more sensitive eyes than humans, but we have no way of knowing exactly what they can see.

Transmission of the different colors depends on the dissolved pigments, on optical properties of pure water, and on suspended materials in the water. In the clearest lake and ocean water blue light is transmitted best, hence the deep blue appearance of many large lakes and seas. The transmission of blue light is affected more than other colors as the amounts of dissolved pigments and suspended material (including one-celled plants) increases. Coastal water appears green, and turbid water appears yellow-brown. All waters are relatively opaque to red light.

Transparency of surface waters is commonly measured by a Secchi (pro-

Percentage of incident light

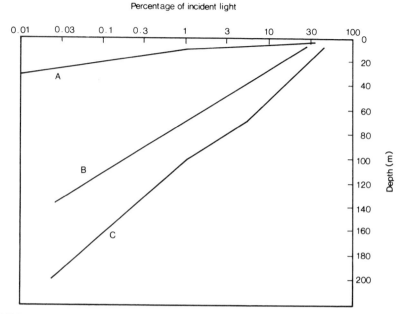

FIGURE 4.13 Light penetration in some ocean waters. (A) Average of some coastal waters. (B) Caribbean Sea. (C) Sargasso Sea, one of the clearest of ocean waters.

nounced sek-key) disk 20 or 30 cm in diameter (Fig. 4.14). It is lowered on a graduated line until it just disappears and then is raised until it just appears; the average of the two depths is called the Secchi disk transparency. This measure of transparency ranges from about 60 m in the clearest ocean water to a few centimeters in the turbid water of river mouths. It is equivalent to the level of penetration of between about 10 and 15% of solar radiation. More precise

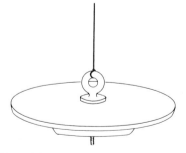

FIGURE 4.14 A Secchi disk used for estimation of the transparency of the surface water layer.

observations and measurements of waters other than the surface layer require optical equipment, but such measurements are rarely made during fishery studies.

The light of the sun is not the only light in the water; many animals and plants produce their own. They are the luminescent forms that are abundant in the sea but rare on land and in fresh water. Many bacteria, phytoplankton, jellyfishes, worms, crustaceans, mollusks, and the fish produce light. They can be seen at night in most warm seas when the water is disturbed by either a wave, a propeller, or an oar. Many of them live in a twilight zone, moving up at night and down at daylight. Many of the fish that produce light also have large eyes, which may enable them to see at unusually low levels of illumination.

The ecological significance of the bioluminescence is not well understood. It is speculated that light flashes may confuse predators, attract prey, or serve to communicate with others, but we really do not know.

4.5 DISSOLVED MATERIALS

All aquatic organisms maintain a constant exchange of dissolved materials between their bodies and the surrounding water. Some materials such as dissolved oxygen and carbon dioxide are indispensable for life itself. Other materials such as the nutrient materials determine the productivity of the water, and the salts of the sea influence greatly the density of the water and pose a condition to which all organisms must adapt.

4.5.1 Oxygen

Next to temperature, the dissolved oxygen content of the water is probably the most common measurement of biological significance. Oxygen is essential to almost all life, so its shortage or absence commonly limits the distribution of plants and animals. Some fish in warm water may be distressed at concentrations less than 5 mg/liter; others at lower temperatures may tolerate less than 1 mg/liter. Unlike the oxygen of the atmosphere, which occurs in a nearly constant proportion of 20.9%, the oxygen dissolved in water varies from none to twice or more the saturation value of about 0.7% by volume or 0.0001% by weight, or as it is commonly expressed, 10 mg/liter.

The amount of oxygen that dissolves in water from the air depends on barometric pressure, temperature, salinity, and proportion of oxygen in the air. The pressure at the surface of the water near sea level is usually close to 1 atm, and the proportion of oxygen varies but slightly depending mostly on the amount of water vapor in the air, so these factors have only a slight influence. Temperature and salinity, however, cause large changes in the saturation levels (Fig. 4.15).

The quantity of dissolved oxygen in water has been determined traditionally

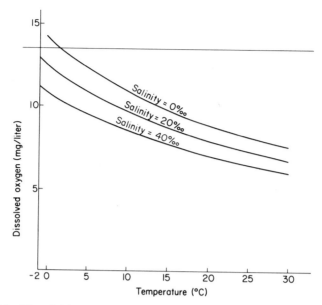

FIGURE 4.15   The solubility of oxygen in equilibrium with water-saturated air at a pressure of 760 mm Hg. [From G. A. Gruesdale, A. L. Downing, and G. F. Lowden (1955). The solubility of oxygen in pure water and sea water. *J. Appl. Chem.* **5**, 53–63.]

by the Winkler titrimetric method, but recently electrical methods for continuous measurement have been developed. Descriptions and instructions will be found in standard reference works.

Varied and sometimes confusing methods of reporting the quantity of oxygen are in common use, including proportion by weight, proportion by volume, and percent saturation (Table 4.8). The proportion by weight is usually used for fresh water and is expressed as parts per million (ppm) or milligrams per liter, with the liter at 4°C, the temperature at greatest density. Oxygen in seawater will be found expressed as milligram atoms per liter (mg-atoms/liter), or as milliliters per liter, with the liter at 20°C. In addition, the percent saturation is used for both seawater and fresh water because temperature and salinity have large effects on solubility. Percentage saturation is useful because it suggests the

TABLE 4.8   Approximate Multiplication Factors to Convert Units of Oxygen Concentration

	To:	ppm, mg/liter	ml/liter	mg-atoms/liter
From: ppm, mg/liter		1.0	0.70	0.0625
ml/liter		1.43	1.0	0.089
mg-atoms/liter		16.0	11.2	1.0

recent biological history of the water. It is computed as 100 times the quantity observed divided by the estimated saturation value for the temperature and salinity of the sample *in situ* but at surface pressure. At sea level the standard atmospheric pressure of 760 mm Hg is used, but for inland waters it is corrected for altitude. The saturation at depth is not reported with consideration of the increased solubility due to water pressure.

Most oxygen in natural waters comes from the air by solution at the water surface; some comes from photosynthesis in the illuminated layer. Oxygen is removed primarily by respiration of plants, animals, and bacteria at all depths; secondarily by transfer from supersaturated surface waters back to the atmosphere; and thirdly by chemical reactions.

It would seem reasonable to assume, as the early scientists did, that the depths of oceans and lakes are without oxygen and life, but such is not the case. Both have been found at the greatest depths that have been sounded, but oxygen is absent in a few places and is scarce enough in many places to limit the habitation by animals.

Oxygen occurs in the deep waters only because these waters were once near the surface. In most lakes all of the waters are near the surface during the seasonal overturn. In the sea oxygen enters the water in polar regions, from which it is carried by the cold water as it sinks to great depths and flows slowly toward the equator. Because of the low temperatures and the scarce life in the depths the oxygen is used very slowly indeed. Water from the depths of the Atlantic Ocean is estimated to have been at the surface about 500 years ago. The deep water of the Pacific is about twice as old.

An outstanding feature of the distribution of oxygen in the sea is the prevalence of a layer with low oxygen content at intermediate depths—usually between 400 and 1200 m. The oxygen concentration is commonly either 5 mg/liter or higher, but over large parts of the Indian and Pacific oceans it is below 1 mg/liter, a level that limits markedly the kinds of animals that can inhabit such waters. Although the layer with low oxygen content (called the oxygen minimum layer) is usually below the level of fishing, in regions of upwelling it may be within 100 m of the surface or even appear at the surface. A few stagnant basins, of which the Black Sea is a major example, contain deeper water that does not recirculate to the surface. Such water may contain hydrogen sulfide and no oxygen. Oxygen is usually present in abundance in all open ocean waters that can be reached by commercial fishing gear.

Severe restriction of the distribution of animals by shortage of oxygen is much more common in fresh water or in estuaries because the decomposable organic materials are much more concentrated either naturally or because of pollution. The deeper parts of such waters may be devoid of oxygen regularly during the summer. Turbulent streams will normally contain water saturated with oxygen, but in their sluggish lower reaches the oxygen may be used almost entirely for the respiration of plants each night and too little may be left for the fish. Critical shortage of oxygen also develops occasionally when exchange with the atmosphere is cut off by ice—a time when photosynthesis is minimal. The rate of respiration by plants and animals is also reduced by the cold but may remain high enough for all of the oxygen under the ice to be consumed, a situation that is called "winterkill."

The surface of layer of quiet water frequently contains more oxygen than the saturation level. Excess oxygen commonly persists as the waters warm during the day but is used in respiration or lost to the atmosphere during the night. Excess oxygen in air may be harmful to land animals, but excess oxygen in the water at concentrations that occur naturally has never been found to be harmful.

4.5.2 Nitrogen

Fixed nitrogen in the form of nitrates, nitrites, or ammonia is essential for life of all kinds (see Section 4.5.5), but only a few nitrogen-fixing bacteria and plants can use the molecular form N_2. It dissolves natural waters at atmospheric pressure in amounts about 1.7–1.8 times the saturation values of oxygen. The amounts utilized by the nitrogen-fixing organisms are such a tiny fraction of the amount available that the reduction in concentration is rarely detectable in ordinary analyses. In fact, the presence of N_2 at all depths of lakes and seas in amounts near saturation levels has been used as an argument that the deep waters were once in contact with the atmosphere.

However, problems with N_2 in water arise when too much is present. Supersaturation commonly causes trouble for aquatic organisms because it leaves solution inside the body to form bubbles that remain in tissues for a long time. When human divers have this trouble it is called the "bends." Supersaturation of natural waters by N_2 occurs in situations that are of special concern to the fisheries scientist. First, emerging groundwater is occasionally supersaturated and unusable without deaeration for rearing of fish. Second, water falling from natural falls or dams carries air below the surface of the plunge basin, where it dissolves under the added pressure of the water above. The extra N_2 may cause trouble for fish downstream from the plunge basin. Third, water warmed either naturally or artificially may become supersaturated and harmful to fish.

4.5.3 Carbon Dioxide and pH

CO_2 is present in natural waters in highly variable amounts because it reacts with the water and other dissolved materials to form either carbonic acid, H_2CO_3; bicarbonate ions, HCO_3^-; or carbonate ions, CO_3^{2-}. Apparently CO_2 is always present in adequate quantities for the biological needs of plants and is never a limiting factor for animals in a natural environment because of abundance (unless pH is also low). The tolerance of animals varies greatly, but in fresh water artificial concentrations greater than about 50 mg/liter generally cause distress or death.

The measure of hydrogen ion concentration, pH, varies in natural fresh waters from a low of 1.7 in extremely acid waters to a high of about 12 in a few soda lakes. Values outside the range of about 4.5–10 are detrimental to fish but rarely occur.

The pH of the open sea, in contrast to that of fresh water, remains almost always between 8.1 and 8.3 in the surface layer. At the depth of the layer of minimum oxygen the pH may be about 7.5; in stagnant basins containing hydrogen sulfide it may be as low as 7.0. The pH is not known to be a limiting factor in the sea for animals and plants of commercial importance.

4.5.4 Salinity

Pure water does not occur in nature. Even rainwater contains dissolved gases and various salts derived from sea spray or dust. The water that is commonly regarded as fresh contains less than 1 g/kg of dissolved solids. "Hard" water contains about 300 mg/kg; exceptionally "soft" water, about 40 mg/kg; average river and lake water, 100–150 mg/kg. Average seawater contains about 35,000 mg/kg, or 35 g/kg, a value that is commonly expressed as 35‰, which is the equivalent of 3.5%.

Salinity is defined as the total amount of solid material in grams contained in 1 kg of seawater when the carbonate has been converted to oxide, the bromine and iodine replaced by chlorine, and the organic matter completely oxidized. Seawater is a mixture of remarkably constant proportions of the halide, carbonate, and sulfate salts of sodium, magnesium, calcium, potassium, and strontium, together with small quantities of other substances and minute traces of many other elements. The proportions between the major constituents are so constant that the total salinity has been estimated simply by determining the amount of chloride (plus bromide), the preponderant anion, and applying the formula

$$\text{Salinity } (‰) = 0.03 + 1.806 \times \text{chlorinity } (‰).$$

Recently, a formula has been developed to relate the salinity to the electrical conductivity, because this property can be measured with great precision.

Such a constant proportion of salts does not occur in fresh water or in enclosed basins with salt lakes such as the Caspian Sea, the Dead Sea, or Great Salt Lake. These may contain preponderantly carbonates, sulfates, chlorides, or any possible mixture of them. The proportion of salts in estuaries where rivers mix with the ocean may be slightly modified by the salts brought in by the river.

The salinity of the open ocean surface varies only from about 33 to 37‰ because of the variations in the amounts of evaporation and rainfall. Salinity tends to be higher in the dry, trade wind belts of the subtropical regions than in the regions of the moist, westerly winds. *Haloclines* (increases of salinity with depth) may exist at the bottom of the mixed surface layer (Fig. 4.10), but below the halocline the salinity varies only slightly with depth, locality, or season. None of the variations in the open ocean is likely to be of direct biological importance.

Vastly different in magnitude and biological significance are the variations in salinity along the shores, where the effects of the river flow and evaporation may be much greater. In estuaries at the mouths of rivers the salinity varies from that of fresh water to full ocean strength according to depth, location in the estuary, tide level, and the seasonal change in freshwater flow. Some coastal waters in arid climates may have salinities greater than the open ocean; 40‰ is found in both the Red Sea and the Persian Gulf; 60–100‰ may be found in some small saltwater lagoons from which water evaporates and into which seawater enters only occasionally.

In coastal waters the differences in salinity and the rapid changes restrict all plants and animals to the salinity zones that they can accommodate (see Section 6.3). The *euryhaline* forms are those that can accommodate the large changes and include most of the animals and plants that customarily live in the estuaries. Some anadromous animals, notably the salmon, migrate when young from fresh water through the estuary to the open sea and return to fresh water when ready to spawn. The *stenohaline* forms are those that tolerate little change and include those that live in either fresh water or the open sea.

4.5.5 Nutrient Elements

In the waters, along with the common halide, carbonate, and sulfate salts of sodium, magnesium, calcium, potassium, and strontium, occur compounds of most of the known stable elements. Most of these occur in minute quantities, but many are important. In addition to the foregoing, nitrogen, phosphorus, silicon, iron, manganese, zinc, copper, boron, molybdenum, cobalt, and vanadium are known to be essential for one plant or another. Others, such as nickel, titanium, zirconium, yttrium, silver, gold, cadmium, chromium, mercury, gallium, tellurium, germanium, tin, lead, arsenic, antimony, and bismuth, are concentrated by aquatic organisms, presumably for some beneficial purpose. In addition, certain vitamins and other organic compounds occur in solution and may be essential for some organisms.

Most of these elements and compounds have either functions or effects that are poorly understood. Some may be limiting factors when too much or too little is present. There is no question, however, about the role of phosphorus, nitrogen, and silicon, which are commonly called the *nutrient elements*. These must be present as soluble salts for plants to grow. Phosphorus and nitrogen go through cycles of incorporation into organic compounds by the plants and animals and subsequent breakdown into soluble salt after discard or death. Phosphorus is the most important to the ecologist, because it is the most likely to be deficient and is usually concentrated, that is, needed by plants in proportions greater than any other element. Nitrogen is available in abundance as a gas. The gas is readily soluble in water, but only certain bacteria and plants, sometimes in association with each other, can "fix" nitrogen into a form of ammonia (NH_3), nitrite (NO_2), or nitrate (NO_3) that can be used by plants. In one of these forms the nitrogen can be incorporated into the bodies of plants and used in amino acids or proteins. The plants may be eaten by herbivorous animals and in turn these are eaten by carnivorous animals. In each step of this conversion the nitrogenous compounds are first digested and then either egested or assimilated. If assimilated, they are either made part of the body or excreted. The parts that are egested or excreted and the dead bodies are decomposed by bacteria back to ammonia, nitrite, and nitrate (see Chapter 7).

Phosphorus is available only from certain rocks and deposits on earth from which it enters the rivers and groundwater as a phosphate ion. In this form it is usable by plants principally in the formation of proteins and fats. Like nitrogen, phosphorus goes through a cycle of plant to animal to bacteria as it is built into organic compounds and broken down again to the inorganic form. Unlike nitrogen, a large part of the phosphorus is quickly adsorbed at the surface of the mud, in shallow waters of seas and lakes. It is restored to organisms by various processes that are not well known. Such deposition on the mud is not part of the cycle in parts of the seas where depth is greater than about 250 m.

Silicon is abundant on earth but not readily soluble. It is not required for nutrition in the strict sense of the word but is needed by certain phytoplankton for their skeletons. Apparently it is recycled rapidly from organic to soluble inorganic form, but the chemistry is not well understood. A shortage has been shown to limit phytoplankton populations, but unlike nitrogen and phosphorus, it is usually available in adequate quantities.

All three of these nutrient elements usually occur in the surface waters of lakes and the ocean in small quantities that are rapidly changeable depending on their utilization by plants. Phosphorus is usually the least abundant; only 10–30 mg/m^3 of inorganic phosphate is available at the surface of many uncontaminated lakes, and only 1–5 mg/m^3 is available in large parts of the ocean surface. Nitrogen as nitrate is more variable in amount; between 0 and 500 mg/m^3 is present in surface waters. Such small quantities and wide variability pose special problems. The quantities may be too small to be detected by ordinary laboratory chemical techniques. The variability from place to place and time to time may require large sampling programs, and interpretation of the results may be difficult.

Much more significant from the standpoint of productivity than the amount of nutrients measurable is the rate at which they are supplied to the surface layers. Phosphorus as phosphate at depths of 100–1000 m in the oceans is present in amounts about 10 times greater than the surface concentration of 1–5 mg/m^3. Even greater quantities of nitrates are usually found in the same intermediate water layers. High concentrations of nutrients also develop in the deeper waters of lakes or in shallow seas, where some nutrients may be adsorbed by the mud or deposited in a layer on the mud.

These benthic nutrients were at one time part of the biological cycles in the surface layers. There is a net downward flux of organic matter that decomposes as it sinks and the elements are returned to inorganic form at some depth in the lake or ocean. They are brought back into the biological cycle in the surface layer by physical processes such as upwelling, turbulence, and convection currents. The places where nutrients are brought up are enormously more productive than other parts of the ocean (Fig. 4.16). In lakes the restoration occurs during the seasonal overturn of the waters (see Section 4.7.4).

High concentrations of the nutrient elements are not always desirable. Phos-

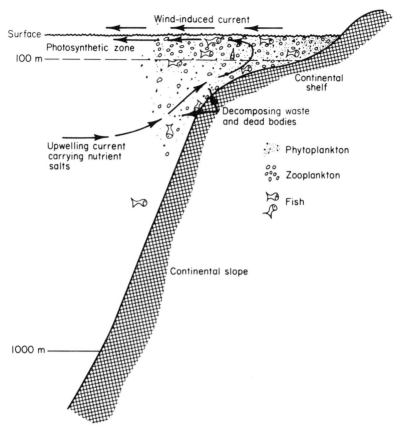

FIGURE 4.16   Diagram of the bio-geo-chemical cycle in upwelling areas of the ocean.

phates and nitrates are abundant in domestic sewage wastes and in the drainage from farmlands. They frequently cause unwanted blooms of plankton in rivers, lakes, and estuaries, so we try to control the amount in our wastewaters. It is an ever-increasing problem to put them where they will produce desirable results.

4.6 PHOTOSYNTHESIS

Carbon dioxide, water, chlorophyll, nutrient elements, trace elements, and the energy of light are converted to a living protoplasm in plant cells with the accessory production of oxygen. In a grossly simplified form:

$$6CO_2 + 6H_2O + \text{energy from sunlight} = C_6H_{12}O_6 + 6O_2.$$

In the absence of light, all cells, either animal or plant, respire with the consumption of oxygen and the release of energy and carbon dioxide:

$$C_6H_{12}O_6 + 6O_2 = H_2O + 6CO_2 + \text{energy}.$$

Because the waters of the earth cover about 71% of its surface, they are potentially the site for about the same percentage of the earth's photosynthesis. Whether the waters do support most of the earth's photosynthesis is not known because the plants are vastly different from those on land, although we guess that photosynthesis occurs at about the same rate per unit area. Practically all of the plants in the waters are the minute, pelagic phytoplankton that drift throughout the surface layer. The attached algae and higher plants that occur only in shallow water along the shores comprise but a minute part of the total weight of aquatic plants, although they may be important in marshes, small lakes, or streams.

The photosynthetic zone in the waters may be determined by observing the distribution of attached plants on the bottom since they can live only where they receive adequate light. Such direct observations are not satisfactory for use in the open water because some of the drifting phytoplankton sink below the photosynthetic zone and may survive there for a time.

Instead, samples of the water containing the phytoplankton are taken from various depths; each is divided in two parts, one of which is placed in an opaque bottle, the other in a clear glass bottle. Both are incubated at the temperature of the water from which the samples were taken, and the clear bottle is exposed to light of the intensity at the depth at which it was taken. After incubation the amounts of photosynthesis or respiration are estimated by measurement of either of the changes in the oxygen content of the bottles or of the changes in the carbon content of the plants. The latter change is measured by adding a small known amount of radioactive carbon (^{14}C) to the bottle before incubation and measuring the radioactivity of the plants after incubation (see Chapter 7).

Such methods have shown that the amount of photosynthesis that occurs in the water is enormously variable. There are many kinds of phytoplankton that occur in different concentrations, and they react differently to light of different colors, to different temperatures, and to different light intensities. In general, however, the maximum amount of photosynthesis occurs in the water layer receiving between 70% and 30% of the light at the surface regardless of the temperature, depth, actual amount of light, and whether the water is fresh or salt. Less photosynthesis occurs in the brightly lighted surface and less occurs in the dimly illuminated depths below (Fig. 4.17).

An important indicator is the *compensation depth*, the depth at which photosynthesis is just balanced by respiration. It is usually found at approximately the depth reached by 1% of the incident light, although it varies with temperature and the kind of phytoplankton present. A convenient approxima-

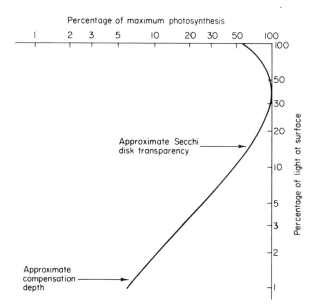

FIGURE 4.17   Diagram of distribution of photosynthesis in the surface layer. [Adapted from W. Rodhe (1965). Standard correlations between pelagic photosynthesis and light. *Mem. Inst. Ital. Idrobiol.* **18**, Suppl., 365–381.]

tion is that the compensation depth is about three times as great as the Secchi disk transparency.

The total amount of photosynthesis under a unit area of surface varies according to light, nutrients, temperature, and kinds of plankton. In the sea the amount of photosynthesis varies markedly from place to place and serves as a useful index of the ability of various regions to produce fish. The range is about 10-fold; the regional minima occur in the clear subtropical oceans and the regional maxima in two kinds of locations of special importance to the fisheries (Fig. 4.18), one on the west coasts of the continents and the other in the subarctic areas.

4.7 CIRCULATION

Currents are always present in the waters, even in seemingly quiet lakes. The currents transport heat, dissolved materials, and solids. The heat influences the climate over the waters and the surrounding land, the dissolved materials sustain the life of the waters, and the solids determine the kind of bottom. Ocean currents flowing from the subtropical latitudes warm great areas of the northern Atlantic and the northern Pacific oceans. Vertical currents bring to the surface not only the cool waters but also nutrient materials from the depths of

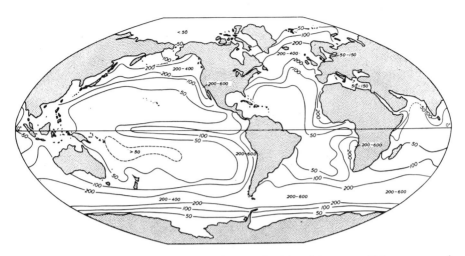

FIGURE 4.18 Estimated annual photosynthetic productivity of the oceans. Units are grams of carbon per square meter per year. [From R. H. Fleming (1957). *In* "Treatise on Marine Ecology and Paleoecology. Vol. 1: Ecology" (J. W. Hedgepeth, ed.). Mem. 67, Geol. Soc. Amer.]

lakes and the sea. All of the currents pick up, carry, sort, and deposit nonliving particles in the water to form the bottoms and the shores.

Organisms adapt either body or habit in many ways to live with the currents. Bodies are shaped for swimming rapidly or slowly. Many animals and plants are equipped with attachments to suit the bottom and velocity of turbulent waters in streams or on wave-pounded beaches. Spores, eggs, and larvae may drift with the currents in the open sea for hundreds of kilometers. Some animals depend on currents to bring food to them. Some whales and fish may ride ocean currents for 10,000 km in their regular migrations.

Of great interest to the fisheries are the currents that concentrate and sustain the animals. The tidal currents bring food to oysters and cause fish to concentrate where they are easy to catch. The great upwellings along the west coasts of the continents and the large plant production sustain the vast shoals of the herringlike fishes. The poleward extensions of the warm currents make large areas of the oceans habitable for many fishes. They even include the current in the stream that sweeps the angler's fly into the fish's mouth.

4.7.1 Causes of Currents

Water currents are the result of a complex of forces, most of which receive their energy from the sun either directly or indirectly. Many of the forces are almost vanishingly small, but their effects are substantial as the balance among them changes. Wind drag on the surface of the water can pile it up slightly along the shore. Atmospheric pressure commonly varies by the equivalent of between 10 and 20 cm of water and occasionally by more than 50 cm. Water changes its volume at the surface as it is either heated or cooled and as it either receives rain or evaporates. The gravitational attraction of the sun and moon pulls the water toward them in regular daily and lunar cycles to cause the tides. All of the resulting paths of water movement are affected by the rotation of the earth. Finally, all water masses in motion have inertia and very little friction, so they tend to have long-lasting oscillatory motions.

The *wind* is the primary driving force of the surface currents in both lakes and oceans. As the wind increases in speed, it accelerates the water by the shearing stress it exerts on the surface, and the surface water in turn exerts a lesser shearing stress on the water layers below. As the wind drags the water across a lake it lowers the water level slightly on the side toward the wind and raises it on the far shore. From the far shore the water will tend to flow back as a gradient current either along a part of the lake less exposed to the wind or in the lower part of the mixed surface layer. In the open sea the wind may push the water for thousands of miles at varying speeds, and the gradient current may be noticeable along only the coast on which the wind blows.

The effects of wind stress on water prevail only in the upper layers—principally in the mixed layer at the surface. In the thermocline layer and in the deeper waters the wind-driven currents can have only a secondary effect through piling up or moving away the upper layer. Here the currents are sustained largely by differences in density.

Coriolis force, or the effect of the rotation of the earth, affects any water in motion. It causes flowing water to be deflected toward the right in the Northern Hemisphere and toward the left in the Southern Hemisphere. It is a force so small that its effect is negligible in rivers and lakes, yet it has a major influence on currents that sustain several of the world's major fisheries.

The strength of the coriolis force, a function of latitude, is greatest at the poles and zero at the equator. It is also a function of depth, being greatest at the surface and least in deep water.

The deflection, according to theory, should be 45° at the surface and increase with depth, so that in a deep layer of water the average deflection is 90°. But the measured deflection at the surface is only 10–15° and it does increase with depth to a substantial average deflection.

This force, acting on currents driven by winds, changes the transport of water to the benefit of the fisheries in several parts of the oceans. It causes offshore movement of shallow waters and upwelling (Fig. 4.16) even when winds blow parallel to the coasts or impinge on them at a slight angle. The result for the fisheries is greatly enriched waters off the west coasts of Europe and North America, where north to northwest winds predominate, and off the west coasts of southern Africa and South America, where south to southwest winds predominate. It also causes upwelling and cooler surface water at the equator, where the currents are driven by the southeast trade winds in the Southern Hemisphere and the northeast trade winds in the Northern Hemisphere.

Changes in density arise at the surface as the water is either warmed or cooled and as the sea either receives rain or evaporates. The changes extend throughout the mixed layer above the thermocline because of wind-caused turbulence. The surface layer will sink when it becomes more dense than the deeper water during the seasonal temperature changes in lakes and the poleward movement of water in the ocean. Once below the surface, a water mass remains almost constant in temperature and salinity; these properties are "conservative" properties since they are changed only slowly by mixing processes. The masses of water that sink in the colder oceans move slowly toward the equator, so slowly that some are estimated to have been at the surface 2000 years ago!

However, neither wind nor density is the cause of currents so obvious to everyone living near where tides occur. Tides along ocean coasts and in estuaries have vertical ranges generally between 1 and 5 m and, in certain places, more than 8 m. These periodic changes in sea level are associated with periodic currents in estuaries and over the continental shelves. Tidal currents are usually much stronger than other movements. Their speeds and direction depend not only on the rise and fall of the tide but also on the depth and configuration of the bottom. Maximum tidal currents may exceed 5 m/s (10 knots) in some narrow straits, and tidal currents of 0.5–1 m/s are found commonly on continental shelves and in estuaries. Such currents on the continental shelves are rotating currents that result in nearly zero transport of water throughout a tidal cycle. They do, however, have a major influence on the bottom composition of the shelves and sometimes enhance the upward transport of nutrient materials through the turbulence that they cause.

Except along the exposed seacoasts and their estuaries, tides and tidal currents have little effect (so far as is known) on either circulation or fertility of the waters. Tides in the open sea seldom vary more than 1 m in height. Tides in the larger lakes are separable from oscillatory motion and tides in the earth's crust only with considerable uncertainty.

In enclosed or nearly enclosed basins, such as estuaries, inland seas, or lakes, the water surface tends to oscillate after having been moved by any force like water sloshing up and down in a tub. Such oscillations are *seiches*. Their amplitude depends on the original force; their frequency on the size and shape of the basin and on the number of harmonics. In simply shaped basins the frequency is predictable from formulas. Lakes of about 10 km in length have been found to have principal oscillatory periods ranging from 12 to 32 min and shorter-period harmonics.

Not only may a lake surface oscillate, but also the layers of different densities may oscillate relative to one another. Such *internal seiches* may be evidenced by a periodic rising and falling of the thermocline. These internal seiches, although not well understood, have a much greater period and amplitude than surface seiches; therefore they probably have a greater influence on bottom sedimentation and in transport of nutrients from the bottom to surface layers.

4.7.2 Waves

As everyone knows, the wind causes waves on the surface of the water, small ones at first and then larger ones as it increases in speed. These waves progress with the wind at speeds much faster than the water currents set up by the wind, which are commonly only about 1 or 2% of the wind speed. The movement of a water particle in a nonbreaking wave is nearly orbital, with a maximum speed in the direction of the wave movement in the wave crest, a maximum speed in the opposite direction in the wave trough, and nearly vertical motion when the water is at its mean level (Fig. 4.19).

The speed of a wave in deep water (not the net current) is related to its wavelength; its speed in centimeters per second is approximately 12.5 times the square root of its wavelength in centimeters. In water shallower than one-half of the wavelength the wave slows down to become a function of depth and when it slows sufficiently it begins to break. Since the speed of the wave varies according to the wavelength, the long storm waves travel faster than the shorter waves. Long-wavelength waves travel thousands of kilometers with little attenuation, so the common state of the ocean's surface is a mixture of *swell*, the long-wavelength waves generated by a distant storm, and *sea*, the shorter waves produced by the local wind.

Most density gradients in natural waters also have waves. Such internal waves are usually found in the thermocline. They tend to be larger than surface waves but travel much more slowly. Some associated with diurnal tides have a period of about 12 hr and an amplitude of between 30 and 50 m. Others have periods of between 15 and 30 min and amplitudes of about 6 m. Their cause is not well known. Sometimes they can be seen on the surface when they cause parallel slicks or bands of alternating rippled and smooth water. They are usually measured by repeated determination of the depth of the density gradient (thermocline) at a fixed point (Fig. 4.20).

Such waves probably enhance the vertical circulation of nutrients in some places, but they are of interest to fishery scientists primarily because they are associated with seiches, which are the cause of the principal subsurface currents in lakes. Waves also cause substantial variation in the depth of the thermocline and interfere seriously at times with the interpretation of echos from echo-ranging equipment.

FIGURE 4.19   Diagram of the direction of water movement in a progressive wave.

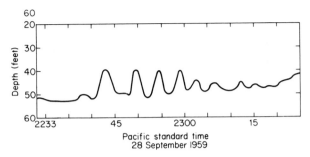

FIGURE 4.20 Small internal waves off San Diego were evidenced by the changing level of the 67°F isotherm. These waves moved shoreward at a speed of about 10 cm/s. [After O. S. Lee (1961). Observations on internal waves in shallow water. *Limnol. Oceanogr.* 6(3), 312–321.]

4.7.3 Measurement of Currents

A major problem for all who are concerned with the circulation of the waters is measurement of the speed, direction, and volume transport of the currents. Speed past a fixed point is measured directly where possible by current meters, some of which may record time and direction of flow. Trajectories may be traced by observing the drift of floating bottles, poles, cards, dye, or buoys. The latter may be equipped with radio in order to be tracked by direction-finding radio receivers. Also, a ship drifts with the surface current, which may be estimated by the deviation of the ship from its projected course.

Such direct methods are used effectively in rivers, lakes, and estuaries where instruments can be easily anchored or where numbers of drifting objects can be easily observed. They are, however, of limited usefulness in the open sea or in large deep lakes where anchoring is difficult and where it is very expensive to follow drifting objects. Here it is possible to use an indirect method to estimate the complete field of motion in a body of water. This is the dynamic or geostrophic method based on measurements of the internal distribution of pressure.

The pressure field cannot be measured directly by pressure meters with adequate accuracy but can be estimated from the density of columns of water ending at the surface. The density is determined from very accurate measurements of temperature and salinity at various depths along the length of the column. The necessary temperature and salinity determinations are obtained by *hydrographic casts* in which a series of collecting bottles with reversing thermometers are lowered on a wire to a depth usually of 1000 m or more. These are closed and reversed when at depth and retrieved with the temperature reading and a water sample from which salinity and other determinations can be made. Such casts are made in a series or network of stations.

This method of estimating currents is based on the assumption that the currents are steady, a state that is in fact approximated to a remarkable degree by the great ocean currents. These are induced and modified by winds, atmospheric pressure, change in density, coriolis force, and gravity, all of which are relatively tiny forces acting on a very large mass with great inertia. The major currents are little affected by tides or other waves.

A full explanation of the dynamic method requires complex mathematics that will not be introduced here. The explanation that follows is a grossly simplified approximation.

The method is based on a measurement of the difference in pressure gradient from which can be computed the difference in current velocity between two surfaces or horizons in the sea. If one of these surfaces has no motion or if its motion is known, then the actual motion of the other surface can be computed. Usually the surface of no motion is assumed to be at a depth of 1000 m, an assumption that has been found to be reasonable in many localities.

The determination of density from temperature and salinity for a water column between a lower surface and a higher surface (usually the air–water surface) permits a determination of the thickness of the intervening layer. This is expressed in terms of the work done per unit mass against gravity, but the unit has been chosen as $D = 10^5$ dyn-cm/g, which is very nearly equivalent to a linear meter. It is called *dynamic meter*. Because only the difference in thickness is needed for the computation of currents, the results are expressed conveniently as dynamic height differences from a standard water column at 35‰ salinity and 0°C, water slightly more dense than that ever found at sea. The differences found in practice are commonly a few dynamic centimeters.

As an illustration of the principle of measuring dynamic heights, consider two points A and B, at which the density of a deep column of water is determined (Fig. 4.21). The less dense water above a surface of no motion is at B; hence the air–water surface at B is higher. The pressure is equal in A and B at the surface of no motion indicated by 0, but above this surface in the B column the pressure is higher than at the same level in A. Since the basic condition is a steady current that maintains the difference in elevation, the water does not flow directly downhill but rather flows in the direction induced by the coriolis force at right angles to the slope.

When the dynamic height is determined at a network of stations it is possible to compute a dynamic topography (Fig. 4.22A) and to estimate the current direction and velocity (Fig. 4.22B). The current direction is clockwise around an elevation and counterclockwise around a depression in the Northern Hemisphere. Thus, if an elevation is on an observer's right, the direction of flow is away from the observer. The velocity is higher along steeper slopes and can be predicted with reasonable accuracy from the gradient of the slope.

Another indirect method has been developed around the measurement of the electric potential produced by the highly conductive seawater moving through the magnetic field of the earth. The gradients in electrical potential (see Section 4.4.5) are measured with a geomagneticelectrokinetograph (GEK). When suit-

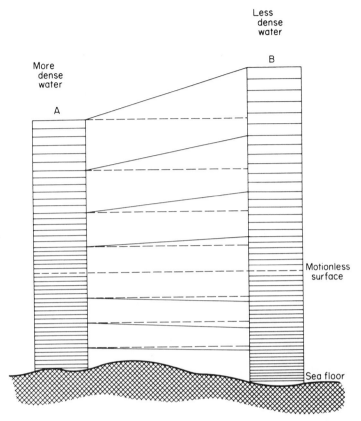

FIGURE 4.21 Diagram of two water columns with a different density distribution above and below a layer of no motion. The solid lines indicate equal pressure; the dashed lines indicate a level surface. The *less* dense water in the upper part of column B is under *more* pressure than the water at the same level in column A.

able measurements are made along different compass courses, the direction and speed of the surface currents can be estimated with fair accuracy.

No method is satisfactory for all purposes. The indirect methods work best in the open sea, where the assumption of a steady state may be approximately correct and where short period changes are less important. The direct methods work best in estuaries, in lakes, and on continental shelves, where tidal and other oscillatory currents preponderate. Measurement of deep currents and vertical currents is especially difficult and expensive.

The direction of water flow is always reported as the downstream motion, that is, an easterly current moves water toward the east. This practice is opposite to that of reporting wind direction, in which an easterly wind blows from the east.

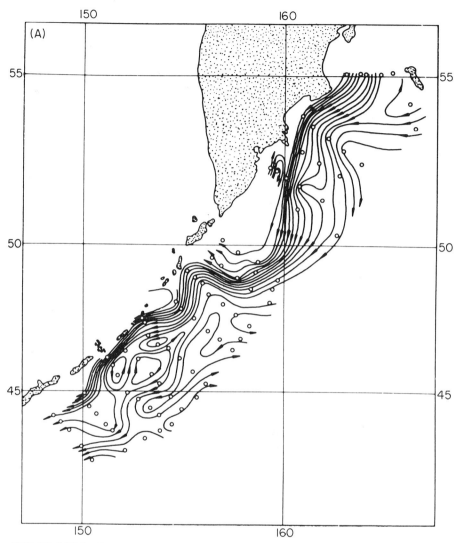

FIGURE 4.22A Dynamic topography of the sea surface estimated from determinations made at the stations indicated by circles. Contours are spaced 2 dynamic centimeters. [From L. M. Fomin (1964). "The Dynamic Method in Oceanography." Elsevier, Amsterdam.]

Volume of flow can be estimated after determination of the velocity over a representative cross section of the current. The simplest situation is in a stream in which volume is estimated from width, average depth, and average velocity. Measurement of the volume of ocean or lake currents is done in fundamentally the same way, but determination of the boundaries and average velocity is much more complicated.

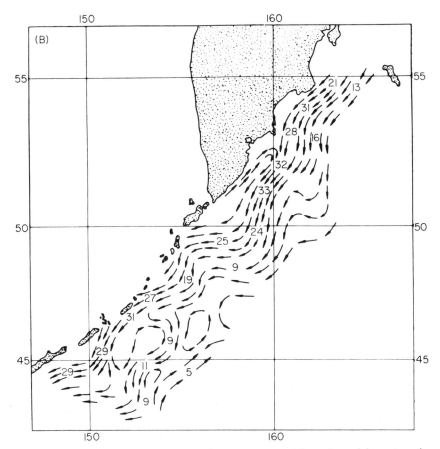

FIGURE 4.22B Currents computed in centimeters per second from slope of dynamic surface. [From L. M. Fomin (1964). "The Dynamic Method in Oceanography." Elsevier, Amsterdam.]

4.7.4 Circulation in Lakes

The surface waters of lakes are moved to and fro by the wind. The resulting currents are always turbulent and depend on many factors, including the shape and depth of the lake, as well as the strength, steadiness, and fetch (length of water stressed by the wind) of the wind. Consequently, they are highly irregular and largely unpredictable. What needs emphasis here is the disturbance of the deeper waters, which reflects in part the disturbance of the surface and which may be significant to the fisheries.

When the wind piles up the water at the downward end of a large lake, the thermocline is also tilted downward at the downwind end. A pileup of a few centimeters at the surface may produce a depression of the thermocline of several meters and the consequent displacement of a large volume of deep water

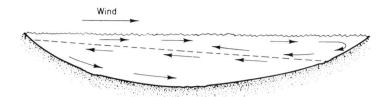

FIGURE 4.23 Diagram of a vertical section of a lake, showing circulation above and below the thermocline (------).

below the thermocline. The pileup includes the warm surface water, the removal of which exposes cooler water at the upwind side, thus enhancing the mixing of the surface layer. The return gradient currents of the surface layer may be located just above the thermocline and will be accompanied by parallel currents in the layer just below the thermocline (Fig. 4.23). Since the lake basin is never symmetrical, the currents all tend to rotate. They are never either laminar or smooth flowing, but turbulent and accompanied by waves, both surface and internal. Sometimes the currents reach surprisingly high velocities; surface current averages about 2% of wind speed. When the wind dies, the currents and waves do not stop but continue to oscillate in various ways. The movement is analogous to the sloshing of water in a tub, but the changes in height of the surface layer are slight, whereas the changes in height of the thermocline are substantial. Such currents are of special importance to fisheries when they displace upward or downward a cold layer, a low-oxygen layer, or a nutrient-rich layer.

Another significant circulatory pattern develops in lakes with large tributaries having a different density from the lake surface because of either temperature or silt. If the inflow is lighter it will spread over the surface; if heavier, it will descend and intrude in a layer at the depth where it finds its density (Fig. 4.24). The latter condition is common in reservoirs with cold tributaries. If the river water includes organic material that uses by the oxygen, the layer of water may be a barrier to the movement of fish.

4.7.5 Circulation in Estuaries

Estuaries are mixing basins of fresh and salt water. The circulation is much more rapid than in either the lakes or open ocean. No two estuaries are alike and the mixing is as varied as the tidal range, the river flow, the shape, size, and depth of the estuary, the wind, and the effect of the coriolis force. Thus, the circulation of each estuary under study needs to be individually determined. If detailed information under varied conditions is needed, it may be necessary to build a hydraulic scale model.

The mixing in each estuary occurs as the fresh water flows *out* on the surface and the salt water flows *in* below and mixes with the fresh water above because

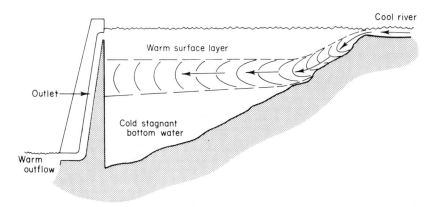

FIGURE 4.24 Diagram of distribution of tributary water in a reservoir during the summer.

of turbulence (Fig. 4.25). The relative inflow of salt water with respect to fresh water varies according to the depth and the amount of mixing, but in large, deep estuaries the saltwater inflow may be between 10 and 20 times as great as the freshwater inflow. The mixture of fresh and salt water flows out on the surface. The actual flushing pattern can be determined for an estuary, and often effluents can be introduced where they will flow rapidly out to sea and not circulate in the estuary.

Another aspect of the mixing is the horizontal and vertical incursion of the fresh- and saltwater layers. The incoming tides may push relatively unmixed seawater far up along the bottom of a river; the outgoing tides may take relatively unmixed fresh water far out on the surface of the estuary. These limits will also be the limits for animals that can tolerate only salt water and for those that can tolerate only fresh water.

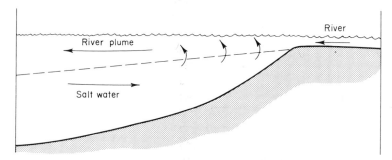

FIGURE 4.25 Diagram of water flow in an estuary above and below the halocline (------). The volume of inflowing salt water is usually several times as great as the volume of inflowing fresh water.

The tides affect not only the area where fresh and salt water mix but also the fresh water upstream. Normally the fresh water rises and falls with the tide upriver to the zone where the riverbed is approximately at the elevation of the high tides. Toward the sea from this zone will be a zone in which the fresh water will just stop flowing at high tide, and still further seaward will be a zone in which the fresh water will flow back and forth with the tides.

4.7.6 Surface Circulation in the Oceans

Knowledge of ocean currents has been important to navigators who have sailed the oceans. They have learned about the dangerous currents and the currents that can help them along. Most of this knowledge has concerned the coastal currents, many of which are caused by the tides. Extensive charts and tables showing the strength and direction of such currents are available.

Not nearly as well known in detail are the nontidal currents of the open ocean, which are driven by wind and differences in density of the water. These include the great ocean rivers such as the Gulf Stream, which have been known to navigators for centuries but little understood and poorly charted. These are the currents that are of importance in the transfer of heat around the globe and to the life in all oceans. They are the subject of this discussion.

The circulation of the world oceans is dominated by the five great gyres: the North Pacific, South Pacific, North Atlantic, South Atlantic, and southern Indian oceans (Fig. 4.26). These are all clockwise in the Northern Hemisphere and counterclockwise in the Southern Hemisphere. All have easterly flowing currents driven by west winds at about 40 or 50° latitude north or south, and westerly flowing currents near the equator driven by the northeast or southeast trade winds. All have currents flowing toward the pole in western oceans and toward the equator in the eastern oceans.

Off these great gyres spin smaller gyres in neighboring gulfs and seas. The Gulf of Alaska, the Tasman Sea, and the Greenland Sea all have connected gyres. More isolated waters such as the Bering Sea, South China Sea, Chukchi Sea, and Gulf of Mexico seem to have independent gyres. Almost every basin in every ocean contains a gyre that may or may not turn in the same direction as the major gyres.

Between the great gyres of the northern and southern oceans are the countercurrents that flow toward the east near the equator. One of these, a surface current in the Pacific at about 8° north latitude, has been well known for centuries. But recently a weaker surface countercurrent south of the equator was discovered, and more important, a powerful submerged current, the Equatorial Undercurrent or Cromwell Current, was also discovered. Both surface countercurrents and undercurrents occur also in the Atlantic and Indian oceans.

The discovery of the Pacific Equatorial Undercurrent by Townsend Cromwell of the U.S. Bureau of Commercial Fisheries illustrates how little is known of subsurface ocean currents and indicates something of the opportunities remain-

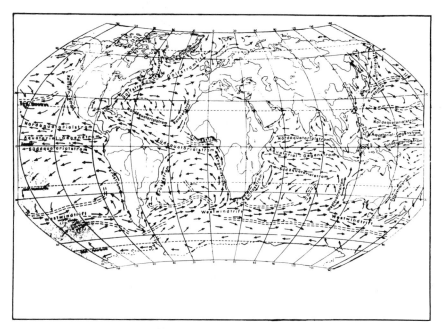

FIGURE 4.26   Major currents of the oceans. [From U. Scharnow (1961). "Ozeanographie für Nautiker." Transpress, Berlin.]

ing in oceanography. Totally unpredicted by oceanographic theory, the presence of the current was suspected from the easterly drift of fishing gear, and its velocity was measured by Cromwell in 1952. Further studies by Cromwell and others revealed a persistent, ribbonlike current about 300 km wide lying only between about 70 and 200 m below the surface, at least 6500 km long, with a speed at the core of about 1.5 m/s (Fig. 4.27). It transports about 40×10^6 m³/s,* a volume half as great as the Gulf Stream or nearly 200 times as large as the Amazon River at flood stage.

The greatest current in the world is the west wind drift; it flows around Antarctica in an unbroken circle to form part of each of the three great gyres of the southern oceans. The transport of water has been estimated to be 150–190 $\times 10^6$ m³/s.

Two of the best-known ocean currents are the Gulf Stream and the Japan Current or Kuroshio. These currents flow northeasterly in the northwest Atlantic and Pacific, respectively. Each of these currents is relatively narrow in the western oceans and then becomes broader as it crosses the ocean. Each brings warmer water from the subtropical ocean to the northeastern oceans to influ-

---

*1×10^6 m³/s is about 35.3×10^6 ft³/s. The unit is called a Sverdrup (sv) by some oceanographers.

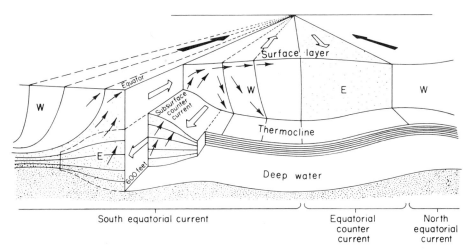

FIGURE 4.27 Diagram of a cross section of the Pacific Ocean about 1500 km long at 150°W longitude showing the major horizontal currents. [From O. E. Sette *et al.* (1954). Progress in Pacific oceanic fishery investigations, 1950–53. *U.S. Fish Wildl. Serv., Spec. Sci. Rep.—Fish.* **116,** 75.]

ence the climate of Europe and northwestern North America. Each also provides fish abundantly—tunas and other subtropical species in the warmer, western parts, herring and bottom fishes in the cooler, eastern parts. Each is connected with the north equatorial currents by diffuse currents flowing south in the eastern oceans.

Two other well-known currents play major roles in the climate and fisheries of the Southern Hemisphere, but they flow northward in the eastern parts of the Atlantic and the Pacific. The Benguela and Peru or Humboldt currents are off southwest Africa and western South America, respectively, and bring cool water northward toward the equator. Both are important in cooling and drying the climate of the coasts. Both support huge fisheries primarily for fish of the herring and anchovy families.

Many other currents are persistent and defined well enough to be named. Each forms but a part of a circulation system and each is significant to the life of the ocean of which it is a part.

No current is a smooth, straight-sided flow, but all tend to be turbulent and all tend to twist, that is, they include some vertical flow as well as horizontal flow. Currents converge, and when they do the less dense water flows over the more dense and the homogeneous surface layer become thicker. Elsewhere currents may diverge, in which case the surface layer becomes thinner and nutrient-rich water rises closer to the surface. Both convergences and divergences may be either semipermanent features of major circulation or temporary phenomena of giant eddies.

Convergences bring together plankton animals of the surface layer. When these plankton animals resist sinking, they accumulate along the convergence zone and provide food for fish. Sometimes a convergence brings together flotsam in a line on the surface, such as is commonly seen in estuaries. Whether marked by debris or not, convergences are boundaries of water masses that may differ substantially in temperature and salinity.

Divergences are especially important for their role in bringing nutrient-rich water closer to the surface. They occur especially off the west coasts of Europe, Africa, and North and South America; that is, they occur on the eastern sides of both northern and southern oceans. As the winds push the water toward the equator, the coriolis force sets the currents to the right in the Northern Hemisphere and to the left in the Southern Hemisphere—offshore in each case. The consequence is transport of cooler nutrient-rich water toward the surface.

Both convergences and divergences occur in the equatorial circulation (Fig. 4.28). The westerly flowing currents tend to diverge exactly at the equator and at the boundary of the north equatorial current and the equatorial countercurrent. They converge at the boundary of the south equatorial current and the equatorial countercurrent.

In the southern oceans two parallel convergences occur in the west wind drift (Fig. 4.29). The most southerly, the Antarctic Convergence, is the zone at which the colder, fresher Antarctic water from the melting ice tends to flow beneath water about 4°C warmer. About 10° latitude farther north is the Subtropical Convergence, the boundary of the intermediate water and the much warmer water to the north. Similar but less well defined convergences appear near the west wind drift of the northern oceans.

Charting the ocean currents and their variations in position and strength remains one of the challenging tasks of oceanography. Major advances have been made during recent decades in gaining a better understanding of the Indian and Arctic oceans. Major difficulties persist, however, such as the great cost of ocean

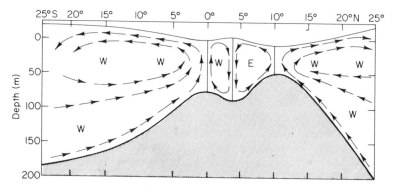

FIGURE 4.28 Schematic representation of the vertical circulation within the equatorial region of the Atlantic. The direction of the currents is indicated by the letters W and E. The water below the discontinuity surface, which is supposed to be at rest, is shaded. (From Sverdrup *et al.*, 1942.)

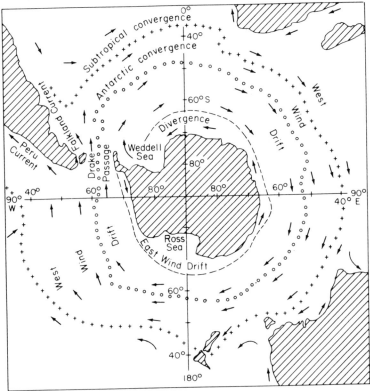

FIGURE 4.29 Approximate positions of the divergence and convergences of the southern oceans. [From G. L. Pickard (1963). "Descriptive Physical Oceanography." Pergamon Press, Oxford.]

observations, the need for deep-water observations that require highly sophisticated equipment, and the great complexity of the current systems.

4.8 SUMMARY: THE OCEAN ENVIRONMENT OF FISH AND FISHING

Let us imagine how it would be to live in a permanent, thick, cold fog through which no eye can see more than 50 m but through which sunlight can penetrate for several hundred meters. Imagine next going to the top of a hill emerging from the fog and being blinded by the bright sunshine. Imagine being able to smell all of the odors from the other organisms and waters that pass you in the veil. Imagine hearing every animal near you and even the echoes of your own noises reflected back to you. Imagine also the sensation of resting with little effort; always with the sensation of weightlessness but always fighting the water to move. Imagine traveling horizontally in the open sea for hundreds of kilometers with no perceptible change in climate but finding intense cold and darkness when descending only a few meters. Imagine the constant risk of sinking down into the dark and cold or rising up to the surface if depth controls ceased to function. Such a fancy gives a remote concept of an animal in the ocean in terms of human senses.

Now let us look at the ocean environment as a whole. Perhaps the outstanding characteristic is the remarkable similarity over long distances of very thin layers that are greatly different from other layers above and below. The pressure, light, heat, oxygen, and nutrient elements all vary greatly with depth, and each is vital to the life in the water. Other features, such as N_2, CO_2, pH, density, and salinity, vary so little with depth that living things are not affected directly, although the slight variations are important for physical reasons.

Another feature is the nearly perpetual motion—mostly horizontal but occasionally and importantly vertical. The currents are always carrying heat, food, eggs, larvae, and the plants and animals themselves. The organism must breast the current or change depth to a favorable current if it is to avoid transportation.

As judged by the scant number of species, the more formidable of aquatic environments are not only the extremely hot, the extremely cold, and the low-oxygen parts of the oceans but also the most variable parts, the estuaries. Here the currents, salinity, and temperature change in tidal and seasonal cycles in ways that few organisms can withstand, and here too most human wastes are discharged to start their final dilution.

REFERENCES

Hutchinson, G. E. (1957). "A Treatise on Limnology," Vol. 1, "Geography, Physics, and Chemistry" (1967); Vol. 2, "Introduction to Lake Biology and the Limnoplankton" (1975); Vol. 3, "Limnological Botany" (1975); Vol. 4, "The Zoobenthos" (1993). Wiley, New York.

Jansen, P. P., *et al.*, eds. (1979). "Principles of River Engineering: The Non-Tidal Alluvial River." Pitman, London.

Sverdrup, H. U., Johnson, M. W., and Fleming, R. H. (1942). "The Oceans: Their Physics, Chemistry, and General Biology." Prentice–Hall, Englewood Cliffs, N.J.

5 | Food Chain and Resource Organisms

Fishery scientists who work to enhance the production of the living resources of the waters must know the animals and plants with which they deal. They must identify them and know the relationships among them. They must recognize the major groups, of which there are relatively few, and know as much as possible about the minor groups, of which there are many thousands. The tasks of identification and classification may be either an essential preliminary to other studies or a major task for a lifetime.

A comprehensive review of aquatic organisms is far beyond the scope of this introductory book, and since the subject here is fisheries, the only general classification presented is the international statistical classification used for reporting catches. This is based on the social importance of the organisms and has little relationship to the evolutionary factors that are the basis for scientific classification. Furthermore, the discussion of food chain organisms, which is essential to an understanding of the productivity of the waters, anticipates the description of food chains in Chapter 7.

5.1 IDENTIFICATION AND NOMENCLATURE

Every scientific study of an animal or a plant should include a correct identification with the scientific name that is based on previously published descriptions of either the organism or related organisms. The scientific name is composed of two parts, the generic and specific names, or three parts if subspecies have been described. The words in the names are latinized regardless of the language or alphabet of the study and are frequently descriptive of a significant feature of the organism. The names and their usage are governed by the International Code of Zoological Nomenclature and by the opinions of the International Commission of Zoological Nomenclature or by their botanical counterparts.

The scientific name of the organism may be followed by the name of the describer of the species and the year in which the description was published. When the species is now in a genus different from that used in the original description, the describer's name appears in parentheses. Examples are the name of the cod of the North Atlantic, *Gadus morhua* Linnaeus[*] 1758, and that of the closely related haddock, which is of a different genus, *Melanogrammus aeglefinus* (Linnaeus) 1758. The names are always underlined in typescript and italicized in print. The generic name is capitalized; the specific name is never capitalized (although botanists formerly capitalized specific names derived from proper nouns). In a second use of the name in manuscript, the genus may be abbreviated (as *G. morhua* in the first example).

Unfortunately, a correct identification is not always easy or even possible. Species are not fixed, unchanging entities, but are stages in the evolution of populations in the generally accepted modern concept. They are not arbitrarily defined entities, but are reproductively isolated, interbreeding populations. The evidence of evolution, reproductive isolation, and interbreeding is often scanty, however, so inferences about species and the practice of identification are commonly based on morphology. Practical and clear definitions of species together with the related definitions of subspecies and genera are given by Milton B. Trautman[†] (pp. 51–52), which he attributes in part to Ernst Mayr, to which I adhere.

> In bisexual animals a species is a natural population, or a group of populations, normally isolated by ecological, ethological, and/or mechanical barriers from other populations with which they might breed and which usually exhibit a loss of fertility when hybridizing. The above definition applies to all except a few bisexual animals. No single definition can be applicable to all animals however, because in the large series of imperceptible and intermediate stages in the evolutionary transition from a subspecies to a species there is no well-defined gap separating the two; hence it is sometimes impossible to know whether two populations are specifically or only subspecifically distinct.

[*]Carolus Linnaeus is the latinized name of a Swedish botanist, Carl von Linne, who published the tenth edition of "Systema Naturae" in 1758, the initial year of the use of the system of binomial nomenclature.

[†]Reprinted from "The Fishes of Ohio," by Milton B. Trautman. Revised edition. Copyright © 1957, 1981 by the Ohio State University Press. All rights reserved. Used by permission of the author, the publisher, and the Ohio Sea Grant Program Center for Lake Erie Area Research. The quotations are not presented in the original order given by Dr. Trautman.

Dr. Trautman emphasized the fluidity of nomenclature by stressing that the foregoing were his concepts at the time of writing the book. He goes on to comment about the changing of names (p. 52).

... It was impossible for the majority of the early describers of fishes such as Rafinesque to review adequately the pertinent literature in order to learn whether their supposedly "new species" had been described, or to compare their specimens with the types of closely related species. As a result, they described many species which had been properly described one or more times previously, and consequently their name for the species, according to the law of priority, had to be synonymized. Some species were thus unknowingly described many times. ... Recent studies of original descriptions of many species have shown that the vague descriptions given by the describers have misled later taxonomists so that they applied scientific names to the wrong species; upon realizing the error another name for these species had to be given. Researchers frequently find a species description in some previously overlooked, obscure publication, which description is of an earlier date than the name applied at present to the species. In that case the law of priority demands that the earliest available name is used, even though the other name has been in usage many years. ...

It was the custom of earliest taxonomists to place many species in a single genus. ... Later when the various species were compared with one another certain differences became apparent. As a result, the large genus was divided into many genera, until in some instances the majority of the genera had become monotypic. ... Still later, when adequate material had been assembled, a monographic study of the group was undertaken, resulting in the accumulation of abundant evidence indicating that a drastic reduction in the number of genera was necessary. ... This swinging from one genus to many, then back to a few when an adequate study of the group was made, is what normally happens.

... The subspecies, or geographic race, is a geographically localized subdivision of a species, which differs genetically and taxonomically from other subdivisions of the species. ... Unlike species, the subspecies comprising a species can freely interbreed without loss of fertility, but mass interbreeding is usually hindered or prevented by some type or degree of geographical, ecological, ethological and/or mechanical barrier or barriers. As indicated under the species, there are many stages in the evolutionary process of species production from an original homogenous group into two homogenous groups or subspecies. Some subspecies are so similar to one another that the two can be separated only through the application of statistical methods, others are so distinct as to be recognizable at a glance. Taxonomists usually consider a subspecies to be valid if 75% or more of one population can be separated statistically from the other population.

... A genus is a systematic unit including one species, or a group of species of presumably common phylogenetic origin, separated by a decided gap from other similar groups. ... The genus is largely a human conception whose function is to group species in order to stress their relationships to each other. Unfortunately all taxonomists do not place the same value upon the various characters, so they are divided into those who consider any clear-cut character to be of generic value, and those who believe that stressing such characters destroys the purpose of the genus, which is to stress relationship. Such diversity of opinion results in disagreement relative to the number of genera, the "lumpers" stressing relationships, leaving subgenera to stress the minor gaps between groups with a genus, and the "splitters" preferring to give generic rank to what the "lumpers" designate as subgenera.

In practice, organisms are identified from their structure, on the basis of their external characteristics as much as possible. The commonly used characteristics of fish are shown in Fig. 5.1. The identification is facilitated by the development of a dichotomous key to the organisms of limited faunal areas (Table 5.1).

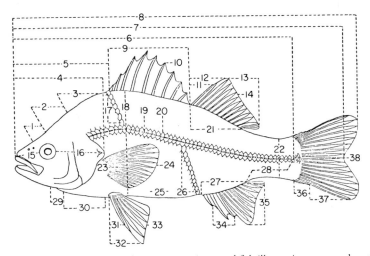

FIGURE 5.1   External anatomy of a common spiny-rayed fish illustrating parts and methods of counting and measuring: (1) interorbital, (2) occipital, (3) nape, (4) head length, (5) predorsal length, (6) standard length, (7) fork length, (8) total length, (9) length of base of the spinous or first dorsal fin, (10) one of the spines of the dorsal fin, (11) one of the spines of the second or soft dorsal fin, (12) height of second dorsal fin, (13) length of the distal, outer, or free edge of second dorsal fin, (14) one of the soft rays of the second dorsal fin, (15) snout length, (16) postorbital head length, (17) scales above the lateral line or lateral series that are counted, (18) body depth, (19) one of the lateral line pores (since in this figure all of the scales in the lateral series are pored the lateral line is complete), (20) one of the lateral scales that, with the remainder, forms the lateral series, (21) length of base of the second or soft dorsal fin, (22) least depth of the caudal peduncle, (23) the pectoral fin, (24) one of the soft rays of the pectoral fin, (25) abdominal region, (26) scales below the lateral line or lateral series that are counted, (27) length of the base of the anal fin, (28) length of the caudal peduncle, (29) the isthmus, (30) the breast, (31) the pelvic spine, (32) height of pelvic fin, (33) one of the soft rays of the pelvic fin, (34) spines of the anal fin, (35) soft rays of the anal fin, (36) the rudimentary rays, (37) one of the principal rays of the caudal fin, (38) the caudal fin. (Reprinted from Fig. I of "The Fishes of Ohio," by Milton B. Trautman. Revised edition. Copyright © 1957, 1981 by the Ohio State University Press. All rights reserved. Used by permission of the author, the publisher, and the Ohio Sea Grant Program Center for Lake Erie Area Research.)

TABLE 5.1   Excerpt from a Dichotomous Key for Classification of Fish[a]

		Family Salmonidae—Salmons
61	(72)[b]	Rays in anal fin, 8 to 12

62	(69)	Teeth on head and shaft of vomer; spots black (sparse in *S. salar*) (spots may be faint if fish has been in salt water for some time)
		Genus *Salmo*—Trouts
63	(66)	No red band along side of body; no red dash below lower jaw
64	(65)	No spots below lateral line
		Atlantic salmon, *Salmo salar*
65	(64)	Large black spot below lateral line, each surrounded by pink or red halo
		Brown trout, *Salmo truta*
66	(63)	Either red band along side of body or red dash below lower jaw
67	(68)	Red band along side of body; no red dash below lower jaw, no teeth on back of tongue
		Steelhead trout, *Salmo gairdnerii*
68	(67)	No red band along side of body; red dash below lower jaw, teeth present on back of tongue
		Coastal cutthroat trout, *Salmo clarkii clarkii*

69	(62)	Teeth on head of vomer only; spots yellow and red, never black (frequently yellow or red spots may be faint if fish has been in salt water for some time)
		Genus *Salvelinus*—Chars
70	(71)	Spots on back, pale yellow; vermiculations on back and dorsal fin absent or weak
		Dolly Varden, *Salvelinus malma*
71	(70)	No spots on back; vermiculations on back and dorsal fin dark green, prominent
		Brown trout, *Salvelinus fontinalis*

72	(61)	Rays in anal fin, 13 to 19
		Genus *Oncorhynchus*—Pacific salmons

[a] Adapted from Clemens and Wilby (1961).
[b] The key is based on the "true–false" method. If statement 61 is false, the reader should look at the number in parentheses, i.e., 72. If the statement is true, the reader should go on to the next statement, i.e., 62.

5.2 CLASSIFICATION

After a species has been established, it is classified with other species into the larger groups of a hierarchy. Ideally this classification reflects the evolutionary history of the groups; the most closely related are those that have had common ancestors most recently. In practice, the evidence for the evolution is usually incomplete or even totally lacking, so the relationships must be inferred from the structure. But organisms always resemble each other in some parts and differ in others, so that deciding on the structure that best reflects the evolution is usually controversial. The classification would be much easier if fossils of all intervening forms were available, but of course they never are. The classifier faces the equivalent of a giant jigsaw puzzle with only a small and unknown fraction of the pieces.

As a consequence, there is never enough evidence to settle many of the questions about relationships. A classification must have a niche for all forms if it is to be useful, so guesses are made about the relations and separations among groups. Only a few people can hope to master a major group in order to revise the classification on the basis of the extensive and scattered scientific literature. Usually a major revision is a task that occupies the productive lifetime of a scientist. Nevertheless, the disagreements are not important to fishery scientists, who are concerned with a few well-known groups. They must accept the differences until better evidence accumulates.

Many modern fish classifiers follow Berg's (1940) classification of fishes for the following reasons: (1) it is recent enough to include most of the evidence available, (2) it is comprehensive, including all known living and fossil forms, (3) it tends to show relationships by lumping groups instead of splitting them, and (4) it uses a reasonably standard system of units and endings. The student will find, however, several other classifications in common use; there are almost as many variations as there are senior taxonomic ichthyologists. In this book, the general classification of fish outlined by Lagler *et al.* (1977) and the list recommended for North American fish by the American Fisheries Society (Robins *et al.*, 1991) are followed; both of these are close to the classification set forth by Berg.

The standard units of the hierarchy are species, genus, family, order, class, and phylum, to each of which the prefixes *super-* or *sub-* may be added (Table 5.2). Endings of *-idae* for animal families and *-inae* for subfamilies are in almost universal use, but endings for other units are more variable. Berg uses a common system of endings of *-oidae* for superfamily, *-oidei* for suborder, and *-iformes* for order. The ending of *-aceae* is commonly used for plant families.

5.3 MAJOR GROUPS

What are the major kinds of aquatic organisms? The student of botany or zoology will easily find complete discussions of the classification of plants and animals, so it is unnecessary to summarize them here. Instead it seems desirable to discuss the classifications and general habits of the most useful groups, because many of these common animals are less important taxonomically than the bizarre forms, which may have no use at all. The groups of plants and animals that are involved primarily in the aquatic production cycle and the principal groups that are harvested by people have been chosen. The latter are

TABLE 5.2   Scientific Classification of Coastal Cutthroat Trout

Group	Name	Meaning
Phylum	Chordata (Vertebrata)	Having notochord
Subphylum	Craniata	Having cranium
Superclass	Gnathostomata	Having jaw mouth
Class	Teleostomi	Bony fish
Subclass	Actinopterygii	Rayed fins
Order	Clupeiformes	Herringlike
Suborder	Salmonoidei	Salmonlike
Family	Salmonidae	Salmonlike
Subfamily	Salmoninae	Salmonlike
Genus	*Salmo*	Salmon
Species	*clarkii*	(Named in honor
Subspecies	*clarkii*	of a Mr. Clark)

Scientific name:
> *Salmo clarkii clarkii* Richardson 1836
>> Genus—always capitalized
>> Species and subspecies—never capitalized
>> Name always in two or three parts, which are underlined in typescript, italicized in print
>> Name and date identify original species describer
>> Name in parentheses if species now in different genus

those identified in the international fishery statistics by the Food and Agriculture Organization (Table 5.3). Many of these useful groups fit the commonly accepted family units of organisms, but others embrace the more inclusive units of the hierarchy. Families or larger groups in the following discussion have been chosen somewhat arbitrarily.

5.3.1 Phytoplankton

The aquatic plants most familiar to us are the larger plants that we see in the fresh waters, in estuaries, and along rocky ocean shores. These provide a large part of the photosynthesis in estuaries, but in fresh waters they are seldom as important. The plants that play the primary role in photosynthesis in flowing streams are the attached algae, which we usually see as "slime" on the rocks. Similar algae occur in the shallow waters of lakes and oceans, but their contribution to productivity vanishes as soon as the depth exceeds the compensation depth (see Section 4.6). Thus in lakes, in the ocean, and even in large quiet rivers, most photosynthesis is accomplished by the unattached, mostly microscopic phytoplankton.

Plankton is commonly defined as the animals and plants that drift in either fresh or salt water. It is classified according to size and ability to conduct photosynthesis: *phytoplankton* when it can conduct photosynthesis and *zoo-*

TABLE 5.3 International Standard Statistical Classification of Aquatic Animals and Plants (ISSCAAP)[a]

Code and division	World catch in 1992 (metric tons)
1 Freshwater fishes	13,054
1.1 Carps, barbels, and cyprinids	7037
1.2 Tilapias and other cichlids	940
1.3 Miscellaneous freshwater fishes	5094
2 Diadromous fishes	
2.1 Sturgeons, paddlefishes, etc.	0.014
2.2 River eels	0.205
2.3 Salmons, trouts, smelts, etc.	1447
2.4 Shads, milkfishes, etc.	0.66
2.5 Miscellaneous diadromous fishes	0.67
3 Marine fishes	82,534
3.1 Flounders, halibuts, soles, etc.	1175
3.2 Cods, hakes, haddocks, etc.	10,543
3.3 Redfishes, basses, congers, etc.	5853
3.4 Jacks, mullets, sauries, etc.	10,498
3.5 Herrings, sardines, anchovies, etc.	20,389
3.6 Tunas, bonitos, billfishes, etc.	4395
3.7 Mackerels, snoeks, cutlass fishes, etc.	3309
3.8 Sharks, rays, chimaeras, etc.	698
3.9 Miscellaneous marine fishes	10,337
4 Crustaceans	
4.1 Freshwater crustaceans	289
4.2 Sea spiders, crabs, etc.	1585
4.3 Lobsters, spiny rock lobsters, etc.	200
4.4 Shrimps, prawns, etc.	2912
4.5 Krill, planktonic crustaceans, etc.	296
4.6 Miscellaneous marine crustaceans	105
5 Mollusks	
5.1 Freshwater mollusks	303
5.2 Abalones, winkles, conchs, etc.	71
5.3 Oysters	1067
5.4 Mussels	1338
5.5 Scallops, pectens, etc.	1004
5.6 Clams, cockles, arkshells, etc.	1696
5.7 Squids, cuttlefishes, octopi, etc.	2758
5.8 Miscellaneous marine mollusks	606
6 Whales, seals, and other large aquatic animals (numbers)	
6.1 Blue whales, fin whales, etc.	545
6.2 Sperm whales, pilot whales, etc.	33,880
6.3 Porpoises	199
6.4 Eared seals, hair seals, walruses, etc.	164,839
6.5 Miscellaneous aquatic mammals	1261
Humpback whales	1
Baleen whales	38
6.6 Crocodiles and alligators	506,918

(*continues*)

TABLE 5.3   (*Continued*)

Code and division	World catch in 1992 (metric tons)
7 Miscellaneous aquatic animals (MT)	
7.1 Frogs and other amphibians	6189
7.2 Turtles and other reptiles	1590
7.3 Sea spiders and crabs	9227
7.4 Sea urchins and other echinoderms	102,374
7.5 Miscellaneous aquatic invertebrates	442,603
8 Miscellaneous aquatic animal products (MT)	
8.1 Pearls, mother of pearl, shells, etc.	7270
8.2 Corals	3191
8.3 Sponges	222
9 Aquatic plants	
9.1 Brown seaweeds	3.2 TMT[b]
9.2 Red seaweeds	1.3 TMT
9.3 Green seaweeds and other algae	20 TMT
9.4 Miscellaneous aquatic plants	657 TMT
(627 TMT from East Asia)	

Overall Distribution
  Marine areas   85%
  Fresh waters   15%

[a]From "FAO Yearbook of Fishery Statistics," Vol. 74 (1992). FAO, Rome.
[b]TMT means thousands of metric tons (1000 MT = approximately 1% of world catch).

plankton when it cannot. Nonliving material such as inorganic silt or organic detritus is excluded. Plankton ranges in size from the smallest of living forms that can be recognized, such as bacteria close to 0.001 mm in diameter, to jellyfish several meters long. *Nannoplankton* consists of the plants and animals that pass through the most finely meshed nets that are practical to use (with openings about 0.05 mm in diameter). These organisms are separated from the water by filtration through membrane filters in the laboratory or by centrifugation. The intermediate-sized plankton, called either *microplankton* or *net plankton*, is retained by plankton nets. Most of the larger phytoplankton is retained by nets with openings about 0.05 mm in diameter but passes through openings about 0.2 mm in diameter. The larger net plankton, including most adult crustacean zooplankton, is retained by nets with openings about 0.2 mm in diameter. The largest plankton, or *macroplankton*, includes animals such as jellyfish and the intermediate-sized crustacea such as euphausids, which may be several centimeters long. It will be obvious that there is no precise point of separation either between the size classes of plankton or between the drifting plankton and the swimming *nekton*, which includes the fish; both are part of the pelagic community of organisms.

Although nannoplankton is extremely difficult to collect and study, it appears that the relative masses of the different groups are inversely proportional to the sizes of individuals. The nannoplankton constitutes the great mass of living material in most waters; the microplankton, a lesser mass; and the macroplankton, the least.

The phytoplankton converts the energy of the sun to organic material and by doing so supports almost the entire web of life in the sea. It can do this, not by virtue of size, as most of the terrestrial plants do, but by virtue of number; vast numbers are spread throughout the photosynthetic zone of all waters. The bottom of the photosynthetic zone is the level reached by about 1% of the sunlight; it is about 150 m in the clearest water, 10 m in typical coastal waters, and less than 1 m in turbid waters. Phytoplankton falls out of the photosynthetic zone and can live in darkness for a time but is rarely found below 200 m.

The phytoplankton is predominantly composed of diatoms (Fig. 5.2) of the class Bacillariophyceae. These are one-celled algae that have a cell wall of silica that is more or less covered by a jelly. Some live individually; some adhere in colonies or filaments. They usually reproduce by simple cell division and occa-

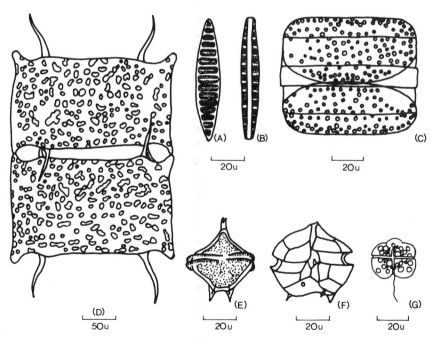

FIGURE 5.2   Some common phytoplankton. (A,B) Top and side views of *Fragilariopsis*, a diatom that is the principal food of the krill of the southern oceans; (C) *Coscinodiscus*, a diatom; (D) *Biddulphia*, a diatom; (E) *Peridinium*, a dinoflagellate; (F) *Gonyaulax*, a dinoflagellate frequently involved in paralytic shellfish poisoning; (G) *Gymnodinium*, a dinoflagellate that occasionally causes red tides.

sionally by a kind of sexual reproduction. Their storage products from photosynthesis are predominantly fats and oils, in marked contrast to higher plants, which store carbohydrates. Not all diatoms are planktonic; some live on the bottom in shallow water, where they may form a thick slime.

Most of the rest of the phytoplankton are flagellated organisms that can be considered either plants or animals. They swim by means of either one or two whiplike flagella but possess either chlorophyll or another similar pigment that enables them to feed as plants. The principal group is that of the dinoflagellates, which have two flagella and a prominent groove around the body (Fig. 5.2).

A characteristic feature of phytoplankton populations is the recurrence of *blooms*. When environmental conditions improve, as with spring warming in temperate latitudes or an increase in nutrients, phytoplankton reproduce very rapidly. One or a few species find some advantage and predominate in the population. They may become abundant enough to color the water or even to make it resemble a thin soup. After gaining predominance, one species commonly gives way to another, and that one in turn to another, so that there is a succession of blooms through the growing season. Such blooms of plant material may radically increase the food for the grazing animals, and these in their turn may flourish and provide increased food for the carnivores. Other blooms of plant material may have an opposite effect and condition the water in a way injurious to animals. An example is a bloom of dinoflagellates called *red tide,* which can kill large quantities of fish or even be directly irritating to people who come into contact with the water.

5.3.2 Insects and Zooplankton

The aquatic stages of insects are important food chain animals of the "slime" zone in shallow fresh waters. Insects are mostly air-breathing animals and a very few live on the surface or carry air down with them. But larvae and nymphs of several orders of insects are completely adapted to benthic aquatic life. These include stone flies, mayflies, dragonflies, caddis flies, and several families of beetles and two-winged flies. These stages of insects have diverse food habits, including plants, animals, and detritus, but they, and the adults when emerging from the water or laying eggs in or near the water, are major food for stream fishes and nearshore fishes in lakes.

But outside the edges of lakes, estuaries, and oceans, in the open waters where phytoplankton are the prevailing plants, almost no insects live. The next organisms in the food chain are mostly zooplankton. This group includes a vast assortment of animals from tiny, one-celled protozoans that may be part of the nannoplankton to crustaceans and jellyfish. It includes a large group of animals that is permanently planktonic as well as a large group of the larval forms of worms, mollusks, crustaceans, and fish that become nektonic or benthic later in life.

Zooplankton is of special interest in the aquatic production cycle because it includes a large proportion of the grazing animals that feed on plant material.

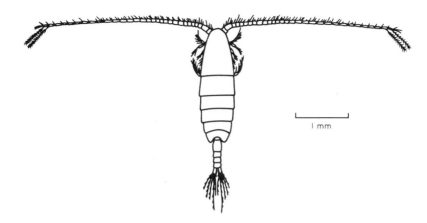

FIGURE 5.3 *Calanus finmarchius*, one of the common copepods of the northern oceans.

These are the larval and older forms that are either small enough to subsist by eating individual phytoplankton cells or possessed of a filter apparatus that will retain large numbers of phytoplankton. Thus, not all plant-eating (i.e., grazing) animals are filter feeders, and many filter feeders do not eat plants, but feed on tiny animals through a filter system that is not fine enough to retain phytoplankton.

The predominant grazing animals of the water are the Microcrustacea, which swarm in the sea as insects swarm on land. Many thousands of kinds exist, but the most important group is the subclass Copepoda. The copepods are mostly of pinhead size, though individuals of one of the more abundant genera, *Calanus* (Fig. 5.3), are about 3 mm long. Like other Crustacea, the copepods grow by molting. The egg is either free in the water or carried by the female, and it hatches into a larva that goes through several molts before maturing. The copepods of the Arctic apparently breed only once a year, whereas the tropical species may have several generations.

Another prominent group of grazing zooplankton consists of tiny shrimplike Crustacea of the high seas, the euphausids. Most of them are less than 1 cm long, but the group includes the krill (Fig. 5.4) ranging from 2 to 7 cm long,

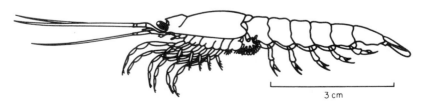

FIGURE 5.4 *Euphasia superba*, the krill of the southern oceans that is the major food of baleen whales.

which comprise the primary food of the baleen whales of the Arctic and Antarctic oceans.

5.3.3 Rhodophyta and Phaeophyta—Red and Brown Algae

The aquatic seed-producing or higher plants live along the sheltered edges of the waters at depths of less than about 3 m. These include rushes, sedges, grasses, and lilies, many of which have been useful to humans through the centuries. They remain useful to small groups of people in many parts of the world, but some of them become serious nuisances at times (e.g., water hyacinth). They are not, however, either a major part of the biomass of aquatic plants or a major part of the commercially important plants.

The most commercially valuable marine plants are the red and brown algae that are commonly known as *seaweeds*.* The seaweeds are attached to the bottom at depths to about 20 m by a holdfast that superficially resembles a root of a higher plant. Above the holdfast they have a stemlike stipe that connects to filaments or blades, in which most of the photosynthesis occurs. The whole plant may be more than 100 m long in some species, such as the giant kelp *Macrocystis* (Fig. 5.5) of the Northeast Pacific. Reproduction of these algae is highly varied but commonly involves a sexual generation that produces gametes and an alternate generation that produces spores. Either or both generations may be conspicuous; in the brown algae the sporophytes are large and the gametophytes are tiny. Both spores and gametes are microscopic, and either or both drift (or swim actively) with the water currents.

The red and brown algae are used directly for human and stock food in many countries; this use is extensive, however, only in the Orient, where many kinds are eaten regularly. Nori, the brown algae of the genus *Porphyra*, is cultivated extensively in Japan. In other parts of the world the seaweeds are valuable principally for their colloids, which are extracted. The brown algae produce algin and the red algae produce agar and carrageenin; all are thickening agents useful in the formulation of drugs, paints, and foods.

5.3.4 Decapoda—Shrimps, Crabs, and Lobsters

The many thousands of species of Crustacea were separated by the earlier naturalists into the Entomostraca, the tiny insectlike forms, and the Malacostraca, the larger forms. This division suits our needs here (even though the Entomostraca have since been divided into several groups) because the Entomostraca are of little direct use to people. Most Malacostraca are bottom living when juveniles or adults and include as a principal group the Decapoda (10 feet), which are taken by the fisheries. These are crustaceans with compound eyes and a carapace that is fused to the thoracic segments. The shrimps and prawns are decapods with long abdomens and with bodies compressed from

---

*Some of the other red and brown algae are unicellular and some of the other red algae form parts of coral reefs. In fresh water another group, the green algae, predominates.

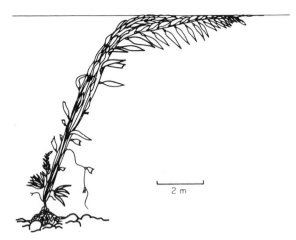

2 m

FIGURE 5.5 A mature plant of the giant kelp *Macrocystis pyrifera*, a brown alga of the Pacific Coast of North America.

side to side. The lobsters and crayfish are larger decapods with long abdomens and with bodies compressed from top to bottom. The edible crabs are decapods with abdomens that are usually short and folded beneath their bodies.

Like all crustaceans, the decapods grow by molting. The molts occur in quick succession during larval stages, when the animals grow rapidly. As the animals mature, the frequency of molting decreases to a common rate of once a year in the larger crabs and lobsters, and perhaps only once in several years in very old animals. In many mature female Crustacea a molt immediately precedes mating, during which sperm are deposited in the female to be used later to fertilize the eggs as they are laid. The newly fertilized eggs of crabs and lobsters are commonly attached to abdominal appendages and carried by the female until hatching (eggs of some shrimp are discharged into the water). After hatching, the larvae of most Crustacea drift or swim with the ocean currents for a time and then descend to the bottom, where they live as juveniles and adults. In crab and lobster fisheries conducted by pots, the egg-carrying females are commonly protected by law and must be released alive.

Many species of shrimps and prawns (the latter word frequently means merely large shrimps) are sought by the fisheries, but two groups are sought above all others. Most important are shrimps of tropical and subtropical waters around the world belonging to the family Penaeidae (Fig. 5.6). The valuable species of these shrimps spawn regularly in the open sea; the larvae pass through several developmental stages while drifting toward land and then grow rapidly for a time as juveniles in estuaries. They then return to the sea bottom to mature in depths of 10–150 m, where they are caught by trawls. The whole life cycle is usually completed in about 1 year. The fisheries for these shrimps are usually bottom fisheries on the continental shelves near or in estuaries.

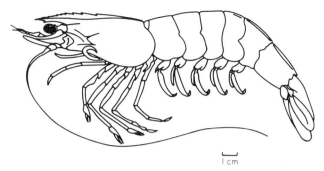

FIGURE 5.6   The brown shrimp, *Penaeus azlecus,* family Penaeidae, one of the principal commercial shrimps of the Caribbean.

The other major group of shrimps is that of the cold-water shrimps of the family Pandalidae. These are found on the continental shelves to 200 m and in the deep esuaries of the temperate and arctic waters of the Northern Hemisphere. The most common species is *Pandalus borealis* of both the Atlantic and Pacific oceans. The shrimps of this family mature first as males and later become females during a life span of 3–5 years. As a consequence, mature females are preferred by the fisheries, and some stocks have been depleted quickly. Other stocks seem to be very large, such as some in the Bering Sea.

Crab fisheries seek only a few of the thousands of species of crabs, most of which are larger species in three families. The Cancridae includes *Cancer,* a genus of edible crabs of the Northeast Atlantic and North Pacific oceans (where one species is the Dungeness crab of western North America). The Portunidae, a family of swimming crabs, includes the genera *Portunus* and *Callinectes* of Europe and North America, which are commonly known as blue crabs (Fig. 5.7).

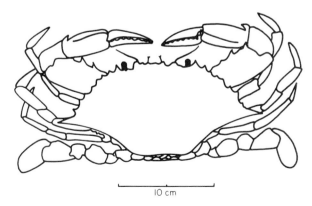

FIGURE 5.7   The blue crab, *Callinectes sapidus,* family Portunidae, of the North Atlantic Ocean.

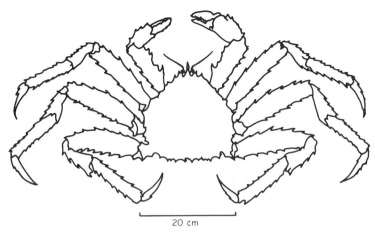

FIGURE 5.8 The king crab, *Paralithodes camtschatica*, of the North Pacific Ocean and the Bering Sea.

The Lithodidae is a family of long-legged crabs in which the posterior pair of legs is degenerate, and hence they appear to be eight-legged crabs. This family includes the genus *Paralithodes*, a species of which is the king crab of the North Pacific Ocean and the Bering Sea (Fig. 5.8). Another member of the family is the centolla of the continental shelves near southernmost South America.

The lobsters belong to two families of large marine Crustacea that live on the continental shelves. The Homaridae (Fig. 5.9) includes the large (to 13 kg) lobsters of the North Atlantic genus *Homarus*, which have very large pinching

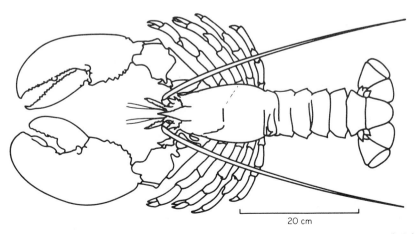

FIGURE 5.9 The lobster *Homarus americanus*, family Homaridae, of the western North Atlantic Ocean. It is very similar to *H. gammarus* of European waters.

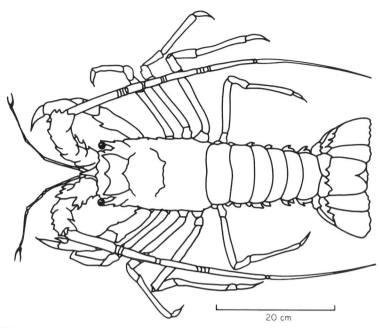

FIGURE 5.10   The spiny lobster, *Palinurus elaphas,* family Palinuridae, from the Mediterranean.

claws on the first pair of legs. The Norway lobster, *Nephrops,* of the Northeast Atlantic and the freshwater crayfish are closely related. The family Palinuridae includes the large sea crayfish or spiny lobsters (Fig. 5.10) of tropical and subtropical waters around the world. These lobsters have no pinching claws on the legs.

5.3.5 Lamellibranchiata—Bivalve Mollusks

The Lamellibranchiata are mollusks with two shells (two values) and a mouth, but no head, and they are bilaterally symmetrical or nearly so. They are all aquatic, and most are marine animals that live on the bottom in the photosynthetic zone. Most of them maintain currents of water through their bodies by means of cilia and feed by separating the microscopic plants and animals from the water current. About 11,000 species have been described, of which only a very few are eaten by humans.

The most diverse group of commercially important species is that of the hard-shell clams, which belong to several families. The taxonomy of these is too complicated to present here, but in general they are symmetrical bivalves with a foot and with two calcareous shells that close tightly together. They burrow in the bottom and must be harvested by digging or dredging. Included are the genera *Venus* and *Cardium* (cockle) of Europe, *Mercenaria* (quahaug)

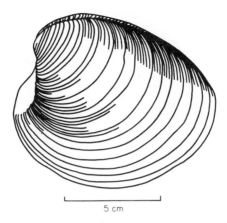

FIGURE 5.11   The northern quahaug, *Mercenaria mercenaria*, family Veneridae, of the Atlantic coast of North America.

(Fig. 5.11), *Spisula* (surf clam), *Arctica* (ocean quahaug), *Saxidomus* (butter clam) of North America, *Venerupis* and *Meretrix* of East Asia, and many others. Most of these are harvested from wild stocks, but some are cultivated, especially *Venerupis* in Japan.

Another group of clams is that of the soft-shell clams of the family Myacidae. They have a gaping, fragile shell. *Mya* (Fig. 5.12), the principal North American genus, lives primarily in mudflats of the intertidal zone on both the Atlantic and Pacific coasts.

The mussels of the family Mytilidae are bivalves that have shells covered with a noncalcareous outer layer and attachments called *byssi*. They occur in large beds. The important genera are *Mytilus* of Europe and *Aulacomya* of South America. *Mytilus* (Fig. 5.13) is abundant in North America but is in little demand, perhaps because of the occurrence of paralytic shellfish toxins in the species of the Pacific Coast. Mussels are cultivated extensively in Europe and to a lesser extent in East Asia.

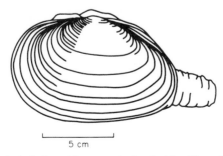

FIGURE 5.12   The soft-shell clam, *Mya arenaria*, family Myacidae, of the Atlantic and Pacific coasts of North America.

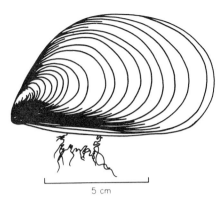

FIGURE 5.13   The bay mussel, *Mytilis edulis*, family Mytilidae, is cultivated in Europe and grows naturally from arctic to subtropical waters on both coasts of North America and Northeast Asia.

The family Ostreidae includes the two important genera of edible oysters, *Ostrea* and *Crassostrea*, which are cultivated extensively in several parts of the world. The family is distinguished by the single adductor muscle, dissimilar shell valves, attachment of the left shell, absence of a foot in the adult, absence of a byssus, and the triangular ligament (Fig. 5.14).

Ostrea is the flat oyster with the left valve that is not deeply cupped and is more or less circular; *Crassostrea* is the larger, elongated oyster with the left valve deeply cupped. The genera have been separated also because of the structure of the larva and the habit of briefly incubating the eggs in *Ostrea*. *Ostrea edulis* is the well-known European flat oyster; *O. lurida* is the Olympia oyster of western North America. *Crassostrea virginica* is the American oyster of eastern North America; *C. angulata*, the Portuguese oyster; *C. gigas*, the Japa-

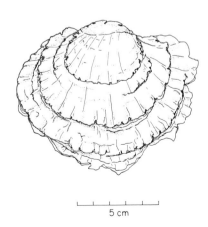

FIGURE 5.14   The European oyster, *Ostrea edulis*, family Ostreidae.

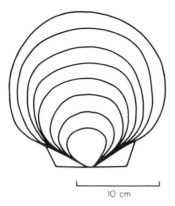

FIGURE 5.15 The giant scallop, *Placopecten magellanicus,* family Pectinidae, of the western North Atlantic.

nese oyster; and C. *curcullata,* the rock oyster of the Indo–Pacific region. A few dozen species of the two genera have been described, but in addition numerous races of some species have been found. They live in the shallow coastal waters and estuaries of the globe outside of the polar regions.

The scallops of the family Pectinidae (Fig. 5.15) are among the few bivalve mollusks that can swim when adult, albeit clumsily. They produce a jet of water by clapping their shells together and can swim for short distances. Commonly they live unattached on the bottom. Usually the scallop is dressed after capture by the fisherman and only the adductor muscle is marketed.

5.3.6 Cephalopoda—Squids and Octopi

The cephalopods are bilaterally symmetrical marine mollusks with a well-developed head surrounded by prehensile tentacles, and they usually have two camera-type eyes. The squid (Fig. 5.16) and cuttlefish have 10 tentacles (Decapoda) and an internal shell; the octopus (Fig. 5.17) has 8 tentacles (Octopoda) and no shell. They include the largest known mollusks; one squid

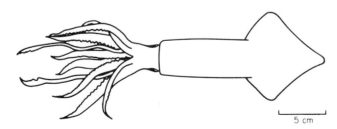

FIGURE 5.16 The common squid, *Loligo vulgaris,* family Loliginidae, of the North Atlantic Ocean.

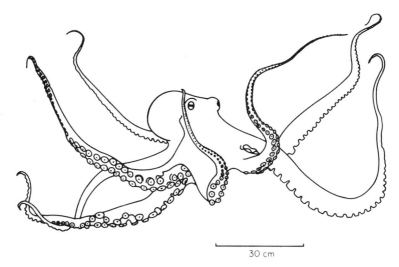

30 cm

FIGURE 5.17   The common octopus, *Octopus vulgaris*, family Octopodidae, of the North Atlantic Ocean.

reaches a length of nearly 20 m. All are carnivorous, feeding on fish, mollusks, and crustaceans.

Octopi live singly, frequently hiding in crevices or holes. They either crawl about with their tentacles or swim by expelling a jet of water. They are caught in some fisheries in earthenware pots placed on the bottom to provide hiding places. The major fisheries are near Japan, Spain, and Italy.

Squids swim agilely, either by jet propulsion or by fleshy fins on the sides of the body. Some species swim in great schools, from which they are caught by trawls. Some schools of squid in midwater on the high seas probably constitute a major underdeveloped fishery resource. The important genera are *Sepia,* the cuttlefish of Europe, and *Loligo,* the widely distributed common squid, or calamares.

5.3.7 Selachoidei—Sharks

Almost everyone who is concerned with the sea has on occasion feared, been fascinated by, or sought out sharks (Fig. 5.18). The 250 or so species are almost all marine and are most abundant in tropical and subtropical waters. A few individuals venture into fresh water, several genera live in temperate waters, and one genus lives in the Arctic. Some kinds prefer to live on the continental shelves and along the coasts, others live on the high seas, and a few inhabit the depths to 1500 m.

Sharks are usually large fishes. However, they range from small adults of 50 cm to the whale shark, the largest fish known, of about 20 m. They are all carnivorous. Some have teeth adapted to crush either crustaceans or mollusks, and others have a filter system designed to capture either small crustaceans or

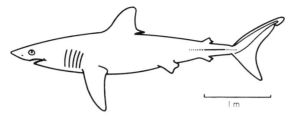

FIGURE 5.18 The porbeagle, *Lamna nasus,* family Isuridae, of the North Atlantic Ocean, a shark sold widely for food.

fish, but many have large teeth suitable for cutting or tearing animals as large as or larger than themselves. The sharks are considered to be more primitive than the bony fishes and are characterized by having between five and seven gill slits, multiple rows of teeth, a cartilagenous skeleton, and a skin armed with placoid scales of the texture of sandpaper. Fertilization is internal. They produce few young, usually fewer than 50 per female. These are either born alive or hatch from a large, horny egg capsule.

The order is divided into about 20 families of modern sharks, and most of the familiar families, genera, and species belong to the suborder Galeoidei. Sharks of this suborder have five gill openings, a nearly cylindrical trunk, an anal fin, and usually two dorsal fins. The nearest relatives of the sharks with any importance as a resource are the skates and rays, of which a few are used for food and oil.

Sharks are justly feared, although the chance of attack in most waters is extremely slight. Most species of sharks are too small, or too sluggish, or have teeth too small to be of any potential danger, but individuals belonging to a few species, notably the white shark *Carcharodon,* the tiger shark *Galeocerdo,* some members of the genus *Carcharhinus,* the lemon shark *Negaprion,* and some of the hammerheads have been known to attack people without provocation. Moreover, sharks have an extraordinarily well-developed sense of smell, which apparently they use extensively when searching for food. They may be attracted to garbage or animal refuse and sent into a frenzy by blood in the water. When in a feeding frenzy, many of the more voracious sharks will attack anything in the water whether or not it is digestible. Such feeding habits also lead sharks to attack fishermen's lines and nets, to which they may be attracted by the captive fish. They can be a fantastic nuisance at times.

On the other hand, sharks are sought in most warmer waters of the world for their flesh, their livers, which are very large and oily, or their hides, which make a fine leather. The flesh of many species is firm and tasty and is readily marketable at good prices. The fins of some species are considered a delicacy; they are preserved by salting and drying for later use in soups. The liver oil of some species is especially rich in vitamin A. Substantial fisheries developed for some of these species around 1940, but they ceased when synthetic vitamin A became available. The hides can be tanned into a tough, fine-grained leather, but they are difficult to remove without damage; the demand for shark leather is relatively small because of the large supply of mammalian leathers and plastics. The flesh of many sharks has a relatively high content of urea, which tends to cause off-flavors rapidly in the preserved flesh and inhibits the canning of the flesh because of the development of ammonia in the cans. Several species of sharks are valued for their high gelatin content, and their flesh is blended with the flesh of bony fish into a variety of fish cakes and sausages. Substantial quantities of the flesh are dried crudely in many of the less developed countries of the world. One species, the mako shark, is prized by anglers because it fights and jumps in a spectacular fashion.

5.3.8 Clupeidae—Herrings

The Clupeidae is the most valuable family of food fishes in the world. It is the principal family of the order Clupeiformes, the group of fish with a form most like the original form of the bony fishes (Teleostei), from which all other bony fishes have evolved. Other families of the order include the Megalopidae (tarpons), Albulidae (bonefishes), and Engraulidae (anchovies). The catch of herrings in 1978 was about 11 million tons, or 15% of the world total. A majority of it was used directly for human food; thus, the herrings comprise one of the most valuable families of fishes.

The Clupeidae are characteristically small (<50 cm), schooling fish with silvery bellies and sides and greenish gray backs. They have no spines in the fins, one short dorsal fin, deeply forked tails, ventral fins on their abdomens far behind the pectorals, deep bodies, and large scales that slip off at a touch. Their flesh is oily, a feature that adds greatly to their flavor and furnishes a valuable oil for industry. They have intramuscular bones in abundance.

The family is considered now to have about 160 species in about 50 genera, most of which are marine and tropical and of little importance. Some live in fresh water; others are anadromous—living in the sea and ascending rivers to spawn. The genera that are of most importance as food are widely distributed around the world and may be extremely abundant in many places. The principal genera will be discussed separately.

Clupea, the sea herring (Fig. 5.19), is the major herring of the temperate and subarctic waters of the Northern Hemisphere. It is found in the Atlantic from Cape Hatteras north to Greenland, east to Spitsbergen, and south to the Bay of Biscay. In the Pacific it is found from Korea to the Arctic Ocean and south to California. The genus is divided into a number of species and subspecies that are not easily distinguishable, but more importantly it comprises dozens of different races, each with distinctive life habits. Some migrate thousands of kilometers annually; others remain in the same bay. Most races spawn in spring; others spawn in summer or fall. Some grow rapidly and rarely exceed 5 years of age; others may live to be 25 years old. Some races may spawn in the intertidal zone; others spawn in 200 m of water. The eggs are adhesive and are

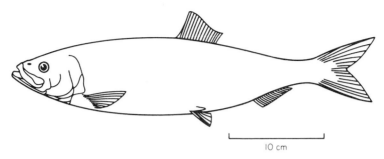

10 cm

FIGURE 5.19   The North Atlantic herring, *Clupea harengus*, family Clupeidae.

laid on the bottom. In some small races the females each produce fewer than 10,000 eggs at every spawning; in others they produce more than 100,000. The sea herrings eat zooplankton with an extraordinary efficiency. They are found where the zooplankton is abundant, whether the plankton is composed predominantly of copepods, euphausids, crab larvae, molluscan larvae, worm larvae, or fish larvae.

In most parts of the world herring are either salted, smoked, or dried. Many small herring are canned and called *sardines* and as such should not be confused with the genera *Sardinops, Sardinia,* or *Sardinella,* which are properly called *sardines.* Considerable quantities are processed for the oil and dried to fish meal. The oil is a valuable base for paint and ink; the meal is used as a vital ingredient of domestic animal and poultry rations.

The genus *Sprattus,* the sprat, a small herring of the Northeast Atlantic Ocean and the Mediterranean and Black seas, is similar to *Clupea* in habit and forms the basis for an extensive fishery where it occurs. The genus *Clupeonella,* also a sprat, supports valuable fisheries in the Sea of Azov, the Black Sea, and the Caspian Sea.

The genera *Alosa,* the shad of Europe and North America, *Ilisha,* the shad of India, and some of the species of *Caspialosa,* the shad of the Black and Caspian seas, are larger herrings that migrate up rivers to spawn. One North American species, *Alosa pseudoharengus,* the alewife, usually enters fresh water only to spawn, but may become landlocked. Recently this species has become the predominant fish in the Great Lakes of North America and apparently could support a large fishery.

Sardinops, the pilchard or sardine (Fig. 5.20), is the second most important genus of the family. Major stocks occur off Japan, the western United States, Southwest Africa, Australia, New Zealand, and western South America. It is a small, rapidly growing herring with a usual commercial size of 10–25 cm and age of 1–7 years. It may reach 30 cm and 15 years, however. *Sardinops* are found primarily in cool, upwelling waters of temperate latitudes that are rich in plankton. They feed on either phytoplankton or zooplankton or both by filtering the smaller forms out of the water and striking at the larger ones. Spawning

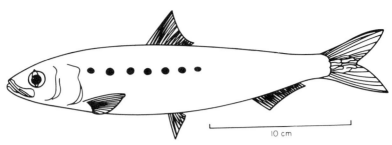

FIGURE 5.20   The Japanese pilchard (Ma-iwashi), *Sardinops melanosticta*, family Clupeidae, one of the important commercial fishes of Japan.

may occur throughout the year or in a certain season in some races. Most stocks migrate a few hundred kilometers each year.

Sardinia, the sardine of the Mediterranean regions, is similar to *Sardinops* in habits. It is a small herring; those caught by the commercial fishery are usually between 10 and 20 cm long and less than 3 years old. It spawns in autumn and winter. It is nowhere as abundant as species of *Clupea* or *Sardinops*, because the Mediterranean waters are relatively barren, but it supports important local fisheries.

Sardinella is the major genus of tropical sardines. It also is a small herring, rarely reaching more than 20 cm long and more than 3 years old. The principal stocks occur in the Arabian Sea, off Southeast Asia, in the Gulf of Guinea, and off Northwest Africa, but one species or another can be found in most tropical coastal waters of the world. It supports many primitive fishermen, who commonly attract the fish to the surface with lights and catch them with either dip nets or enveloping nets.

Brevoortia, the menhaden, occurs off eastern North America from the Gulf of Mexico to the Gulf of Maine and along eastern South America off southern Brazil and northern Argentina. It is a medium-sized herring measuring, on the average, between 30 and 40 cm in length, and living up to 10 years. Unlike most other herrings it feeds almost entirely on diatoms and the smaller zooplankton, which it obtains through a filtering system that is one of the more efficient known among fish. It has been reported that an adult can filter 25 liters/min.

The menhaden is too bony to provide good food and is useful to the fisheries for oil and fish meal. In recent years, the landings of menhaden in the United States have been greater in volume than those of any other species taken by the United States fishermen.

General   When one compares the world distributions of all the herrings (including those minor genera that were not discussed) and the anchovies (to be discussed next) with the biologically enriched areas of the oceans, it appears that almost every enriched part of the ocean has an abundance of one species from one family or the other. *Clupea* and *Sprattus* occur in the cold waters of the north (not the south), *Sardinops*, *Brevoortia*, and *Engraulis* occur in the temperate and subtropical waters, and *Sardinella* occur in the tropical waters. These fish are the principal converters of energy from phytoplankton and zooplankton to fish flesh; they are the predominant small pelagic fish of the world.

It is interesting that few of these fishes are known to occur in abundance in the enriched areas of the southern oceans, but this dearth may be due to a lack of exploration. It is interesting also that rarely are two species from these families equally abundant in the same area. When two species occur along the same coast, they do so either at different depths or at different seasons; one species predominates and seemingly excludes the other, but this problem is one of population ecology, and it will be discussed in a later chapter.

5.3.9 Engraulidae—Anchovies
The total production of the anchovy family, which had been nearly 20% of the world's total around 1970, declined to about 4% in 1978 with the collapse of some major stocks. Part of the catch is dried into meal for animal food, and part is cured or canned for human food.

The anchovy family includes only about six genera and 40 species. These are widely distributed in tropical, subtropical, and temperate waters. Anchovies resemble herring superficially but have a larger mouth and a rounded belly. They are small fish; few exceed 20 cm in length and most live 3 years or less. Like the herrings, they feed predominantly on the tiny animals that constitute the bulk of the zooplankton, such as the copepods, larval crabs, and larval mollusks.

Almost all of the commercial anchovies of the world belong to the genus *Engraulis* (Fig. 5.21). The major stock is found off Peru and northern Chile; lesser stocks are found in the western North Pacific, off Australia, in the Mediterranean and Black seas, and in the Atlantic off Europe.

5.3.10 Salmonidae—Salmons and Trouts
The family Salmonidae includes some of the most exciting fish of the world, either to eat or to catch for sport. Their flesh is usually richly flavored and

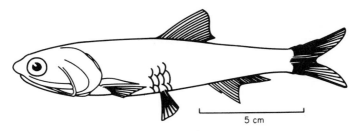

FIGURE 5.21 The Peruvian anchovy, *Engraulis ringens*, family Engraulidae, the species that once supported the largest fishery in the world.

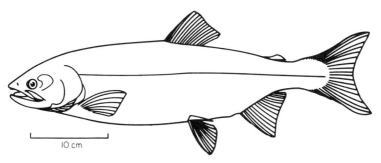

FIGURE 5.22 The pink salmon, *Oncorhynchus gorbuscha*, family Salmonidae, the most abundant of the Pacific salmons.

commands a premium price. Some offer a spectacular display when hooked on angling gear, and anglers may pay hundreds of dollars per day for the privilege of seeking them. Most species inhabit clean, well-oxygenated, colder waters of the Northern Hemisphere. Several species have been introduced into suitable waters in the Southern Hemisphere.

The family includes three important genera. *Salmo* contains the Atlantic salmon and the rainbow trout, brown trout, and steelhead trout. *Oncorhynchus* (Fig. 5.22) includes the six species of Pacific salmon, one of which occurs only in East Asia and the rest of which are found around the North Pacific rim from Japan to California. *Salvelinus,* the genus of the chars, has the brook trout, lake trout, Dolly Varden trout, and the Arctic char. Another less widely known genus is *Hucho,* a large predatory trout (to 1.5 m) of the Danube River. The family is characterized by the presence of an adipose fin, fins without spines, and smooth (cycloid), small scales. Closely related fishes are the whitefish of the family Coregonidae (sometimes named in the Salmonidae as a subfamily) and the smelts of the family Osmeridae.

Many members of the Salmonidae family have an unusual tolerance for either fresh water or salt water. The Atlantic salmon, Pacific salmon, steelhead trout, and Dolly Varden trout are commonly anadromous; they spawn in fresh water, move to sea in the juvenile stage to feed, fatten, and mature, and return to fresh water to spawn. Some of the other species, such as brown trout and brook trout, may go to sea occasionally. On the other hand, all of the anadromous species can spend their lives in fresh water, although usually they do not grow as rapidly or reach as large a size as they do in the sea. A recent spectacular example of freshwater life for an anadromous fish was the successful introduction of the coho salmon, *Oncorhynchus kisutch,* into the Great Lakes of North America.

The anadromous habit of many species and the ability of individuals to return to the home streams in which they were born has attracted much attention. Some species undergo spectacular color changes as they prepare to spawn. Individuals of some species make regular migrations for thousands of kilometers at sea; indeed, the Atlantic salmon from Norwegian home streams may feed off Greenland, and the Pacific salmon from Japanese home streams may feed in the Gulf of Alaska.

All of the members of this family are carnivorous. When young, their diet usually consists of either small crustaceans or insects in fresh water and small crustaceans in the sea; as they become older, commonly their diet includes a great proportion of small fish.

Most of the members of the Salmonidae produce relatively few (<5000) large eggs. Commonly the females lay their eggs in a nest or redd that they build in gravel through which water is circulating. The males fertilize the eggs and then the females cover them with gravel. After an incubation period of between 30 and 150 days the eggs hatch into relatively large, agile larvae. The larvae of several species, notably the rainbow trout, will accept artificial food as soon as they begin to feed; therefore, such species are very adaptable to aquaculture.

5.3.11 Cyprinidae—Carps and Minnows

The Cypriniformes are unique in having a number of small bones that connect the swim bladder to the hearing apparatus in the skull. Most of them are freshwater fish; some occur on all of the continents. They include the common characins of Africa and America in the family Characidae, many families of catfishes of the suborder Siluroidei (see Section 5.3.12), and the carps and minnows of the family Cyprinidae.

The Cyprinidae contains more species than any other family of fish. They are freshwater fish of Africa, Asia, North America, and Europe, but unless artificially reared they are not found in Australia or South America. They possess no teeth on the jaws but normally have well-developed teeth on the pharyngeal bones. Their fins are usually without spines. Usually they possess a single dorsal fin. Ordinarily the bodies are covered with smooth cycloid scales, though occasionally they may be naked.

The biology of the cyprinids is highly variable. A few live in cold waters and require a large supply of oxygen, but many prefer warm waters and some tolerate an amazingly low quantity of oxygen, as little as 0.5 mg/liter. Most female members of the family deposit large numbers of eggs; these are fertilized externally and usually are not cared for. Most cyprinids do not make long migrations from feeding areas to spawning grounds. Larval cyprinids usually feed on zooplankton, but the older animals eat many kinds of food. Most juveniles and adults feed on bottom animals. Some adults live on other species of fish; some are herbivores and feed either on phytoplankton or detritus, and a few even feed on higher plants.

Hundreds of species of the Cyprinidae are small fish. These include the minnows, daces, chubs, and shiners. These feed predominantly on insects and zooplankton throughout their lives and in turn provide forage for larger predatory fish.

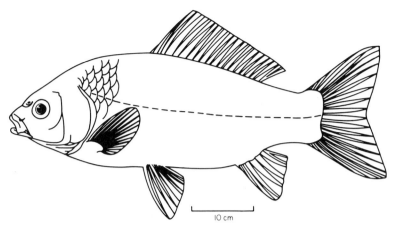

FIGURE 5.23   The Asian carp, *Cyprinus carpio*, family Cyprinidae, a species introduced over much of Europe and North America and extensively cultured in many parts of the world.

A few species of the family provide highly important food fishes in Europe, Asia, and Africa. These include the roaches, the breams, the carps, and the tench. The wild carp of the genus *Cyprinus* (Fig. 5.23) lives in many of the sluggish rivers and lakes of Europe and Asia, where it is subject to intensive fishing. The species has been bred for thousands of years and several domesticated varieties have been developed (see Section 11.2.1). Other genera of Asian carps of the family Cyprinidae are sought likewise in the wild and grown in captivity; these include *Catla, Labeo, Barbus,* and *Cirrhina.*

Another important genus is the crucian carp, *Carassius.* This fish is closely related to the wild carp. It is extremely hardy, grows to a length of 45 cm, and serves as a food fish in parts of Europe and Asia. A form of *Carassius,* the goldfish, has been selectively bred into hundreds of varieties for aquarists.

5.3.12 Siluroidei—Catfishes

The order Cypriniformes includes the suborder Siluroidei, to which belong about 28 families of catfishes. Most of them lack scales, but some have spines or plates. Usually they have a head with several pairs of barbels and a body flattened from top to bottom. Many have a single strong spine in some of the fins.

The catfishes are of special importance as food fish because they have few bones in the flesh, and most are especially tolerant of poor water with small quantities of oxygen. Some, such as the walking catfish, can even live on land in moist vegetation for hours at a time. Most of the species build nests; commonly the males guard the eggs and young. Most are predatory bottom feeders and they feed on invertebrate animals or other fish. Some species are well adapted to freshwater aquaculture.

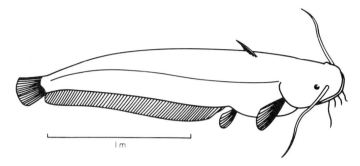

FIGURE 5.24   The European catfish, *Silurus glanis*, family Siluridae, one of the largest freshwater fishes.

The suborder includes the Old World catfishes of the family Siluridae (Fig. 5.24), which may reach 2.5 m in length and 300 kg in weight. Others are the Bagridae of Africa and Asia, the Clariidae of Africa and Southeast Asia, and the Ictaluridae of North America. Several diverse families occur in South America, some of which are spiny or armored. Two families, Bagridae and Plotosidae, include species that live in the sea.

5.3.13 Scomberesocidae—Sauries

The sauries are little-known, small (<45 cm), pelagic fishes of tropical, subtropical, and temperate parts of the open ocean. Their closest relatives are the halfbeaks, family Hemiramphidae; flying fishes, family Exocoetidae; and needlefishes, family Belonidae. They spawn in the open sea, and their eggs and larvae are pelagic. Their food consists of zooplankton, small crustaceans, and small fishes. The principal commercial fisheries are in the subtropical and temperate waters of the Northern Hemisphere.

The Atlantic saury, *Scomberesox* (Fig. 5.25), is distributed widely in the Atlantic as far north as Newfoundland and northern Norway. It occurs also in the Mediterranean and Black seas. The Pacific saury, *Cololabis*, inhabits the waters of both the Asiatic and American coasts, north to Sakhalin and Alaska. The flesh of these sauries is moderately oily, very tasty, and in great demand when available. They are taken with floating gill nets or purse seines but are

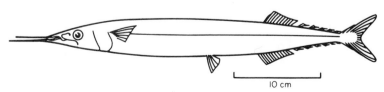

FIGURE 5.25   The Atlantic saury, *Scomberesox saurus*, family Scomberesocidae.

caught most successfully at night by dip net after having been attracted to the surface by lights.

5.3.14 Gadidae—Cods and Hakes

The gadoids, or codlike fishes, include only about 70 species, but they make up one of the most important families of food fishes in the world. World production in 1978 was about 10.4 million tons, an amount that reflects the long, slow growth of the fisheries from prehistoric times. The gadoid fish populations tend to be relatively stable in production.

The Gadidae are characterized by lacking spines in the fins, by having ventral fins located far forward on the belly, and by being physoclistous (i.e., when an air bladder is present it has no connection with the gut). Most of the family have either three dorsal fins and two anal fins or second dorsal and anal fins with very long bases. They also lack intramuscular bones, a feature that makes them easy to eat.

The family is usually divided into three subfamilies. The Gadinae includes *Gadus*, the cod (Fig. 5.26); *Odontogadus*, the merlang of Europe; *Melanogrammus*, the haddock; *Pollachius*, the pollock or saithe; and *Theragra*, the walleye pollock. The Merlucciinae contains only the genus *Merluccius*, the hake (Fig. 5.27). The Lotinae, much less commercially important, has *Brosme*, the cusk; *Urophycis*, another hake; and *Lota*, the single freshwater genus.

The commercially important genera are all marine. All live in temperate or polar waters predominantly in the Northern Hemisphere with the exception of *Merluccius*, which is abundant off South Africa, Argentina, and Chile, as well as off Europe and both coasts of North America. All live and feed on or near the bottom of the continental shelf or continental slope except *Merluccius* and *Theragra*, which live at times near the surface or at intermediate levels. All migrate along routes suited to their reproductive and feeding needs—some of them up to 2000 km annually. Most of them spawn in late winter or early spring, each female producing many millions of eggs. The eggs are broadcast in

50 cm

FIGURE 5.26   The Atlantic cod, *Gadus morhua*, family Gadidae, a species of major importance to many countries.

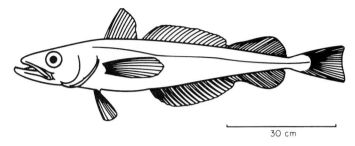

FIGURE 5.27 The Pacific hake, *Merluccius productus*, family Gadidae, an abundant species off the west coast of North America.

the ocean to drift with the currents. The larvae and young, up to a few centimeters long, are also pelagic; many drift up to 3 months and hundreds of kilometers before descending to the bottom.

Like many larger fishes of cold waters, most gadoid fishes grow relatively slowly and live many years. Cod reach sexual maturity between 4 and 15 years, depending on the race, and some may live 30 years. Most of the gadoids in commercial catches are between 2 and 10 years old.

All of the gadoids feed on other animals, principally on invertebrates when small and on fish when large. Starfishes, sea urchins, mollusks, crustaceans, and fish are common foods.

5.3.15 Percoidei—Perchlike Spiny-Rayed Fishes

The order Perciformes includes a large proportion of the total species of fish. These fishes usually have two dorsal fins, of which the first is spiny. Their pelvic fins are usually located under or in front of the pectoral fins.

This morphologically diverse order is divided among about 20 suborders, of which three are especially important for either food or angling. One of these, the Percoidei, includes about 100 families that are most like the original type of the order in that they have spines in the fins. They inhabit both marine and fresh waters ranging from tropical to subarctic, but a large proportion of the marine species are found in the coastal zones of either tropical or subtropical waters. The species of major importance are mostly those in the following eight families, which are grouped together here because they are similar in general appearance and habit. (The other especially important suborders are the Scombroidei, the tunas and mackerels, and the Cottoidei, which includes the rockfishes of the family Scorpaenidae.)

The sea basses and groupers are the best known of the more than 400 species of the family Serranidae (Fig. 5.28). They are usually robust with large mouths, small teeth, ctenoid scales, and either a round or a truncate caudal fin. Most species live in shallow tropical or subtropical seas, but a few occur along temperate shores or in fresh water. Most are predatory on other fishes or

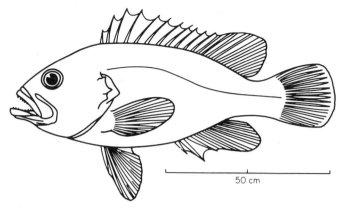

FIGURE 5.28 A grouper, *Epinephelus guaza*, family Serranidae, from the Mediterranean.

invertebrates. Some may exceed 400 kg in weight. Most are excellent to eat, and many are sought by anglers. Some of the smaller tropical species can change color rapidly and are therefore called "chameleons of the sea."

The snappers of the family Lutjanidae (Fig. 5.29) include about 250 species. Most are large, voracious, and brightly colored fish of shallow tropical or subtropical seas. They have large mouths with canine teeth, robust bodies, ctenoid scales, and usually truncate or forked caudal fins. Most are highly valued for either food or sport, and some for both. The name *snapper* has been applied because some species snap their jaws tightly after being caught.

The croakers and drums of the family Sciaenidae (Fig. 5.30) comprise about 150 species, most of which live in shallow tropical or subtropical seas. They are usually somber in color and in some tropical fisheries are the major fishes taken from the bottom at depths of 10–80 m. Many species are sought for food; a

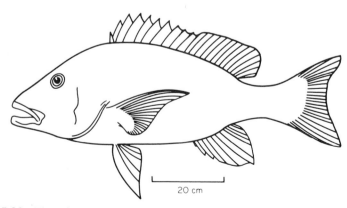

FIGURE 5.29 The red snapper or pargo, *Lutjanus aya*, family Lutjanidae, one of the excellent food fishes of the Caribbean region.

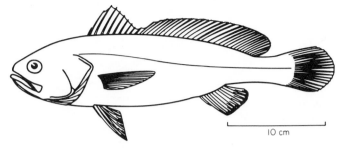

FIGURE 5.30 The yellow croaker, *Pseudosciaena manchurica*, family Sciaenidae, one of the most abundant food fish of the East China Sea.

few are sought for sport. The name *croaker*, or *drum*, has been applied to many of the species because they have a resonant air bladder with a vibratory mechanism that produces fairly loud croaks or grunts. These fishes usually have a medium-sized or small mouth, a second dorsal fin with a longer base than the first, and ctenoid scales. All are carnivorous.

The porgies of the family Sparidae (Fig. 5.31) include more than 100 species. These occur in seas ranging from tropical to subarctic. A few species attain a large size; some are prized by anglers. Some species of moderate size comprise the major commercial stocks on tropical continental shelves between 60 and 200 m. Usually they have small mouths, compressed bodies, and forked caudal fins. All are carnivorous.

The jacks, scads, and pompanos of the family Carangidae (Fig. 5.32) include about 200 species. They inhabit seas ranging from tropical to temperate. Most

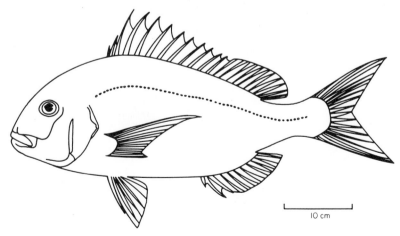

FIGURE 5.31 The Japanese Ma-dai, *Chrysophrys major*, family Sparidae, one of the most delicious fish in the world.

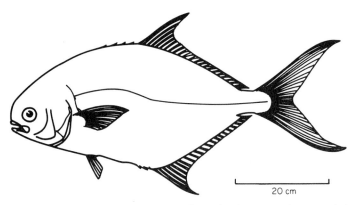

20 cm

FIGURE 5.32   The pompano, *Trachinotus carolinus*, family Carangidae, one of the choicest food fish of the western tropical Atlantic.

are good eating fishes. Some are highly prized as game fishes, for example, the amberjacks of the genus *Seriola* and the cavalla or kingfish of the genus *Caranx*. Some are schooling fishes, such as the genus *Trachurus*, which includes the maasbanker of South Africa and the jack mackerel of the North Pacific Ocean, and they support substantial commercial fisheries. The members of the family are characterized by compressed bodies, small, smooth scales or none at all, forked caudal fins, two separate spines in front of the anal fin, and, in some species, a ridge of bony plates along either side of the caudal peduncle. All are carnivorous and robust, swift swimmers.

The perches of the family Percidae (Fig. 5.33) are freshwater fishes of the Northern Hemisphere. The common perches of the genus *Perca* of North America, Europe, and Asia and the walleyed pikes of the genera *Stizostedion* of North America and *Lucioperca* of Europe and Asia are important fish for food and sport. Commonly the members of the family have a medium to small

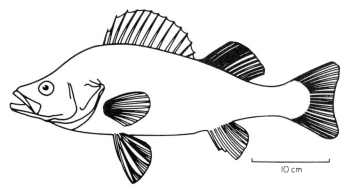

10 cm

FIGURE 5.33   The yellow perch, *Perca flavescens*, family Percidae, of fresh waters in Europe and North America.

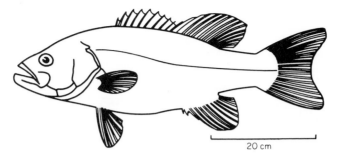

FIGURE 5.34 The smallmouth black bass, *Micropterus dolomieu*, family Centrarchidae, a choice sport fish of North America.

mouth, separated dorsal fins, and ctenoid scales. They are all carnivorous and feed on either small fishes or invertebrates. A great majority of the American species in the family belong to the subfamily Etheostominae, the darters. These are commonly brightly colored fish less than 10 cm long that live in running water on stony bottoms.

The sunfishes and freshwater basses of the family Centrarchidae (Fig. 5.34) occur in fresh water in North America. Only about 30 species are recognized, but many are well known to anglers in the lakes and slow rivers from Canada to Mexico. All prefer moderate to warm waters with either sand or mud bottoms and abundant vegetation. All species use nests. The males build the nests and guard the eggs and the young. All species have more or less flattened bodies, ctenoid scales, and small to medium mouths. They feed principally on either small fish or invertebrates.

Similar fishes of the family Cichlidae live in the fresh waters of Africa and South America. These include the genus *Tilapia* (Fig. 5.35), an important food

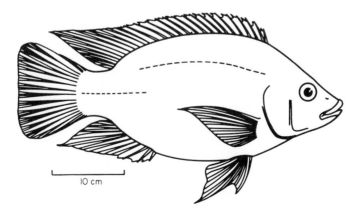

FIGURE 5.35 The cichlid *Tilapia mossambica*, family Cichlidae, from fresh waters of Africa, a species of great promise for aquaculture.

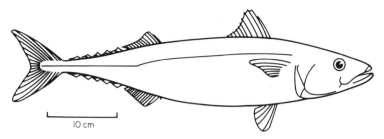

FIGURE 5.36   The Atlantic mackerel, *Scomber scombrus*, family Scombridae.

fish that is of increasing use in aquaculture. The family also has many small colorful fish with unusual breeding habits (e.g., mouthbrooding) that are highly prized by aquarists.

5.3.16 Scombridae*—Mackerels and Tunas

Most scombrids are schooling marine fishes of surface waters ranging from tropical to temperate. They may migrate over the continental shelf or far outside its edge; some tunas make transoceanic migrations. All species are prolific and most broadcast their eggs in the surface layer of the open sea. Most scombrids have rich, oily flesh and are excellent eating. Some of the larger species are highly prized game fish.

The scombroid fishes are robust, torpedo-shaped, voracious fish. Some have small, inconspicuous scales; others are partly naked. The mouth is large and the color is usually blue or green on top and silvery white underneath. Their dorsal and anal fins are followed by finlets, and their caudal peduncles are reinforced by one to several keels on the sides.

The more abundant mackerels are of the genus *Scomber* (Fig. 5.36) and are found in most temperate waters of the world. They support major fisheries on both sides of the North Atlantic and North Pacific oceans, in the Black Sea, and in a few places in the Southern Hemisphere. They range in size up to about 60 cm, congregate in great schools, and migrate toward colder water during the spring and summer. A closely related genus that supports important fisheries is *Rastrelliger* of the Indian Ocean.

Larger (to 200 cm), swifter, more voracious, yet still slender mackerellike fish are the Spanish, or king, mackerels of the genus *Scomberomorus*. These fish inhabit tropical and subtropical waters. They are prized as choice food and game fish but seldom are taken in large quantities because they tend to migrate either as individuals or in small schools.

Three genera are intermediate in characteristics between the mackerels and

*Taxonomists agree that the mackerels and the tunas are closely related and commonly refer to them as the "scombroid fishes." They disagree, however, as to whether the groups should be subfamilies, families, or even suborders. Here they are considered subfamilies of the family Scombridae.

50 cm

FIGURE 5.37   The yellowfin tuna, *Thunnus albacares*, family Scombridae, of the tropical and subtropical oceans.

tunas: *Auxis,* the frigate mackerels; *Euthynnus,* the little tunas; and *Sarda,* the bonitos. These fish range in size from small to medium (to about 110 cm). *Euthynnus (Katsuwonus) pelamis,* the skipjack tuna, is an extremely valuable species found around the world in tropical and subtropical waters, but the other species in these genera are only locally important.

The large scombrids, the big tunas of the genus *Thunnus,* are the objects of major fisheries around the world in waters ranging from tropical to temperate. The largest species, the bluefin, *T. thynnus,* reaches a length of 3.5 m and a weight of more than 800 kg. It tends to be more abundant near the coasts on both sides of the Atlantic and Pacific in subtropical to temperate waters than elsewhere in its distribution. Recoveries of tagged individuals have shown transatlantic crossings. The other species of temperate waters is the albacore, *T. alalunga.* It seldom exceeds 120 cm in length but is especially valuable for canning because of its white flesh. This species has been shown to make transpacific migrations. The species that prevail and support large fisheries on the tropical and subtropical high seas around the world are the yellowfin tuna, *T. albacares* (Fig. 5.37), and the bigeye tuna, *T. obesus.* These species reach a length of about 180 cm.

5.3.17 Scorpaenidae—Scorpion Fishes and Rockfishes

Many of the more than 250 species of fish in the Scorpaenidae are considered to be among the ugliest of the fishes. They are distinguished by large, heavy, bony heads with spines and a heavy bone extending along the cheek from below the eye. They have small teeth and ctenoid scales. Most species range in size from small to medium, rarely exceeding 100 cm in length.

All of the scorpaenids are marine fish occurring from the Arctic to the tropics in waters ranging from shallow to deep. Most species prefer rocky areas, hence the name *rockfishes.* Many of the species are brilliantly colored and many have fleshy flaps on their fins, which effectively camouflage them among rocks or reefs. All are carnivorous. Many species, including those of

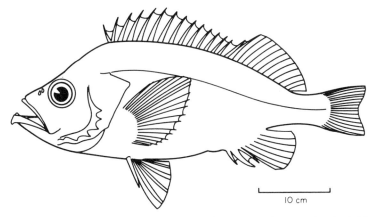

FIGURE 5.38 The rosefish, *Sebastes marinus*, family Scorpaenidae, a major food fish of the North Atlantic.

greatest commercial importance, are ovoviviparous. The eggs are fertilized internally, and the eggs and larvae are contained within the females until the larvae are spawned.

The family is commercially important for two reasons. Many species have venom sacs at the bases of their fin spines and can inflict painful wounds when handled. Two genera of the tropical Pacific, *Synanceja** and *Pterois*, produce wounds that may be fatal to humans.

Despite the ugly appearance and venomous equipment of some, almost all of the larger species are choice eating. Of most value are the three species of *Sebastes* (Fig. 5.38), the ocean perch or rosefish of the North Atlantic, and the more than 60 species of *Sebastodes*, the rockfish of the North Pacific. The species of both of these genera are the objective of extensive trawl and longline fisheries.

5.3.18 Pleuronectiformes—Flatfishes

The more than 500 species of the order Pleuronectiformes are asymmetrical fish when adult. Their bodies are flattened from side to side. They swim upright when in the larval stage and for a time thereafter, but usually swim on one side later. As they begin to swim on one side, an eye migrates to the "top" side and the top side becomes colored, while usually the "bottom" side remains white. They are anatomically similar to the perchlike fishes but commonly lack spiny rays in their fins.

The order is usually divided into seven families, of which only three are of commercial importance: Pleuronectidae (Fig. 5.39), the righteye flounders; Bothidae (Fig. 5.40), the lefteye flounders; and Soleidae, the soles. The English

---

*Synanceja is placed in another family by some recent authors.

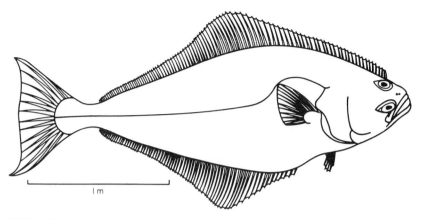

FIGURE 5.39   The halibut, *Hippoglossus hippoglossus*, family Pleuronectidae, of the North Atlantic.

common names are not consistent with the family affiliations, however. The name *sole* is applied to the thinner species of all three families. The names *dab, flounder,* and *plaice* are used for the medium-sized species, and the name *turbot* is applied to the larger, more robust species of the first two families listed. The halibuts, the largest of the flatfishes (reaching more than 200 kg and 3 m), are in the family Pleuronectidae.

The flatfishes live in closer association with the bottom than most other fish. They occur from the tropics to the Arctic, but the more important species are in

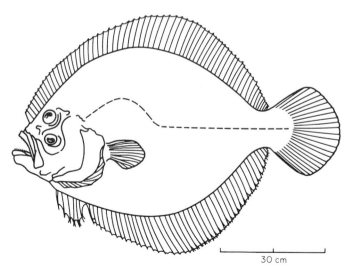

FIGURE 5.40   The turbot, *Rhombus maximus*, family Bothidae, of the Northeast Atlantic.

temperate and arctic waters of the Northern Hemisphere. They are all marine, but a few species migrate up rivers into fresh water. They live either on the continental shelf or on the upper part of the continental slope. They feed mostly on either invertebrate animals or small fish. Commonly flatfishes rest on the bottom, changing their color and color patterns to match it. They spawn on the bottom, but the eggs and larvae are usually pelagic, drifting with the ocean currents. The postlarvae are pelagic only until they turn on their sides, when they descend to the bottom. Many species are highly esteemed for their nearly boneless, delicately flavored flesh. They are caught usually by trawls and are associated frequently with codlike fish.

5.3.19 Cetacea—Whales

All of the larger marine mammals have been sought for food, oil, or pelts by coastal peoples in many parts of the world, but in recent years the large cetaceans, the whales, have been the species of most value. These include the largest of the toothed whales, the sperm whale, and several species of rorquals, one of the groups of baleen whales. The right whales, another group of baleen whales, and many of the numerous species of porpoises and dolphins are taken in minor quantities.

The rorquals are the most slender and fastest of the baleen whales. They have short baleen plates and a small dorsal fin. They include the famous blue whale (the largest living animal, which surpasses a length of 30 m and a weight of 120 tons), the fin whale, the sei whale (Fig. 5.41), and the humpback whale. These mammals have a life span of about 30 years, during which time each female produces about 10 large young, each measuring about one-fourth the length of a mature animal.

The right whales are large, slow baleen whales with large jaws and long baleen plates. They are no longer abundant.

The baleen whales bear their young either in tropical or in subtropical waters. Food is usually scarce in these waters; hence, most baleen whales migrate toward the poles in the spring to feed and fatten during the summer.

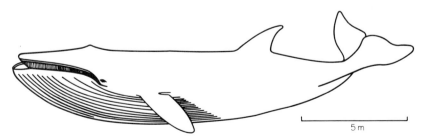

FIGURE 5.41 The sei whale, *Balaenoptera borealis*, family Balaenopteridae, now the most abundant of the baleen whales.

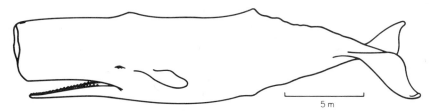

FIGURE 5.42   The sperm whale, *Physeter catadon,* family Physeteridae, the largest of the toothed whales.

The small crustaceans on which they feed, such as krill or small fish, are abundant in summer in the far northern or far southern oceans. They return to breed and have their calves, and they commonly fast during the winter in warmer waters. On the other hand, the sperm whale (Fig. 5.42) feeds largely on squid and it finds food abundant in some tropical waters as well as in the higher latitudes. Unlike the baleen whales, it is more widely distributed in tropical, subtropical, and temperate latitudes; only a few solitary males go to the edges of the ice—both north and south.

The whale fisheries of the world developed near land along which the whales had regular migratory routes and where men in small boats were able to catch some of the slower species, the right whales. Whaling spread farther out to sea from the fifteenth to the seventeenth centuries as men took to sailing ships to seek whales for oil and whalebone. Arctic whaling for right whales became a major pursuit in the seventeenth and eighteenth centuries and resulted in a depletion of the stock by the middle of the eighteenth century. Tropical whaling, especially for the sperm whale, developed in the eighteenth and nineteenth centuries but declined after 1850 for economic reasons (not due to a scarcity of sperm whales).

Throughout these centuries most of the fast rorquals escaped the whalers, but, with the development of fast catcher ships and harpoon guns in the early twentieth century, these species became catchable. The fishery expanded and severely reduced first the Arctic and then the Antarctic stocks. The blue whale is now nearly extinct; the fin whale is very scarce. The recent whale fishery of the world focused on mainly sei whales and sperm whales.

REFERENCES

Berg, L. S. (1940). Classification of fishes, both recent and fossil. *Tr. Inst. Zool. Akad. Nauk. Gruz. SSR* **5**, 87–345. (In Russian. Reprinted in 1947 with English translation. Edwards, Ann Arbor, Mich.)

Clemens, W. A., and Wilby, G. V. (1961). Fishes of the Pacific coast of Canada, 2nd ed. *Bull. Fish. Res. Board Can.* **68.**

Food and Agriculture Organization (FAO) (1992). "Yearbook of Fishery Statistics," Vol. 74. FAO, Rome.

Lagler, K. F., Bardach, J. E., and Miller, R. R. (1977). "Ichthyology." Wiley, New York.

Robins, C. R., *et al.* (1991). "Common and Scientific Names of Fishes from the United States and Canada." Amer. Fish. Soc., Bethesda, Md.

6 | Biology of Aquatic Resource Organisms

Increasingly, fishery scientists are being asked questions such as the following: How does a fish find a lure? What temperature, salinity, and oxygen concentrations limit the distribution of an aquatic animal? How does an animal navigate on the high seas? What is the best place and time to catch a given species? How and why do fish form schools?

Answering such questions requires an understanding of how the individual organism lives in the aquatic environment, how it senses its surroundings, and how it breathes, eats, excretes, reproduces, and develops. Every organism performs these functions in ways to which it is suited by structure and habit; therefore, the biologist studies each species separately. Such studies are essential preliminaries for analysis of exploited populations (Chapter 8).

In this chapter a few pertinent functions of certain aquatic resource organisms are summarized, with emphasis on the behavioral aspects that are useful in answering questions like those posed here. Many other aspects of the biology of the diverse organisms of the waters, such as structure, organization, integration, and evolution, which are appropriate for more advanced courses in biology, are omitted. Ecology will be discussed in the next chapter, and examples of the application of biological knowledge to major fishery problems of the day will be postponed to Part III.

6.1 SENSES

Every skilled angler knows how easily a trout is alarmed by seeing or hearing a person. Every purse-seine fisherman knows the frustration of having a school of fish take alarm and escape the net before it can be closed. All fishermen approach their quarry knowing that their presence can be sensed through vision or hearing or in some other way. When close enough, fishermen may try to lure the fish to bite or enter a net. They may employ either a visual stimulus, such as an artificial lure or light, or a chemical stimulus, such as a natural bait or sexual product. They may frighten fish into their nets. In most of their strategy, fishermen depend on overcoming the quarry's senses and behavior.

6.1.1 Light Reception

Many of the simpler invertebrate animals can sense light and will orient themselves to light from one direction. The light-sensitive organs range in complexity from light-sensitive spots to camera eyes with lenses capable of focusing an image on a retina. Crustacea have compound eyes with many tubelike units, each of which receives light from a different direction; these animals probably receive a mosaiclike image. Many bivalve mollusks have simple eyes with lenses; some have more complex eyes.

Camera eyes occur in the cephalopod mollusks and in most fish. Most such eyes have mechanisms for focusing on either near or distant objects and for controlling the amount of light. The light-controlling mechanism may be either an iris near the lens or movable cells in the retina that put pigment in front of the visual cells.

Even with the best of eyes, the vision of underwater animals is severely restricted by the particles in the water. The limits are about the same as those of the Secchi disk transparency, which are about 60 m in the clearest of ocean waters and only a few centimeters in turbid coastal waters. Some of the best eyes among fish are probably possessed by the tunas; they live in the clearer ocean waters and communicate among themselves with visual signals. Tuna living in experimental tanks were able to see brightly illuminated black and white stripes that subtended an angle of as little as 4 feet of arc (Nakamura, 1968). Under the same excellent conditions, a net twine 1 mm in diameter could not be seen at a distance of more than 1 m.

Vision may be better when the eye of the animal can perceive colors, but the poor transparency of water restricts color vision at the red end of the spectrum. Water best transmits the blue-green part of the spectrum (499–500 μm). Fish eyes appear to be most sensitive to these colors, and the bioluminescent organs produce the brightest light within this range.

The image of an object in the air, when seen from below the water surface, is severely restricted by the refraction of light at the surface. When the surface is disturbed, as it almost always is, the image is broken into many moving parts. Even with a calm surface, the image of everything above the water from horizon

to horizon is restricted to a cone, or Snell circle, subtended by an angle of 97.6°
(Fig. 6.1). Objects near the horizon are visible but greatly distorted.

Most fish, especially the pelagic carnivores that pursue relatively large prey,
rely on vision to locate their food. They feed mostly during the day, seldom at
night. But some fish are attracted by light, and fishing is conducted at night
with lights in many parts of the world. Usually a powerful lamp is hung over
the side of a stationary ship, either below or above the surface, until the fish are
congregated in such a way that they may be encircled by a net. Several species of
herrings, anchovies, mackerels, tunas, and sauries (all schooling fishes) are
taken by this method. Perhaps most spectacular is the technique used in the
fishery for sprat in the Caspian Sea; the fish are attracted by lights to the mouth
of a suction hose and merely pumped aboard. It is not known whether fish are
attracted directly by light or by their ability to see food in the lighted area, but
in either case they are responding to a visual stimulus.

Animals that either school (see Section 6.4) or orient to each other probably
rely principally on vision; some of these animals, such as the tunas, communicate with visual signals. Some animals develop sexual differences in color and
can change color rapidly during mating, perhaps as a way of communicating.

Migratory aquatic animals (like numerous terrestrial animals) may orient to
the sun. Evidence of this behavior was gathered from observations in the field
and from experiments in training fish to respond to a visual cue for a reward.
Fish that had been moved from their home and marked with small floats or
sonic tags were found to follow a consistent compass direction when the sun
was shining and to become confused when the sun was obscured. Trained
captive fish were observed to follow a definite compass course using the sun for
guidance and compensating for the change in the azimuth of the sun as it

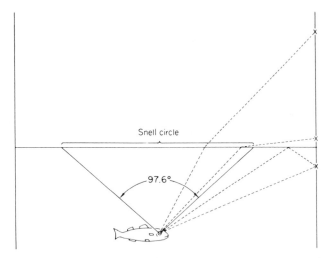

Snell circle

97.6°

FIGURE 6.1 Diagram of the field of vision from below the water surface.

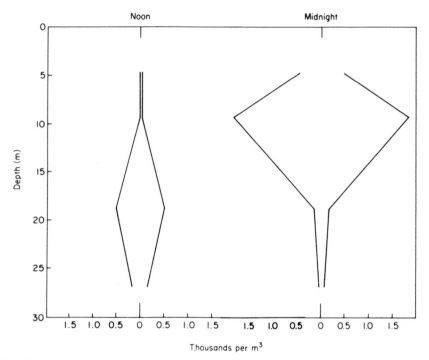

FIGURE 6.2 Vertical distribution of calanoid copepods in Lake Nerka, Alaska, during a summer day. [Data from D. E. Rogers (1968). A comparison of the food of sockeye salmon fry and threespine sticklebacks in the Wood River lakes. *Univ. Wash. Publ. Fish.* (N.S.) 3, 1–43.]

passed through the sky. Some further evidence indicates that fish can also change their orientation according to the maximum altitude of the sun (Hasler, 1966). Such orientation implies not only a remarkable ability to measure the angle of the altitude of the sun from below the surface of the water, but also a fairly accurate sense of time, both daily and seasonal.

Another response to light is evidenced by the extensive daily vertical migrations of many plankton animals. This behavior was discovered by some of the first people to tow plankton nets when they caught much more plankton near the surface at night than during the day (Fig. 6.2). More observations on the prevalence of the phenomenon were collected by the operators of sensitive echo-sounding equipment; they obtained records of strong echoes from numerous layers that rose toward the surface in the evening and descended in the morning. They called them "deep scattering layers" (Fig. 6.3). At first the cause of the echoes was a mystery, but later it was found to be layers of plankton of many kinds, some of which were attracting fish.

6.1.2 Chemical Reception

In terrestrial animals, taste receptors come into direct contact with a substance and smell receptors detect substances at a distance; both senses require

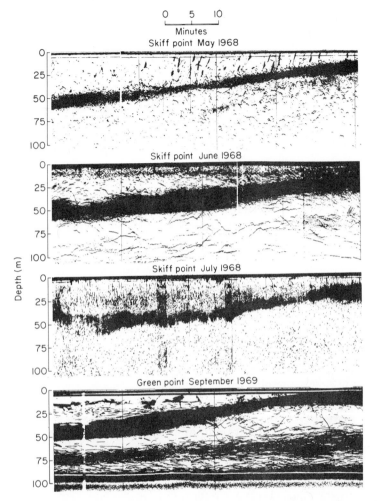

0 5 10
Minutes
Skiff point May 1968

Skiff point June 1968

Skiff point July 1968

Green point September 1969

Depth (m)

FIGURE 6.3 An echo sounder on a drifting vessel in Puget Sound repeatedly recorded the vertical migration of euphausids during the hour before sunset from depths of about 50 m up to a depth of about 25 m. (Records by R. T. Cooney, Dept. of Oceanography, Univ. of Washington, Seattle, Wash.)

that the substance be dissolved in water of the mucous membrane. In aquatic animals the substances are dissolved in water at all times, but it is useful to retain the terms *taste* for contact reception and *smell* for distance reception, even though the organs may not be in the mucous membrane in the mouth or nasal area.

Taste and smell may be the primary senses by which many aquatic animals locate food, homes, mates, and enemies, even if the animals have good eyes. In some fish, apparently the ability to smell is at least equal to that of the most sensitive of terrestrial animals. It is put to good use in waters where the visi-

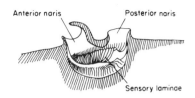

FIGURE 6.4   A longitudinal section of the olfactory organ of the haddock *Melanogrammus aeglefinus*. [Redrawn from R. H. Burne (1909). The anatomy of the olfactory organ of teleostean fishes. *Proc. Zool. Soc. London*, pp. 610–663.]

bility is only a few centimeters. Usually taste and smell organs are associated with the feeding mechanisms, but in many animals, including fish, they are associated with sensitive chemical receptors located in other parts of the body.

Such senses are probably especially useful to aquatic animals because most waters have odors resulting from animals and plants, some of which are detectable even by relatively insensitive human noses. Many algae cause odors that are troublesome or even offensive when the water is used for either drinking or swimming. Many live fish have characteristic odors, and some coastal fishermen can detect schools of fish in the water by the odor.

The olfactory organs of most fish are cavities lined with sensitive cells through which water is circulated. Some of their olfactory organs and the associated parts of the brain are so large that the fish would be expected to have an excellent sense of smell. An example is the haddock (Fig. 6.4).

Salmon use the sense of smell for homing. This was established by moving adult salmon from their spawning streams to locations some miles away and surgically interfering with various senses. Those with their eyes covered quickly found their way back to the stream from which they had been taken, but those with plugged olfactory organs became lost. In other experiments, salmon with normal olfactory organs found their own home stream even though they had been mixed with salmon in another spawning stream, whereas those with their olfactory organs plugged did not return. Clearly, the home stream had a distinctive odor that they could recognize.

Salmon can detect the odor of enemies. A spectacular escape reaction will be caused in a group of resting adult salmon in a stream if a bear wades some distance upstream. The salmon will swim wildly downstream when they come into contact with the water in which the bear has walked. The same reaction would be obtained if a person were to rinse his hands in the water upstream from the salmon.

Finding and sorting food is probably the predominant role of the taste and smell senses. Many experiments have shown that aquatic animals react to extracts of food. They sense the food by their olfactory organs; some also sense the food through special receptors on fins and barbels. The ability of sharks to detect blood in the water and their subsequent ferocity in a feeding frenzy are well known.

Chemical receptors play a major role in the spawning of shelled mollusks. The oyster is an example. The male oyster may be stimulated to spawn by rising temperature, by the sperm of other oysters of the same species, or by a wide variety of organic substances. The females are more specific; they spawn only when stimulated by sperm from the males. An entire oyster population will spawn soon after the first male spawns, because his sperm will start a chain reaction with other males and they in turn will stimulate the females.

Fishermen have long used bait to attract fish, but the possibilities for taking advantage of the special chemical sensitivity of fish and other aquatic animals are only beginning to be scientifically studied. There will be many applications for effective attractants and repellents that can be used in catching, guiding, herding, or protecting aquatic animals.

6.1.3 Temperature Reception

Most mobile animals seem to prefer certain temperatures; each species is commonly found only in waters with temperatures within a certain range. Many migrate seasonally according to the warming and cooling of the water, and good fishing is frequently found in the temperature gradients of oceanic fronts. One may assume that marine animals can detect temperature differences and respond accordingly.

However, the role and method of temperature reception are not well understood. Fish respond to temperature changes as small as 0.03–0.07°C when the changes occur fairly rapidly, but how such responses can guide migrations is not clear. Nor is it clear how temperature changes are perceived; some evidence indicates that taste organs and the lateral-line system are involved.

6.1.4 Mechanical Reception

All of the higher aquatic animals respond to mechanical stimulation; some are extremely sensitive to touch. The use of touch in sensing their surroundings is universal among animals and needs no further emphasis, but one related function is common only to most fish. This is the ability to sense water currents and low-frequency vibrations (less than about 200 Hz) through the lateral-line system. This is a system of furrows or canals with frequent openings to the exterior that occurs along the trunk and on the head. It has specialized sensory cells at frequent intervals (Fig. 6.5).

The lateral line appears to be a set of distant touch organs that allow the fish to sense the currents set up by the movements of nearby animals and the presence of nearby objects by reflected currents from his own movements. The distribution of sensory cells along the line permits identification of the direction of a point source of stimuli. The lateral line apparently supplements the sense of hearing at low frequencies. In fact, there is no good dividing line between high- and low-frequency reception, and there is some confusion about near-field and far-field effects.

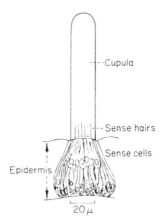

FIGURE 6.5 A sensory unit of the lateral-line system. These units are located either on the surface of the body or in lateral-line canals. [After S. Dijkgraaf (1962). The functioning and significance of the lateral line organs. *Biol. Rev.* **38**, 51–105, Cambridge Univ. Press, New York.]

6.1.5 Pressure Reception

All swimming or drifting aquatic animals must regulate their depth and must be expected to respond to pressure changes. Animals having an air sac, such as fish with an air bladder, have an organ that will change in size according to pressure. Surprisingly, many invertebrates that lack such organs are also pressure sensitive. Some crustaceans as well as some fish have been shown to respond to pressure changes of only 10 mbar, the equivalent of a 10-cm change in the depth of water. It has been postulated that pressure receptors may be developed either around changes in electric potential with pressure across the respiratory surfaces of crustaceans or with differences in compressibility of skeletal and soft tissues. Pressure receptors may be combined with mechanical receptor cells such as those in the lateral-line canals of fish.

6.1.6 Sound Reception

One might expect the aquatic animals to depend on hearing for information more than the terrestrial animals, because sound travels much faster and with less energy loss in water than in air. Such does not seem to be the case; however, the role of sound reception is not well understood and involves some activities that are rare among terrestrial forms. Although the majority of fish and many crustaceans produce sounds of one kind or another (see Section 4.4.4), only a few fish and the marine mammals appear to produce sounds for communication.

The simplest hearing organs in some crustaceans are mechanical receptors, which are small, projecting, hairlike organs that are sensitive to deflection. Similar structures occur in the lateral-line canals of fishes. Such skin receptors are sensitive only to low-frequency vibrations (up to about 200 Hz) or to water currents. An inner ear, or labyrinth, that is similar to the mammalian ear enables fish to detect frequencies higher than about 50 Hz.

The ability of fish to hear sounds of different frequencies has been studied in behavioral experiments in which fish have been rewarded by responses to various frequencies. These studies have shown that the most sensitive frequency range for most fish is 300–800 Hz, with about 2000 Hz as the upper limit. Members of the Cypriniformes, a group of freshwater fish with a special connection between the ear and the air bladder, can detect frequencies up to about 13,000 Hz (the upper limit for humans is about 10,000 Hz).

But the hearing of most fish is not very useful to them, because of their own deficiencies and the sound-transmission characteristics of the aquatic environment. They have little ability to discriminate sound intensities or frequencies, which is an ability that is highly developed in birds and animals that communicate in air. Communication under water is more difficult for two reasons. Sound is transmitted about five times as fast as in air, so it is more difficult for an animal to obtain directional information, and most waters have a high level of natural ambient noise.

When fish do communicate by sounds, they are usually either feeding or breeding. The sounds may be helpful in locating mates, in protecting nests against predators, or in telling others about food. The few attempts to attract fish by sounds were successful only when the natural sounds of food or breeding were imitated. Some fish react strongly to unusual sounds with an escape reaction of leaping or diving away from the sound. A few fish appear to produce sounds for the purpose of echo ranging.

Vastly more sophisticated is the auditory capability of the marine mammals. Some of these produce and hear sounds up to 150,000 or even 170,000 Hz. They have a substantial vocabulary for communication among themselves and make extensive use of echo ranging. Furthermore, they can locate the direction of a sound, since their two ears are acoustically isolated.

Despite the ability of most fish to hear, and despite the alarm reactions of some fish, the use of sound to repel fish has been notably difficult. Experiments with frequencies from 67 to 70,000 Hz and intensities high enough to be described as underwater thunder failed to cause young trout to do more than give an initial start at the commencement of the sound (Burner and Moore, 1962). Other workers who studied the effects of sound on fish entering trawl nets were unable to show that the sound of nets and towing wires has a significant effect either by helping the fish avoid the nets or by attracting them to the nets. Still others were able to make sounds that tended to attract the fish toward the bottom, where they could be caught more easily by trawls. Perhaps the sound would have to be one that the fish associates with danger.

6.1.7 Electrical Reception

Most of the studies about the effect of electricity on fish have been either attempts to understand the use of electric organs by the species of fish that produce electricity or attempts to guide and catch fish by electricity. Little has been done to identify how electricity is received, although it was shown that some special electrical receptors appear to occur in the lateral line of the electric eel, *Electrophorus*. Certain sensory and motor nerves in fish are direct electrical receptors, and presumably the entire body will receive electricity.

The production of electricity occurs in a few species of both freshwater and saltwater fish of diverse families. These include some marine torpedoes and rays (elasmobranchs), some freshwater Mormyridae of Africa, and the Electrophoridae and Gymnotidae of South America—the latter families being related to the minnow and carp family, Cyprinidae. The electric organs are modified muscles. The most powerful of these fishes appears to be the marine torpedo; its maximum discharge is between 3 and 6 kW. The torpedoes and rays use their powerful electric organs to stun prey, whereas the freshwater fishes use their much weaker electric organs for orientation as well as to stun prey. Some of the latter have a remarkable ability to sense differences in the resistance of the water around them. Small wires or nonconductive pieces of glass tubing bring quick responses when moved near them. They can detect differences in electric potential of as little as 0.02 μV/cm.

Salmon have been found to move predominantly downstream with the ocean currents in their oceanic migrations. Since the electric potential in such ocean currents (ranging from 0.05 to 0.5 μV/cm, see Section 4.4.5) is more than that detectable by some electric fish, it has been postulated that salmon might utilize the electric field of the ocean for orientation during their long migrations.

When electric currents through the water are greatly and continuously increased, fish display a characteristic series of reactions. When the current is alternating in polarity, at first the fish is uneasy, then it gives a "minimum response" (a powerful twitch of its tail), then it twitches violently, and then it becomes paralyzed. When the current is direct, the fish becomes uneasy, then faces the positive pole and swims toward it (electrotaxis), and then becomes paralyzed (electronarcosis). When the direct current is interrupted at a suitable frequency (but not alternated in polarity), the fish will swim vigorously toward the positive pole during electrotaxis.

The levels of electric potential at which fish behave in the foregoing manner vary according to temperature, electrical resistance of the water, kind of electric current, and size and species of fish, but the general levels are somewhere between 0.01 and 1.0 V/cm. The responses occur at lower levels of electric potential with lower temperatures, in seawater, with alternating current, and with larger fish.

Such behavior of fish in tanks has suggested to many workers the exciting possibility of forcing fish, by electricity, to swim into nets or away from water intakes. Unfortunately, such hopes have been dashed by the difficulty of maintaining a uniform electric field outside the laboratory and by the huge power required in seawater. As fish enter any electric field some may be repelled before being compelled to swim toward the positive pole; others may be narcotized before reaching the positive pole. Further, the effect of the electric field varies with the orientation of fish relative to the field and the resistance of nearby things, such as the ship hull or the bottom. Nevertheless, considerable success has been achieved by the use of interrupted direct current and by the combination of electricity with other methods of concentrating fish, such as lights or nets.

6.2 RESPIRATION

The respiration of animals includes the exchange of gases either at the surface of the body or in internal passages, the transportation of gases through the body, and the exchange of gases between the circulatory system and the cells. The gases are the needed oxygen, the waste carbon dioxide, and the inert gases, such as nitrogen, that just ride along.

The exchange of gases at the surface of aquatic animals takes place through the integument, gut, gills, lungs, and air sacs. In tiny animals and in the eggs and larvae of larger animals, the integument itself is adequate, but in larger animals larger surfaces are needed. In the more efficient systems, these surfaces are supplied abundantly with blood in order to allow the rapid diffusion of gas to and from the external medium. Most fish, mollusks, and crustaceans have gills in internal chambers through which water is pumped to increase still further the efficiency of the exchange. Some have auxiliary systems for exchanging gases directly with the air either when the water is low in oxygen or when the water recedes during tidal or annual cycles. A surprising number of fishes that live in tropical fresh waters can utilize air when living in water almost lacking in oxygen.

Oxygen consumption varies enormously with species, size, activity, season, and temperature. Fast-swimming animals use much more oxygen during active periods than sluggish animals use during periods of rest. Large animals use less per unit of weight than small animals do. Consumption increases with temperature up to some critical level and then falls off, but many animals compensate for seasonal changes in metabolism and may actually consume more oxygen in winter than during the rest of the year.

The fish that require the greatest amount of oxygen are the active species that live habitually in well-oxygenated water. An example of a marine species is the mackerel; it has a relatively large gill surface and its blood will carry more oxygen than that of most fish. It needs to swim almost continuously in order to ventilate its gills adequately, even though the water in which it lives is always almost completely saturated with oxygen. Freshwater species with large oxygen requirements include the trouts. These fish live habitually in the oxygen-saturated water of cold streams; they cannot long withstand water with less than 5 mg/liter of oxygen. At the other extreme (excluding the air-breathing fishes) are the less active bottom species of freshwater fish, such as carp and catfish. These fish can survive with dissolved oxygen of only 0.5 mg/liter.

Expression of such minimum levels of oxygen is an oversimplification, however, because of the synergistic action of many factors. Increasing temperatures require an increase in the amount of oxygen required to saturate the blood. Increasing carbon dioxide reduces the ability of hemoglobin in the blood to carry oxygen and therefore requires higher environmental dissolved oxygen for equivalent survival. Other related factors are the activity of the fish, its general health, and its acclimatization to the unusual temperature and gas conditions. The last is especially important, because many fish have a substantial ability to adjust to conditions that are less than favorable.

Respiration in the shelled mollusks occurs simultaneously with feeding (see Section 6.5). As water is pumped through the open shell and over the gills, gas exchange occurs and food particles are filtered out. Usually the water circulated is greatly in excess of that needed for respiration alone, so the amount of oxygen removed is relatively small.

6.3 OSMOREGULATION

The membranes of aquatic animals that pass gases and digested food between the environment and the body also pass water and salts. Most aquatic animals maintain a reasonably constant proportion of water and salts in their bodies. When this proportion is different from that of the surrounding water, it requires some mechanism for regulation.

Most invertebrates in the sea are *isosmotic,* or have nearly the same salt concentrations as the environment. Sharks and rays tend to have a slightly higher concentration of salts (including urea) than seawater, but the fluids of most *teleosts* (bony fish) have much lower salt concentrations than seawater. In fresh water the fluids of all fish have a much higher salt concentration than the water. Freezing-point depressions of several fluids are tabulated below.

Seawater	-1.8 to $-2.1°C$
Shark blood	-1.9 to $-2.4°C$
Marine teleost blood	-0.7 to $-1.0°C$
Freshwater teleost blood	-0.45 to $-0.60°C$
Fresh water	-0.0 to $-0.1°C$

Marine teleosts, which lose water osmotically across the gills and must get rid of salt, tend to drink large quantities of seawater and produce only small quantities of urine. Freshwater fish, which gain water osmotically and must conserve salt, drink little or no water and secrete large quantities of very dilute urine. The salt–water balance is maintained by the kidneys, gills, and intestine.

The different salts from the water do not remain in the body in proportion to their abundance in the environment. Most animals are able to concentrate some ions while rejecting or excreting others. Marine animals usually reject magnesium and sulfate, whereas freshwater animals usually concentrate sodium, potassium, and chloride ions.

The mechanisms for salt control and ionic adjustment of most animals are so inflexible that the animals are *stenohaline,* that is, they are restricted either to salt water or to fresh water. Marine fish usually tolerate salinities less than the salinity of the open sea but higher than the salinity of their blood, whereas freshwater fish usually tolerate any salinity less than their blood salinity. Relatively few estuarine animals and migratory fish, such as salmon, can adjust to the change from fresh water to salt water, or vice versa.

6.4 SCHOOLING

It is no accident that most of the world fisheries depend on concentrations of fish rather than on fish scattered through the water. Fortunately for those who seek them, many mobile aquatic animals aggregate or school in ways that make them much easier to catch in certain places or at certain times. They school in order to breed, not only to bring the sexes together but also to release the eggs or larvae in a favorable place. They commonly school in order to feed, either because the food is localized or in order to feed collectively.

They also school because of other kinds of mutual attraction that are largely independent of environmental circumstances. Biosocial aggregations vary from casual, temporary, and irregular groupings of a few animals to very large numbers of a single species of nearly the same size, spaced uniformly, swimming parallel, and acting concomitantly, a behavior that is called *polarized* (Fig. 6.6). The variety of kinds of aggregations and the variety of causes have led to substantial disagreement on the meaning of terms and to unwieldy classification of types of schools. Some of the senior behavioral scientists who are studying fish schools now restrict the term *school* to groups that exhibit polarized behavior. Other kinds of groups are called *aggregations, shoals,* or *clusters.*

The survival value of schooling probably varies among the animals concerned. There are obvious values in finding mates and in herding food. Some fish have been shown to experience less stress when living together than when living alone. A school may give the individual a greater chance to survive because the school is harder for a predator to find, and having found it, a predator can consume only its capacity before it becomes satiated and loses the school. A school may intimidate smaller predators, whereas a solitary individual may not.

Regardless of the biological benefits of aggregating or schooling, many commercially important species do it. The herrings, anchovies, mackerels, and

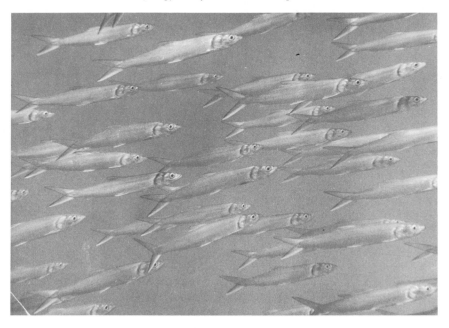

FIGURE 6.6   A polarized school of machete, *Elops affinis*, in the southern Gulf of California. (Photo by E. S. Hobson, Bureau of Sport Fisheries and Wildlife, U.S. Dept. of Interior, Washington, D.C.)

tunas school (in the scientific sense) near the surface of the water; the cods and many perchlike fishes aggregate near the bottom.

6.5 FOOD AND FEEDING

Every animal requires energy for living—for growth, maintenance, and reproduction—which it must obtain from its food. Each starts life with a bit of food received from a parent, but it soon needs to feed itself. It must continue to feed itself with suitable food regularly or die. The regularity must suit the animal's ability to find and ingest food and to store energy. For some, feeding is almost continuous; for others, feeding is in circadian (i.e., daily or tidal) or seasonal cycles. But for most animals, feeding is the dominant activity of their entire lives, because their need is constant and the food is usually scarce.

Understanding this activity is useful to all scientists who are concerned with any aspect of fisheries. If they should want to improve the catch, they either would need to develop better baits or learn about feeding behavior. Should they want to develop a rational method of exploiting a population, they would need to know how food is a limiting factor and how it may be divided among competing animals. Should they culture animals, they would need to study intensively the nutritional requirements of the animals in order to obtain the best growth at the least cost.

6.5.1 Feeding Mechanism and Prey

Not only does every animal feed, but almost every free-living animal and plant is food for something else. No matter how tiny, large, swift, camouflaged, armored, spiny, or poisonous it is, some other animal can find it, catch it, overcome its defenses, and consume it. As consumers, animals in general have an almost infinite adaptation for organic food, not only in their jaws and dentition, but in their food detection techniques and digestive apparatus. For example, the mouths of fish of different species are adapted to cut, grasp, suck, stab, rasp, scrape, dig, crush, and filter. Some invertebrates can bore through wood and the hardest of molluscan shells. The adaptations are usually consistent, from detection of prey through capture and mastication to digestion. An animal that feeds by filtering plant materials will probably have a small gullet, enzymes suitable for digesting carbohydrates, and a large intestinal surface. An animal that catches and gulps large fish will probably have a large gullet, a large and unusually acid stomach, and a small intestinal surface.

Usually the larger and more highly organized animals feed on smaller and more primitive organisms, but there are many exceptions. Some fish can swallow animals larger than themselves; some jellyfish eat fish; some starfish eat mollusks; some fish eat birds or mammals.

At the start of life, all animals differ from adults in their capacity to find, capture, and ingest food. Most animals have completely different kinds of diets at different stages in their lives.

Aquatic animals find their food by using a variety of senses, particularly smell and sight. Many predators are assisted by the characteristic odors of their prey. It is well known that many fish and crustaceans are attracted by either the body fluids of injured animals or pieces of flesh. Having approached a prospective prey, the foraging animals must depend largely on vision for the final attack. Other sensory systems used by certain animals include methods of detecting noise, movement of the water, changes in electrical resistance, and echo ranging.

The importance of vision in the feeding of many animals is indicated by the prevalence of circadian rhythms in feeding. Many foragers feed more vigorously at dawn and evening than at night and noon, although they may do so as a result of the behavior of the prey.

Most animals in temperate climates have seasonal rhythms in their feeding that also are caused by combinations of circumstances. Commonly, the amount of food needed is greater in warm water than in cold, because at lower temperatures digestion and metabolism are retarded. At temperatures near freezing some species cease to feed. The spawning season for some animals is a period of fasting. Perhaps the most common cause is a seasonal cycle in food abundance. Many animals feed intensively during the plankton blooms, the emergence of certain insects, the spawning periods of crabs and fish, or other seasonal phenomena, each of which may last only a week or two.

No simple classification of feeding systems is completely satisfactory. Some animals are herbivorous, feeding entirely on plants; some are carnivorous, feeding entirely on animals; but many are omnivorous, feeding on both animals and plants either at the same time or at different times in their lives. The spectrum of habit ranges from the stationary filterer to the speedy forager and includes the parasite.

6.5.2 Determining the Diet of Aquatic Animals

Because it is rarely possible to observe an aquatic animal feeding under natural conditions, its habitual food or diet is determined by an examination of the stomach contents of a suitable sample of animals. Usually stomachs are easy to obtain from fishermen or from animals collected for other purposes. They are usually preserved for later examination in a laboratory. There it is a straightforward though not often an easy task to identify each organism in the stomach and either weigh it or determine its volume.

More difficult is the task of analyzing and interpreting the data. Complications arise because animals feed irregularly, they sometimes regurgitate the most recently eaten items during the process of capture, food items digest at varying rates, and identification of digested food items may depend on certain teeth or skeletal remains.

The objectives of dietary studies are to understand how animals live and grow, what foods may influence their abundance and distribution, and the relative quality of feeding conditions. Usually such studies are part of life-history studies of such matters as distribution, migrations, growth, and reproduction.

Despite the difficulties of interpreting the data on number and weight (if volume is taken, it is assumed to have a specific gravity of 1) of food items in stomachs, a great deal can be learned. The counts of different food items may be compared directly with each other, but in this method extra emphasis is given to items with parts highly resistant to digestion and to the tiny items; for example, a thousand crab larvae may equal the food value of only a single fish. A much better method is to express the frequency of occurrence of each item among the stomachs examined by showing the percentage of stomachs containing it (Table 6.1). Such a percentage shows the number of individuals in the sample that ate the food item and hence a rough index of the availability of the food. A still better method is to use the volume of each item, since the emphasis given to items with durable parts is less. The volume of each item may be reported either as a percentage of the aggregate volume of all stomachs or as an average of the percentage in each stomach—a method that gives each stomach equal weight, regardless of the volume of contents.

When a representative sample of the food available is obtained along with the stomachs, it is possible to estimate a *selectivity coefficient*, or *forage ratio*,

TABLE 6.1 Example of a Food Analysis[a]

Food organisms[b]	Number of organisms	Percentage of stomachs in which occurred	Percentage of total volume
Crustacea	85,140	66.9	24.8
Squid	3642	55.4	26.2
Fish	5333	70.4	46.7
Flying fishes	23	1.9	3.3
Jack, *Decapterus*	46	2.2	8.4
Pomfret, *Collybus*	449	13.8	3.4
Tuna, *Katsuwonus*	19	1.5	5.1
Unidentified fish	2439	48.2	8.0

[a]Food organisms found in the stomachs of 1097 yellowfin tuna captured in the central Pacific, 1950 and 1951. [After J. W. Reintjes and J. E. King (1953). Food of the yellowfin tuna in the central Pacific. *U.S. Fish Wildl. Serv., Fish. Bull.* **54**(81), 91–110.]
[b]Table is greatly abbreviated; all organisms were identified as accurately as possible, and many more were listed.

for each food item. This ratio is taken by dividing the percentage in the stomach by the percentage in the environment. A value higher than 1 indicates selection; lower than 1 indicates avoidance. The difficulty of obtaining a satisfactory sample of the food in the environment with either nets or grabs limits the determination of this coefficient to animals feeding entirely on small plankton or small bottom invertebrates.

The total volume of all items in a stomach, expressed as a percentage of the weight of the animal, gives an *index of fullness*. The index is useful for comparing the relative feeding activities at different times and places.

Special studies of captive animals in which the total weight of food eaten is related to the gain in weight of the animal yield an estimate of the *nutritional coefficient*, or *conversion ratio*. The nutritional coefficient is the number of units of food required to produce a unit gain in weight on a net weight basis. It varies with size and activity of the animal and with the kind of food. For fish, the value varies from 2 to 40 and a value of about 10 is common. Refined coefficients with various names, which are based on the dry weight of diet components (i.e., proteins), are also used.

6.5.3 Examples of Food and Feeding Habits

Stationary Bottom Filterer: The Oyster The sessile animals of the bottom cannot forage but must filter their food from the water as it flows by. They increase the efficiency of the operation by increasing the size of the filter, by pumping water through an enclosed filter, and by sorting the nourishing from the nonnourishing particles and the harmless from the harmful.

Adult oysters pump a surprisingly large amount of water over their gills, which serve not only as respiratory organs but also as filters and sorters. Large

American oysters, *Crassostrea virginica*, have been found to pump amounts varying from 10 to 20 liters/hr when active, and they may pump up to several hundred liters per day, although their periods of activity are highly variable. Water is pumped over the gills through the open shell by the action of the cilia, which cover the gills and much of the body. The valves of the shell are not moved to pump the water. Water is drawn into the ventral region, passed by diverse routes over the gills, and exhaled posteriorly (Fig. 6.7). The particles in the water become entangled in the mucus on the gill surface and are pushed toward the mouth. They are sorted either along the way or at the labial palps near the mouth; the food is passed into the mouth; the waste is pushed out as pseudofeces. This filter system will remove particles as small as 2 μm from the water.

The food of oysters is not satisfactorily known because a large proportion of it is the naked and minute nannoplankton that is too delicate to examine easily, even before ingestion by an animal. Despite disagreement about details of their food, it is generally known that the oysters, like most other bivalve mollusks, consume mostly phytoplankton (see Chapter 7 on ecology), and, hence, they are relatively efficient in transforming the basic product of the water into food that can be consumed by humans.

Carnivorous Bottom Forager: The Cod   The cod might well be selected as the bottom-dwelling fish with the least specific food habits. It has a mouth that is about average in size and power. Its jaws and the roof of its mouth are

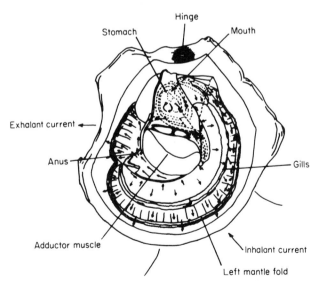

FIGURE 6.7   Ciliary water currents during feeding of the European oyster, *Ostrea edulis*. Right shell and mantle removed. [Redrawn from C. M. Yonge (1926). Structure and physiology of the organs of feeding and digestion in *Ostrea edulis. J. Mar. Biol. Assoc. U.K.* **14**, 295–386. Reprinted with the permission of Cambridge University Press.]

equipped with numerous rows of small, pointed teeth suitable for holding animals but not for tearing or crushing. Its gill rakers and gut have no unusual features to suggest other than an ordinary food habit.

As soon as the larval cod consumes its stored food, it must feed regularly on particles of the proper size. At first the food is small phytoplankton, but soon the pelagic larva is large enough to eat tiny crustaceans such as copepods, barnacle larvae, and crab larvae. As it grows, it becomes a bottom forager and turns to larger crustaceans such as shrimp or euphausids, to miscellaneous bottom invertebrates such as tunicates, brittle stars, sea cucumbers, squid, clams, or mussels, and to fish in great variety. It eats almost any small fish, including young cod. After the cod reaches a length of about 50 cm, fish is the predominant item in its diet.

Such a varied diet indicates a ready adaptability to changes in the abundance of food. Apparently cod can utilize whatever animal food is caught easily and are found commonly to eat different diets in different places at different times of the year.

This readily changeable food habit is typical of many carnivorous foraging fish with ordinary dentition in either fresh waters or the sea and for either pelagic or bottom-dwelling species. Most of them tend to feed more on invertebrates (insects in fresh water) than on fish when small and more on fish than on invertebrates when larger.

Giant Filter Feeders: The Whalebone Whales As the size of an animal's food becomes smaller relative to the size of the animal, more energy must be expended to find and capture the food relative to the energy provided if the food items must be taken singly. Food particles each weighing less than about 0.1% of an animal's daily ration may not supply enough energy unless they can be taken in quantity by some kind of a filter system.

The largest animal filter system is possessed by the whalebone whales, which have up to 300 baleen plates on either side of their mouths. These plates may be up to 4 m long in right whales and up to 1 m long in rorquals. They are arranged about 1 cm apart in a row around the upper jaw and are fringed on the inner edge with hairs. These hairs form a matlike strainer that prevents particles larger than about 2 mm from getting out. The whales operate this apparatus by swimming along with their mouths open and occasionally swallowing or by striking at a dense aggregation of food. The food of the whalebone whales in the Antarctic is mostly krill, a euphausid crustacean that varies in length from 2 to 5 cm. In the Arctic, however, krill is not as abundant, and their food consists of large copepods, other crustaceans such as crab larvae, or schooling fish such as smelt, mackerel, and herring.

It is interesting that a large part of the whale's food is also made up of filter feeders. The krill, like most of the small crustaceans, is equipped with rows of hairlike projections on its forward legs that strain and move particles toward its mouth. It feeds principally on diatoms larger than about 0.04 mm. Fish of the sizes and species taken by whales are partial filter feeders. They feed principally on small crustaceans but may include larger colonial phytoplankton in their diet, especially when small.

6.6 REPRODUCTION AND EARLY DEVELOPMENT

Most aquatic organisms spend much of their lives and energies reproducing. After a brief juvenile stage, they develop sperm or eggs, spawn, recover, and repeat the process in a cycle that continues until senility or death. The sperm and eggs are usually produced in such quantities that they represent a large expenditure of energy. The spawning process may be preceded by migrations or nest building and followed by care of the young. The postspawning recovery is usually coincident with or immediately followed by preparation for the next spawning.

The eggs and larvae start a life that is completely different from the life of the adults. The young of all sessile animals are mobile for a time. Commonly, the number of young is related to their fragility. The pups, or calves, of marine mammals number one or two at a time and are usually able soon after birth to feed and live much like the adult. At the other extreme, the eggs produced by an oyster or a shrimp number in the millions and hatch into feebly swimming larvae, completely unlike the adults. Most such larvae drift with the currents, in which they are easy prey for other animals.

The cycle of breeding activity is closely correlated with feeding activity, and these two activities account for most of the migrations and aggregations of the aquatic animal resources. Both the process of developing sperm or eggs and the process of recovering from spawning require extra energy and good feeding conditions. Some animals store large amounts of energy in addition to that in the gonads, which is used while migrating to spawn or while fasting during spawning.

Fishery scientists are commonly concerned with many aspects of the reproductive process. With wild animals they need to understand the reproductive process either to catch them easily or to protect them if they are unduly vulnerable, or to explain the great fluctuations in abundance due to failure of the young to survive. With domesticated animals they will try to control the environment to make the process as efficient as possible. They will be concerned with preventing waste of animals in poor condition as a result of breeding activities or of young being cared for by their parents.

6.6.1 Maturity

The cold-blooded animals with which we are concerned do not attain a definitive size or social behavior that may be used to define maturity, as birds or mammals do. The period of life in which sexual maturity is attained varies greatly; a few species are mature at birth, whereas others mature only once near the end of their lives. Most of the higher aquatic animals mature at an age equivalent to between one-fifth and one-third of their maximum life-span. Within a single species, the better-fed individuals living in warmer waters are the first to mature, and frequently males will mature at a younger age than females.

Some differences in the definition of *maturity* will be found among authors; some use the term to designate animals at all times after they are ready for their first spawning, whereas others restrict the definition to animals during the time that they are actually spawning. In this volume the term in the first sense is used.

The onset of maturation is indicated by the rapid development of the gonads. Among juvenile fish, the gonads are slender strands of tissue in the dorsal part of the abdominal cavity. As the gonads increase in size, the individual eggs in the ovary become visible to the naked eye and the testes change to a pale reddish color. When the gonads attain full size they seem almost to fill the body cavity; ovaries commonly make up between 10 and 25% of body weight (Fig. 6.8), whereas testes make up between 5 and 10%. At this stage the fish are considered to be *ripe* and will soon spawn naturally. Soon after the gonads reach maximum size in some species, such as salmonids, the ova and the sperm (called *milt*) can be expressed from the body by light pressure on the abdomen and fertilization can be accomplished artificially.

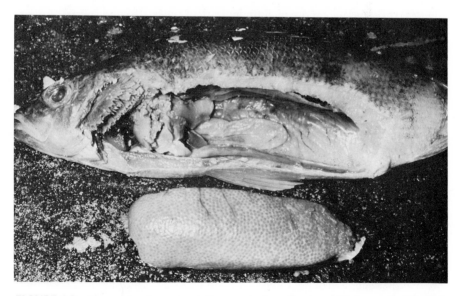

FIGURE 6.8 The ovaries of a mature yellow perch, *Perca flavescens*, fill almost all of the body cavity. (Photo by Professor L. S. Smith, School of Fisheries, Univ. of Washington, Seattle, Wash.)

6.6.2 Fecundity

Fecundity among egg-laying animals refers to the number of eggs being readied for the next spawning by a female. *Relative fecundity,* the number of eggs per unit of weight, is commonly used as an index of fecundity. *Total fecundity* is the number of eggs laid during the lifetime of the female.

Fecundity is more or less inversely related to the size of eggs and to the care given to eggs and larvae. Some sharks may have eggs between 6 and 9 cm long and produce only about 12 at a time. Scarcely more than this number of eggs are produced by some of the tropical fishes, but they are tiny and incubate either in their mouths, in special pouches, or in nests. Near the other extreme is the case of the animals that broadcast tiny, fertilized eggs to drift with the plankton. Most of the important marine fishes have this habit; many of the larger cods or flounders produce more than 1,000,000 eggs at one spawning. The largest number of eggs is produced by the sessile invertebrates, such as oysters, which may produce tens of millions of eggs at one spawning.

Fecundity is related to the hazards faced by the eggs, larvae, and young. Most animals maintain a reasonably constant population size. In bisexual animals, with a sex ratio of one to one, each female may be expected to produce enough eggs during her entire life to result in an average of two more adults. The species that have survived the process of evolution have been only those that have produced enough eggs and given them enough care to maintain their numbers (see Section 7.15).

One characteristic of the more fecund species is especially important to the fisheries: they fluctuate more in abundance than the less fecund species. Apparently the survival from many spawnings is very poor, whereas the average survival of two adults is maintained by an occasional highly successful spawning. Among fish that spawn annually, the progeny of such a spawning is called "a dominant year class," and among mollusks, "a successful set of spawn." Frequently, fishery scientists are asked to explain whether such fluctuations in abundance are due to natural causes or to the effects of fishing. Such species also offer them the tantalizing challenge of discovering and controlling the limiting factors in order to obtain a successful spawning every year.

6.6.3 Secondary Sexual Characteristics

Most mature fish and crustaceans develop external sexual differences in addition to the primary sexual characteristics, the gonads. These may help either humans or prospective mates to recognize the sexes; they are called *secondary sexual characteristics* and may be either accessory to spawning or not. The accessory structures include the intromittent organs of some male crustaceans and fish and certain fin or mouth structures that may be used in the care of either eggs or young. Other sexual differences (which are more common among freshwater fishes) include differences in size, shape of body, shape of fins, color, and presence of tubercles. Color differences between the sexes and special nuptial colors for both sexes are most striking among the salmonids and many small freshwater tropical fish, which are prized by aquarists. Perhaps the least sexually differentiated fish are the marine species that broadcast their eggs in the water. The sex of most of the cods, herrings, tunas, and mackerels cannot be recognized outside the spawning season except by dissection.

6.6.4 Sex Reversal

In most vertebrates, sex is determined at the time of fertilization and is fixed for life, but sexual change in midlife occurs in a substantial number of marine tropical fish species. The change is usually once during life and may be from male to female or from female to male. Hermaphrodites, which fertilize their own eggs, are very rare. Fish from several orders are involved, but most of them are from Perciformes.

Sex reversal (inversion) also occurs in some invertebrate resource animals. In many species of oysters, the oyster matures first as a male, slowly becomes a female, quickly becomes a male again, and then continues to alternate in sex throughout life. In colder climates oysters may change sex once a year; in warmer climates, several times. A single change occurs in some species of *Pandalus,* the northern shrimp. They mature first as males, remain so for a year or two, become females, and remain so for the balance of their lives.

These habits have several implications for the fisheries. Some of them are closely associated with aggregative behavior during spawning, when only one sex may be vulnerable to capture. And if the larger *Pandalus,* the females, are preferred by the fishery, their capture to the exclusion of males may seriously destabilize the sex ratio. Furthermore, the possibilities of manipulating the sex of domesticated fish through hormonal control, for example, may provide opportunities for increasing aquacultural production.

6.6.5 Time of Spawning

Among aquatic animals, spawning occurs at a time after the adults have been able to feed well enough to have extra energy for the developing gonads and when they can escape predators. Spawning and incubation must also precede a period of favorable food production that is sufficient to sustain the young. The timing for each race of animals has been determined by evolutionary adaptation; those animals that survived have originated from spawn released under favorable conditions.

For many aquatic animals in the temperate zones, spawning appears to occur in advance of the spring blooms of plankton by an interval about equal to the time required for the embryo and larva to consume the energy stored in the egg. Usually the spawning time is in the late winter or spring, but salmon in the Northern Hemisphere require between 3 and 9 months to pass through the egg and prolarval stages and commonly spawn between July and December. Much variability exists, however, even within the same species. This is especially prevalent for species in tropical and subtropical areas, where environment is relatively uniform throughout the year.

Regardless of the variability among animals, some kind of schedule is followed by each race. The schedules are controlled in most aquatic animals, at least in part, by temperature and length of daylight. These factors may operate separately or jointly. In experiments, spring-spawning animals spawned sooner when either temperature or the period of daylight was increased artificially; autumn-spawning animals spawned sooner when either temperature or the period of daylight was decreased. Other factors that may operate are tides, salinity, and floods.

The effects of such exogenous factors are, however, always subject to the restraints of the animal's own endogenous cycle. For example, an unusually warm spring may hasten spawning for an animal, but only if the gonads have developed adequately already—a process that may have started many months before.

6.6.6 Spawning Behavior

In addition to a schedule, every bisexual aquatic animal has a spawning ritual. Each must locate and recognize a member of the opposite sex of the same species, or recognize the other's sexual products, and then release its own ova or sperm at the right place and time. The rituals and schedules are as varied as the number of races of animals; indeed they are important aspects of the adaptation of the races to their environment. For each race of each species, however, they are relatively invariable. The procedures and sequences must be followed if the spawning is to be successful.

A few species engage in courtship during the later stages of gonad development. This may include pairing off, displays of colored fins, and aggressive behavior between members of the same sex. Such behavior is more common among bottom-dwelling freshwater fishes than among the pelagic marine species, many of which congregate in large spawning shoals.

Commonly, fertilization among fish occurs outside the body, in the water. It is accomplished by the nearly simultaneous release of sperm and ova close together: usually the ovum is fertilized within seconds of being laid. Fertilization by the sessile mollusks also occurs in the water. Both sperm and ova are released to drift with the currents and find each other, either in the mantle cavity or free in the sea (see Section 6.1.2).

Fertilization takes place internally in many crustaceans and in a fair number of fish. The male has an intromittent organ that is used to place the sperm in the female's body. Fertilization may be effected immediately, or the sperm may be stored for a period. Usually the energy for the developing embryo and larva is supplied by the yolk of the egg, whether fertilization is internal or external. In a few species of fish, the female's body will provide additional energy through some kind of placentalike structure.

Some freshwater fish and bottom-dwelling marine fish care for their young for a time. Most nest builders guard their nests and young; some carry the incubating eggs and larvae about in their mouths or in special pouches. Many crustaceans carry their eggs and smallest larvae attached to abdominal append-ages on the female. Marine mammals usually bear one or two living young after a gestation period of as much as 15 months.

6.6.7 Incubation

When one sperm enters the ovum, usually other sperm are excluded; incuba-tion starts immediately and continues until the egg hatches. The single-celled blastodisk divides and differentiates to form an embryo. During this exceed-ingly complicated and delicate process, the energy of the yolk is used according to the coded information in the genetic material to form the new animal, and energy is obtained and carbon dioxide and other wastes are eliminated through the egg membrane.

The incubating eggs of fish usually develop a prominent pair of dark spots at about the middle of the incubation period; these are the eyes of the embryo. These are called *eyed eggs* by fish culturists. The development of the eyes in trout and salmon eggs coincides with the end of a period of extreme sensitivity to shock that starts soon after fertilization.

The period of incubation or gestation varies among species of fish from a few hours to more than a year; the longer periods are required by fish that spawn in the autumn at high latitudes. The eggs of many of the common marine food fish hatch within 1–2 weeks. The period for each species varies inversely according to temperature within the tolerable range of temperatures. For example, trout eggs will tolerate incubation temperatures between 2 and 15°C. Within this range, the eggs of a particular race require about the same number of degree days. When the race requires about 400 degree days, the egg hatches in about 40 days at 10°C, and in about 80 days at 5°C.

6.6.8 Larval Development

The moment of hatching is the beginning of a hazardous period in the life of the aquatic animal. It marks no obvious change in the structure of the larval fish, which merely is freed from its confinement in the shell, but it creates the need for a profound physiological and behavioral change. The tiny animal must shift its respiratory and excretory systems from an exchange through the egg fluids and membranes to an exchange directly with the water through its own gills, skin, and kidneys. It must maintain its own osmotic balance without the help of the egg membrane. It becomes a swimming object and thus is more likely to attract the attention of predators than the motionless egg. It must find, catch, and digest its own food as soon as its supply of yolk is used up.

The larva begins a new life that is not only entirely different from its life as an embryo, but also entirely different from the life it will lead as a juvenile or adult. It probably does not resemble the adult: it does not eat the same food, and it has a much higher metabolic rate. It leads a completely different life—probably not even in the same place as the adult. Furthermore, its life is far from static because, relatively, it is growing faster than it ever will grow later in life, changing in structure and habit every few hours or days as its organs develop.

Larval development continues until the animal reaches the juvenile stage, when it more or less resembles the adult. A larval fish, while still using its stored yolk, is called either a *prolarva* or a *sac fry*. After it has absorbed the yolk, it is called either a *postlarva* or an *advanced fry*. The larval forms of invertebrates are much more varied and numerous. Usually bivalve mollusks have such an indistinctly structured egg shell that no real hatching occurs. While drifting in the water, the embryo develops into a feebly swimming larva called either a *trochosphere* or *trochophore*, from the Greek "trochos," wheel. The name refers to the wheellike rings of cilia. A second stage and much better organized swimming larva is the *veliger*, from the Latin "velum," veil, and "genere," to carry. A crustacean egg usually hatches first into a free-swimming *nauplius*, an unsegmented, egg-shaped larva with three pairs of appendages. A juvenile stage is attained after the next molt in some species, but not until after several intervening stages in others. Two of the common intervening larval forms are called *zoea* and *megalops;* others are named after other adult crustaceans that the larvae happen to resemble.

6.6.9 Examples of Reproduction and Early Development

The Oyster   The ova are fertilized within the mantle cavity of the female in *Ostrea;* in the open sea in *Crassostrea.* In either location, development proceeds for a few days until a ciliated velum has developed. The velum propels the animal and sweeps food into its mouth. At this stage it is a trochosphere. It is about 0.2 mm long and capable of feeding on plant cells no larger than about 0.01 mm. It soon becomes a better organized, veliger larva, and after between 1 and $2\frac{1}{2}$ weeks, depending on temperature, it develops a foot and begins to crawl. It has also developed a shell and become heavier. It can search on the bottom for a place to attach itself, since it can still swim. After more or less exploration, it *sets* by attaching itself on its left valve by means of an adhesive cement. Once it is attached, the velum and foot disappear with 2–3 days, and the shell rapidly enlarges. The *spat,* or young oyster, metamorphoses to an animal generally resembling the adult when the shell measures only about 1.5–2.0 mm in diameter (Fig. 6.9).

This process of reproduction and the factors controlling it have been studied intensively because it is fundamental to oyster culture. An important factor is temperature; after oysters are mature and in water near 20°C they can be induced to spawn. They commonly spawn in several pulses during the summer.

The oyster grower tries to put spat collectors in the water at just the right time to collect the naturally spawned spat and avoid the larvae of other animals, which foul the collectors if they are submerged continuously. Sometimes government agencies provide a forecast service of setting prospects for oyster growers based on temperature and quantitative samples of oyster larvae in the water.

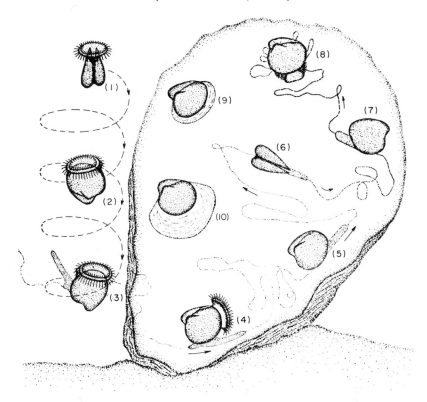

FIGURE 6.9   The setting process in the American oyster *Crassostrea virginica*. (1) and (2), Swimming larvae with protruded velum; (3) and (4), searching phase with protruded foot; (5) to (7), crawling phase; (8), fixation to substratum; (9) and (10), spat, 1 and 2 days old. The size of the larvae and spat relative to the old oyster shell is greatly exaggerated, the former being about 0.3 mm long at stage (1). [From H. F. Prytherch (1934). The role of copper in the setting, metamorphosis, and distribution of the American oyster *Ostrea virginica. Ecol. Monogr.* 4(1), 47–107.]

Penaeid Shrimp   Most of the edible shrimps, crabs, and lobsters either live in the sea or migrate between the sea and the estuaries. Most of them produce weakly swimming pelagic larvae. The larvae may drift with the ocean currents for a few days or a few months depending on species. Most species spawn at a single season of the year in a place in the ocean from which the young will drift to suitable nursery grounds in shallow water.

The penaeid shrimps, the common commercial shrimps of the tropics and subtropics, generally make a migration from shallow water to deeper water to spawn. In most species, the male deposits the sperm capsule on the abdomen of the female to fertilize the ova when they are released by the female. Several hundred thousand semibuoyant eggs are laid by each female. These hatch in about 1 day and then go through a series of about 10 molts as they grow gradually and transform into tiny shrimplike animals. These stages drift with the currents back toward the estuaries, which are the nursery grounds. Here, when the water is warm, the young shrimp grow rapidly, and after a few months they become adults. The adults start back to sea to spawn; usually they spawn only once during their lifetime (Fig. 6.10).

Atlantic Mackerel: Scomber scombrus   Many of the marine food fishes produce large numbers of small eggs that they broadcast in the sea to drift with the currents. Many of these species also experience spectacular changes in abundance due to variation in the survival of the young. The Atlantic mackerel, an important food fish of western Europe and eastern North America, is one of these species, selected as an example because of the extensive studies of the early history reported by Sette (1943), from which this account has been taken.

The mackerel are schooling fish. They reach a maximum length of about 55 cm but mature generally at a length of about 35 cm and at an age of 2 years. They live in the open sea, mostly over the continental shelf, but are not depen-

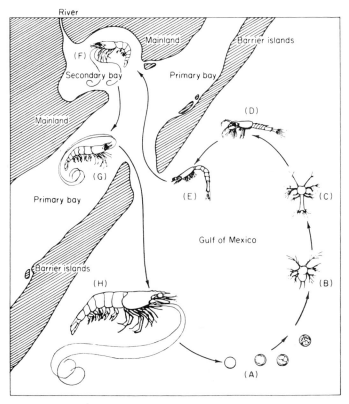

FIGURE 6.10   Life history of a shrimp. (A) Shrimp eggs; (B) nauplius larva; (C) protozoea; (D) mysis; (E) postmysis; (F) juvenile shrimp; (G) adolescent shrimp; (H) mature adult shrimp (not drawn to comparable scale). [From A. W. Moffett (1967). "The Shrimp Fishery in Texas," Bull. No. 50. Texas Parks and Wildlife Dept., Austin, Texas.]

dent either on the coastline or on the bottom at any stage in their lives. They winter in depths ranging from 100 to 200 m along the outer part of the continental shelf, and they move toward the surface, toward the shore, and toward the North Pole during the spring on a feeding and spawning migration. They are almost always found in water temperatures between 7 and 20°C.

The mackerel fishery is pursued during the late spring, summer, and early autumn, when the fish are nearer the shore and the surface. It has been an important fishery for centuries in Europe and since early colonial days in North America. Its changing abundance has plagued and puzzled the fishermen, however, who find mackerel in enormous numbers in some years and in scant supply in other years. North American production has varied from 6000 to 106,000 tons annually.

Approximate data on the reproductive process are as follows: Spawning begins in mid-April in the southern part of the North American range and in mid-July in the northern part (Fig. 6.11), but it continues for only 3–4 weeks in any one locality. An average female produces 400,000 ova. The fertilized egg is 1.2 mm in diameter, drifts in the upper 10-m layer of the sea, and hatches in 4 days. The newly hatched, feebly swimming larva is 3.2 mm long. It absorbs the yolk sac in 5 days, develops fins in 26 more days, when it is 10 mm long, and then reaches the adult stage at a length of 50 mm, 40 days later (Fig. 6.12).

These findings were established or verified by Sette as a preliminary to determining the rates of growth and mortality of the eggs and larvae. This he did from an extensive series of quantitative net samples taken repeatedly over the entire range of the United States stock and throughout the period of spawning and larval development in 1932. After identifying and sorting the mackerel from the numerous other species in the catches, he was able to find *homologous* groups (i.e., eggs fertilized in a brief period in a limited area that could be identified in later catches when further developed). After determining the change in length of each homologous group, he was able to determine the rate of growth. Then, from the numbers of mackerel taken in the same samples, he was able to estimate the actual numbers of eggs and larvae in the sea at different stages in their development.

The rate of larval growth was found to be logarithmic; it proceeded at about 4%/day in length and 12.5%/day in weight! The number of eggs laid was estimated to have been about 64×10^{12} by a spawning population of about 1×10^8 individuals. The larvae died during most of the developmental period at rates of 10–14%/day, except as the larvae were developing fins, when the rate was 30–45%/day! Only between 1 and 10 mackerel reached a length of 50 mm out of each million eggs spawned!

These observations were the record of the failure of a year class due probably to two possible causes: the failure of the newly hatched young to find enough of the special food they needed, or the drift of the young with unusual currents away from their accustomed nursery areas. Sette concluded, on the basis of circumstantial evidence, that the second factor was probably the predominant

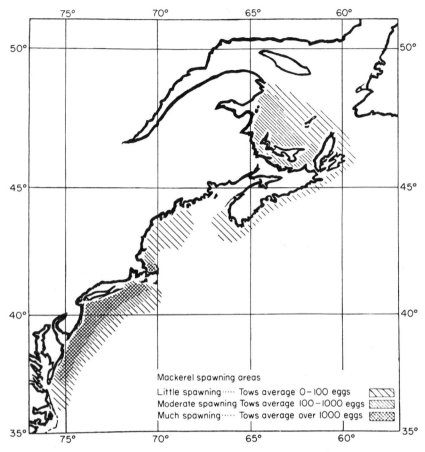

FIGURE 6.11   Mackerel spawning areas along the Atlantic Coast of North America. Relative intensity is indicated by the average number of eggs caught in plankton nets. (From Sette, 1943.)

cause of the failure of the 1932 year class, although the scarcity of plankton may have been a contributing factor.

Sockeye Salmon: Oncorhynchus nerka   The reproduction of salmon, trout, and chars has been studied more extensively than that of almost any other fish because it is spectacular and easy to observe, and it is easy for humans to intercede for the purpose of rearing the young artificially. Almost all of these fish migrate from feeding to spawning grounds, where they become brightly colored and develop sexual differences. The female digs a pocket, or redd, in the gravel in which she lays the eggs and covers them after fertilization by the male. The eggs hatch and the larvae consume the stored yolk while the eggs are protected by the gravel; then the larvae emerge to begin feeding. The sockeye salmon was chosen as an example of a species with this type of repro-

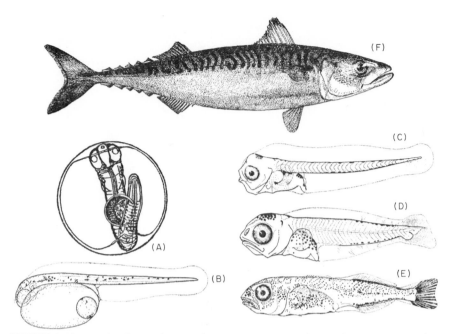

FIGURE 6.12   Mackerel, *Scomber scombrus*. (A) Egg; (B)–(E) larvae of 3.5, 4.6, 7.8, and 15 mm in length, respectively; (F) adult, 40 cm. [From H. B. Bigelow and W. C. Schroeder (1953). Fishes of the Gulf of Maine. *U.S. Fish Wildl. Serv., Fish. Bull.* 53(74), 318.]

duction because it is typical of the group with respect to its egg laying and larval development, but it is more spectacular than most in its migrations and nuptial colors.

The sockeye salmon spawns only once—at the end of a life ranging usually from 4 to 6 years in length and migrations of thousands of kilometers in the ocean. It begins to migrate from the high seas toward its home stream in late spring, traveling at speeds of between 30 and 50 km/day through the open sea, to its estuary, and up its river to its lake system, which is its nuptial and nursery area. Its gonads have begun to develop rapidly while it is still at sea, and it stops feeding as it approaches the estuary. Soon after entering fresh water it begins to change color from bluish silver to bright red in the body and green in the head. It finds its way through the lake to its spawning beach or stream and spawns either in late summer or in autumn.

The female usually digs a redd, then mates with an attending male (Fig. 6.13), and covers the eggs with gravel a few minutes later. She usually deposits her eggs in four to six clusters of about 500 eggs each. The preferred spawning site is in a pool at the head of a riffle or in a riffle through which water flows, with gravel small enough and loose enough to be moved around by the female and with little fine sand or silt that would prevent circulation of the water or the emergence of the larva.

FIGURE 6.13 A pair of spawning chum salmon, *Oncorhynchus keta*, releasing a cloud of sperm. Their spawning behavior is almost identical to that of sockeye salmon. (Photo by A. H. Hamalainen, Univ. of Washington, Seattle, Wash.)

The eggs usually hatch either in late autumn or in early winter; the larvae remain in the gravel subsisting on their yolk sacs (Fig. 6.14) until early spring, when they wriggle up through the interstices of the gravel. They emerge and seek food and shelter along stream banks and lake beaches for several weeks, and then they move out into the open water of the lakes, where they lead pelagic lives. They continue living in the lake until the following spring, or even until the second following spring, when, at a length of between 5 and 10 cm, they obey their urge to migrate down the river to the sea.

6.7 AGE AND GROWTH

Growth is commonly considered to be a gradual increase with time in size or mass or some kind of a living unit. It may be applied to a part of an organism, to a whole organism, or to a population of organisms. It is complicated by the differentiation of parts of organisms, which may grow at different rates or even negative rates (degrowth), and by the death of parts of populations. It is further complicated because the use of energy for growth is not greatly different from the use of energy for maintenance of the unit, for the repair of wear and tear, or for the excretion of products. A more exact definition is "Growth is the addition of material to that which is already organized into a living pattern." In terms of energy equivalents it may be expressed as

$$\Delta W/\Delta t = R - T,$$

in which ΔW is the growth, R is the total energy obtained from the rations, and T is the total energy expended in metabolism, all during the unit of time Δt.

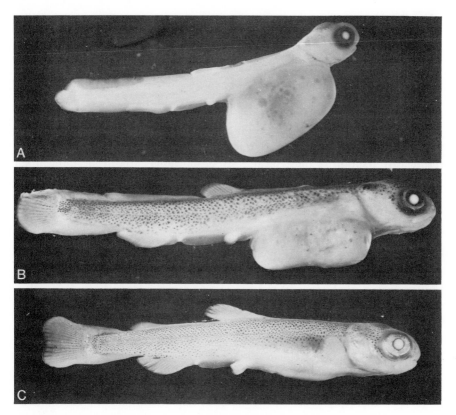

FIGURE 6.14 Young sockeye salmon, *Oncorhynchus nerka*. (A) Soon after hatching; (B) middle of sac fry period; and (C) beginning of postlarval period. The last stage is the time of emergence from the gravel. The mean lengths are about 17, 22, and 25 mm, respectively. [Photos by J. C. Olsen (1964). Fisheries Research Institute, Univ. of Washington, Seattle, Wash.]

This section deals with the growth of aquatic organisms, and Section 7.1.5 deals with the growth of aquatic populations.

Few topics are of more fundamental importance to the fishery scientist than the age and growth of aquatic life. Questions such as these are constantly recurring: How old is this animal? How much will it weigh next year? At what age does it mature? At what rate does it grow at age x? When is it the fattest? When will the progeny of these parents be large enough to catch? Answers to such questions are basic to understanding the life of individuals as well as a necessary preliminary to understanding the fluctuations in resource populations.

Two kinds of growth are found among living aquatic resources: the stepwise growth of the crustaceans, and the "continuous" growth of the mollusks and the vertebrates. The difference is not really too important for most purposes, however, because much of the growth of crustaceans at molting is due to the addition of water to energy already stored, and much of the so-called continuous growth of mollusks and invertebrates varies cyclically in rate with the seasons and schedules of activity.

The growth of almost all of the aquatic resource animals is asymptotic, that is, each species in each environment has a characteristic ultimate size, which it approaches by growth throughout life. This type of growth is common to all cold-blooded vertebrates and, surprisingly, to aquatic mammals as well, which do not reach maximum size at about the age of maturity, as many of the terrestrial mammals do. An important characteristic of asymptotic growth is the relative fixation of the typical body proportions early in life rather than during adolescence and maturity, as in the terrestrial mammals.

6.7.1 Maximum Age and Size

Age and size of aquatic resource animals are more or less correlated. The oldest among the animals tend to be the largest, but there are many exceptions. The long-lived fishes tend also to be those with the following characteristics: they are phylogenetically primitive, sluggish in their movements, bottom- or shallow-water inhabitants; they have accessory respiratory devices; and they are adaptable to extreme fluctuations in oxygen concentration, temperature, and salinity. Examples are sturgeons, sharks, and carp. The short-lived fishes are those with opposite behavior and adaptability, including tunas, salmon, and capelin.

The maximum age of wild aquatic animals is not well known because those that die naturally are seldom seen, and the evidence of age in the oldest of the animals is frequently obscure. The oldest fish on record (albeit a questionable one) is a wild sturgeon whose age was estimated from the rings in its fin rays as 152 years. Many fish have been kept for long periods in aquaria; the records go to a sturgeon kept in the Amsterdam Aquarium for 69 years and a carp kept in the Frankfurt Aquarium for 38 years. Numerous species of larger fishes have been kept for more than 20 years. Some invertebrates also may live a long time: northern lobster, 50 years; king crab, 20 years; and Pacific oyster, 40 years. Apparently, such animals live longer than whales, which start life as relatively large infants, become sexually mature between 3 and 8 years, and survive for a maximum of about 40 years.

The largest living animal, the blue whale, now nearly extinct, is said to reach a length of about 30 m. One blue whale, 27 m long, was weighed carefully at a whaling station and found to weigh 119 tons.

The largest fish is the whale shark; it reaches a length of about 15 m. The largest bony fish is probably either the black marlin or the blue marlin; both exceed lengths of 4 m and weights of 700 kg.

The largest of the individuals of numerous species of fish caught by anglers are of special interest as trophies. The principal marine trophies are registered with the International Game Fish Association of Fort Lauderdale, Florida. The association sets standards for the tackle that may be used, checks reports, and annually publishes lists of record fish for men and women and for various strengths of line. The all-tackle record fish prior to 1968 was a white shark, *Carcharodon carcharius*, of 1208 kg (2664 pounds); the largest bony fish was a black marlin, *Istiompax marlina*, of 707 kg (1560 pounds). Records of North American freshwater fish are kept by *Field and Stream* magazine of New York. Regional and local records of trophies are also kept by many clubs and government fishery departments.

6.7.2 Patterns of Growth

The growth of any animal is accompanied and influenced by many factors, including both the endogenous events of its schedule of development from embryo through maturation to senility and the exogenous changes in its environment. Many of these factors operate independently in ways that are not well understood. Some influence the change in size of the whole animal; others influence the shape of the animal. Commonly, they must be studied empirically (i.e., by actual observation of the growth of a given species).

Endogenous Factors In any organism, a basic pattern of growth is inherited from the parents. Inheritance of rapid growth in trout was shown by breeding experiments in which rapidly growing adults were selected as parents in successive generations. The progeny of such selected trout in 1 year attained the length usually attained in 3 or 4 years.

The growth and differentiation of animals is controlled by hormones. Their role in aquatic resource animals has not been studied as extensively as for some terrestrial mammals. The pituitary gland produces growth hormones; fish have been shown to respond when fed purified extracts of either fish or mammalian pituitaries.

Another factor that contributes to a slightly increased growth in fish with large eggs is the relative amount of yolk. Larvae from eggs with more yolk average slightly larger than those from eggs with a small amount of yolk, and they retain their size advantage for some time.

The larval period is characterized by rapid rates of growth and sudden changes in food and function, and it is commonly also a period of rapid changes in body shape (Fig. 6.12).

After the animal begins to look like the adult, the endogenous changes in growth are associated with the varying proportions of the energy input used for growth and other needs. The juvenile animal uses energy primarily for growth and secondarily for maintenance. Later, energy is diverted to the maturation of the gonads and to preparation of the body for any fasting during spawning. After spawning, when the bulk of the gonads is gone, the animal typically requires some time to regain its average body shape and replenish any depleted stores of energy. A cycle of varying partition of energy among growth, maintenance, and gonads continues until senility. At this time most of the energy input is used for maintenance and very little is used for growth.

Some species have sexual differences in growth. In many flatfishes the females are heavier for a given length, grow faster, and live longer than the males. An example is the yellowtail flounder (Fig. 6.15). In other species, such as the common dolphin, the male grows faster than the female. Such sexual differences occur also in some crustaceans; the male northern king crab is larger than the female.

Exogenous Factors The growth of any animal is influenced by a complex of environmental factors, so that different stocks frequently grow at different rates (Fig. 6.16). Both the quantity and quality of food are probably the most important, but temperature too has a major effect at times. The combined effect of these two factors is usually marked in the fresh waters of temperate or arctic regions that freeze in winter. When the water is near 0°C, both metabolic activity and growth are minimal (Fig. 6.17). The variation in food supply is not restricted, however, to the colder areas; most waters everywhere have blooms of plankton, swarms of insects, or schools of larvae, and these provide food supply for gorging and fattening (Fig. 6.18). One of the effects of the abundance of food on growth appears to be linked to the relationship between the energy cost of the food and the energy supplied. Small particles and hard-to-find food require the expenditure of more energy for grazing or search, and therefore they reduce the proportion of energy input that is available for growth.

Growth may change when the animal is migrating, when extra energy is required for locomotion, when food is changing, or when the animal chooses not to feed. Some fish stop feeding entirely during the last part of a migration to the spawning ground.

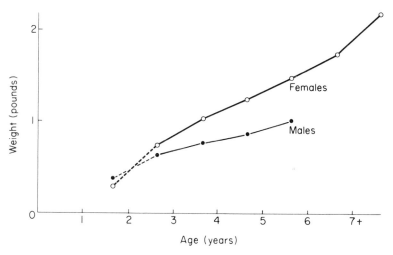

FIGURE 6.15 Attained weight of male and female yellowtail flounder, *Limanda ferruginea*, from southern New England. Data are shown for the autumn, when effects of gonad development are minimal. [From W. F. Royce, R. J. Buller, and E. D. Premetz (1959). Decline of the yellowtail flounder (*Limanda ferruginea*) off New England. *U.S. Fish Wildl. Serv., Fish. Bull.* **59**, 169–267.]

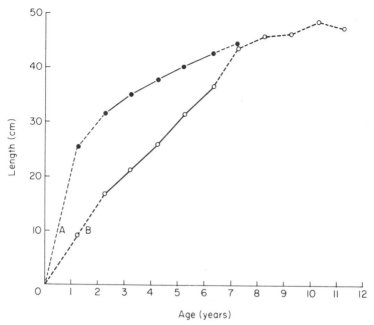

FIGURE 6.16 Annual growth in length of female yellowtail flounder, *Limanda ferruginea*, from two stocks off eastern North America. Dashed lines are used where data are few. Curve A, southern New England stock (Royce *et al.*, 1959); curve B, Nova Scotian stock (Scott, 1954). [From W. F. Royce, R. J. Buller, and E. D. Premetz (1959). Decline of the yellowtail flounder (*Limanda ferruginea*) off New England. *U.S. Fish Wildl. Serv., Fish Bull.* 59, 169–267; and D. M. Scott (1954). A comparative study of the yellowtail flounder from three Atlantic fishing areas. *J. Fish. Res. Board Can.* 11(3), 171–197.]

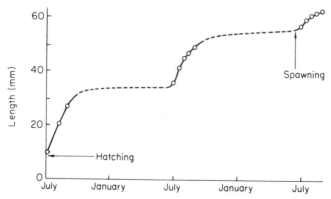

FIGURE 6.17 Growth in length of threespine stickleback, *Gasterosteus aculeatus*, in a subarctic lake in Alaska. Dashed lines show presumed growth during winter. [From D. W. Narver (1966). Pelagial ecology and carrying capacity of sockeye salmon in the Chignik Lakes, Alaska. Ph.D. Thesis, Univ. of Washington, Seattle, Wash.]

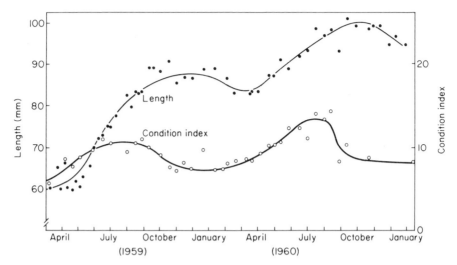

FIGURE 6.18   Growth and condition of oysters held on trays in Willapa Harbor, Washington. Curves are fitted by eye. Decreases in length are caused by breakage of the shell edges during handling. Condition index is dry weight of meat divided by volume of shell cavity × 100. [Data from K. K. Chew (1961). The growth of a population of Pacific oysters (*Crassostrea gigas*) when transplanted to three different areas in the state of Washington. Ph.D. Thesis, Univ. of Washington, Seattle, Wash.]

The amount of living space per animal may affect growth in ways that are not well understood. Crowded animals may suffer from an accumulation of body wastes, a low supply of oxygen, or a small opportunity to get food. Any one of these factors would depress the rate of growth. Even when an abundance of food and water is supplied, the animals still may not grow as rapidly as less-crowded animals. At the other extreme, animals either in isolation or in low concentrations may not grow as rapidly either. An optimal degree of crowding may enhance growth, presumably as a result of obscure social or chemical factors.

Length–Weight Relationship   Every animal grows in both length and weight; the relationship between these is of both theoretical and practical importance. Comparison of the relationship indicates changes in body shape or in the condition of the animal. Further, it is necessary for a fishery scientist frequently to estimate the total weight of an animal of given length or the meat weight of an invertebrate for a given shell dimension. Consequently, determining the length–weight relationship is a standard procedure.

It is obvious that length and weight measurements used for a comparison must be obtained in the same way, but this is a problem of such complexity that it deserves a note here. The length of a fish is commonly measured in three ways: standard length; fork length, or length to the tip of the central rays of the caudal fin; and total length (Fig. 5.1). The first is used regularly for taxonomic purposes, the second is the common measure of the fishery scientist, and the third is used occasionally. Moreover, fish may be measured by the fishery scientist while alive, or immediately after death, or later during rigor mortis, or still later after rigor mortis. Fish are also commonly measured after preservation in formalin. It is important that measurements to be compared be the same measurements obtained under similar conditions or that they be adjusted. Live fish and fish in rigor tend to be shorter than relaxed fish after rigor. Preserved fish are also shorter, because of shrinkage of the tissues. Weight measurements must also be used with caution because most animals lose fluids after death and add fluids after preservation in formalin or alcohol. Changes in both length and weight up to $\pm 5\%$ may occur because of such factors.

With any set of lengths and weights the relationship is usually of the form

$$w = al^b.$$

A line may be fitted easily to the data when they are transformed to (with logs to base 10)

$$\log w = \log a + b \log l.$$

This equation yields a straight line (Fig. 6.19).

When the exponent $b = 3$, the animal is growing without change in shape or specific gravity, that is, isometrically. Among fish, an exponent of 3 is the exception; however, values from 2 to 3.5 are commonly found.

Condition Factors   A number of indexes of fatness, or condition, of aquatic animals have been developed. A direct index of fatness has been obtained by chemical analysis of fat content, which is expressed as a percentage of body weight. A closely related index is the specific gravity of individual fish. This may be obtained if great care is used to eliminate all air bubbles. Those with lower specific gravity are fatter. Other direct indexes of fatness are based on the fat that surrounds the viscera. This may be judged subjectively or removed and weighed. A condition index of oysters is the ratio of the dry weight of the meat to the volume of the shell cavity $\times$ 100 (Fig. 6.18).

Other indexes are obtained by computation of the length–weight relationship. This may be shown as the estimated weight for a given length of fish (Table 6.2). Another index in common use for studies of freshwater fish is based on the cubic relationship.

$$w = Kl^3$$

in which K (usually multiplied by 100,000) is called the *condition factor*. The factor varies, of course, with the units of measurement; K is used commonly for grams and millimeters, C for pounds and inches. Either will vary also according to which length measurement is used.

6.7.3 Allometric Growth

Even after the larval period, at the end of which the shape of most animals becomes similar to that of the adult, some changes in the body proportions continue to occur. Commonly they are minor changes, such as those in length

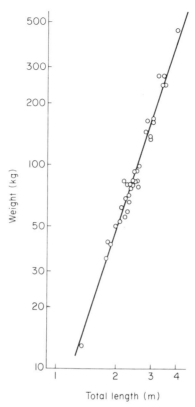

FIGURE 6.19   Length–weight relationship of Pacific blue marlin, *Makaira nigricans*. [Data from W. F. Royce (1957). Observations on the spearfishes of the Central Pacific. *U.S. Fish Wildl. Serv., Fish. Bull.* 57, 495–554.]

TABLE 6.2   Seasonal and Sexual Differences in the Weight (in Pounds) of Male and Female Yellowtail Flounders (*Limanda Ferruginea*) at the Mean Length of 35.869 cm[a]

	Calendar quarter	Male	Female	Difference[b]
	1	0.914	0.955	0.041
	2	0.826	0.945	0.119
	3	0.822	0.872	0.050
	4	0.882	0.953	0.071
Average		0.892	0.933	0.041

[a]From W. F. Royce, R. J. Buller, and E. D. Premetz (1959). Decline of the yellowtail flounder (*Limanda ferruginea*) off New England. *U.S. Fish Wildl. Serv., Fish. Bull.* 59, 165–267.

[b]Note that females are always heavier and differ most greatly in weight from males during the second quarter, which is spawning season.

of fins and fatness of body and the temporary changes associated with the ripening of the gonads. These changes are usually called *allometric growth*, although some grosser changes may be called *heterogonic growth*, such as the differentiation of right- and left-hand claws in crustaceans. The lack of change and the continuation of proportional growth is called *isometric* or *isogonic growth*.

Such changes in body proportions are of special interest to the fishery scientist who tries to differentiate stocks of the same or closely related species on the basis of body proportions. He or she must separate any differences due to growth from those due to either environment or inheritance. This can be done by comparing animals from different regions at identical lengths (Fig. 6.20). The yellowfin tuna, *Thunnus albacares*, of the example, shows substantial variation in fin height due to locality in addition to great variations due to body

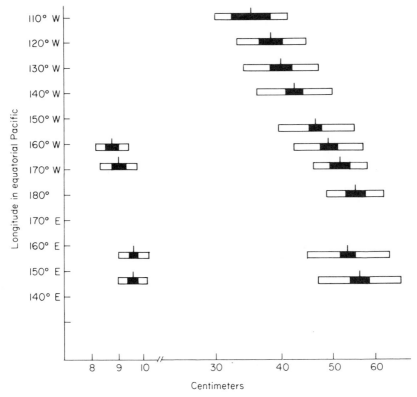

FIGURE 6.20   Height of anal fin of small (65 cm) and large (140 cm) yellowfin tuna, *Thunnus albacares*, from the Pacific equatorial region as estimated from regression statistics. The center line indicates the mean; the solid bar, ± two standard errors of the mean; the hollow bar, ± one standard deviation from regression. [From W. F. Royce (1964). A morphometric study of yellowfin tuna *Thunnus albacares* (Bonnaterre). *U.S. Fish Wildl. Serv., Fish. Bull.* 63(2), 395–443.]

size. Note that if the anal fin grew isometrically after the tuna was 65 cm long, the fin would be only about 20 cm high when the tuna was 140 cm long, instead of 30–65 cm.

Body proportions tend to vary more with growth than the numbers of body parts, such as fin rays or vertebrae. The latter, called *meristic characters*, are more satisfactory for racial studies, but even they vary with growth occasionally.

6.7.4 Rate of Growth

The essential way of expressing any kind of organic growth is to describe its rate during the life of the unit. This problem is one that has attracted the attention of many scientists who have sought general laws of growth by fitting mathematical functions empirically to growth data and by deducing other mathematical functions on theoretical grounds. As may be judged from the examples of growth given here and the discussion of the factors influencing growth, the rate may be the result of many factors operating independently and erratically. An adequate mathematical expression of the seasonal rate alone is extremely difficult, but even if the growth rate were assumed to be uniform throughout the year, the variations in yearly increments among species and populations would remain extremely complex. No generally applicable law of growth has been discovered yet.

One may say, however, that the growth of living aquatic units is almost always exponential; the rate is a function of the size already attained. Thus, it is multiplicative or logarithmic rather than arithmetical. Early in the life of a unit, the growth typically accelerates, and later it slows. The curve of weight against age is at first concave upward, then it flexes and becomes concave downward as it gradually approaches an asymptote (Fig. 6.21). Even so, the rates are usually exponential; the early period positively, the later period negatively. Usually, too, the inflection is early in life, so the curve is an asymmetrical sigmoid with a long right-hand limb (see Section 8.9).

6.7.5 Age Determination

It is essential to determine the age of animals for determination of the rate of growth as well as for the analysis of populations, which will be discussed in Section 7.1.4. Fortunately, this can be done for most aquatic resource animals with relative ease. Three methods are available; the marked animal method and the length frequency method are used occasionally, and the annual ring method is used routinely. Each method has its special virtues and limitations, so they frequently supplement one another.

The *marked animal method* requires that animals whose age is known initially be identified by a mark, such as a tag or by isolation. When growth data are desired, the size can be recorded at appropriate times, but the data must be interpreted with caution because the growth is almost certainly affected by the marking, handling, or isolation. The growth data may be less

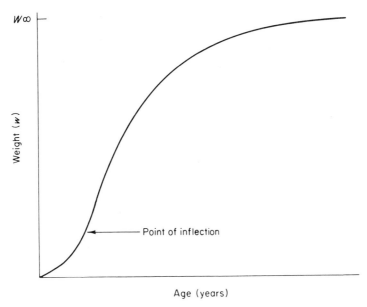

FIGURE 6.21 Typical curve of growth in weight of aquatic animals. Point of inflection is usually before half of the life span, thus an asymmetrical sigmoid curve results.

important, however, than evidence about the characteristics of the annual rings, which can then be identified in other unmarked animals.

The second, or *length frequency method*, is used occasionally and is especially helpful for age determination of crustaceans and young animals or for determinations of seasonal growth. It is sometimes named for C. G. J. Petersen of Denmark, who introduced the method in 1891. This method is useful for broods of animals that have been spawned during a single, short period and that grow individually at nearly the same rate. When they are measured, the length frequency data form a mode or series of modes that can be identified. If, for example, collections were made of young fish, monthly, after their known birth date, it would be found that no other individuals of the same species would be near the same size, and the mode of this homologous group could be followed during successive collections (Fig. 6.22). With other data, a multimodal distribution may be found in which two or three successive year classes can be identified.

The third, or *annual ring method*, which is by far the most useful, depends on the record left in the hard parts by irregular growth and metabolism. The irregularities include the daily cycle of feeding, the occasional changes in food and activity such as spawning, and the annual cycle of changes with the seasons. The daily cycle, surprisingly, leaves a series of rings in bony parts of some fishes, which can be seen under high magnification. These have been found, in

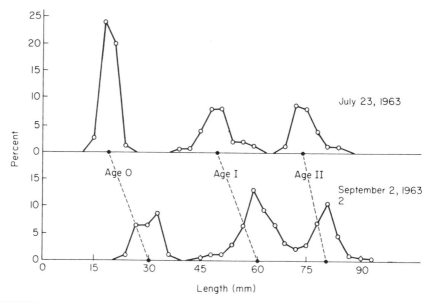

FIGURE 6.22 Percentage length frequencies of pond smelt from Black Lake, Alaska, in mid- and late summer. (Unpublished data. Fisheries Research Institute, Univ. of Washington, Seattle, Wash.)

some cases, to provide a record of age in days of larval fish and a record in days of growth during the year in older fish.

Annual rings, or *annuli*, as they are commonly called, are crowded or irregular rings laid down during annual periods of slow growth, hence another common name, *checks*. They are usually prominent enough to be seen under low magnification or even with unassisted eyes. Checks may also be caused by injury, spawning, or periods of irregular feeding and, with careful examination, their cause can usually be identified. This record, in hard parts that grow at a known rate relative to the size of the body, can also provide an estimate of body size at the time that each check occurred throughout life.

Many hard parts show checks, and the part to use may be chosen according to which is easiest to collect or examine, which has the clearest marks, which has the most consistent size relative to body size, or which may be collected without harm to the animal. The parts in most common use are the scales of fish (Fig. 6.23). Also useful from fish are the otoliths (ear bones) (Fig. 6.24), vertebrae, opercular bones, and fin rays. Similarly useful are the shells of mollusks (Fig. 6.25), the teeth of seals (Fig. 6.26), whalebone, and the earplugs of whales. Unfortunately, crustaceans have no permanent hard parts.

When the part is a fish scale, it may be examined directly, or a plastic impression may be made of the surface. Either the scale itself or its impression may be examined under a microscope, or preferably, its image may be projected

FIGURE 6.23 A scale of a mature sockeye salmon, *Oncorhynchus nerka*, from Bristol Bay, Alaska. Shown are (F) the focus of the scale; (1) one annulus near the end of freshwater life; and (2, 3, 4) three annuli during saltwater life. (Photo by T. S. Y. Koo, Fisheries Research Institute, Univ. of Washington, Seattle, Wash.)

on a screen (Fig. 6.27). When the part is a bone, tooth, or fin ray the checks may appear inside, and it may be necessary to saw it into a thin section and polish this section for examination. Sometimes thin bones can be made translucent by chemical treatment and examined directly.

After any necessary preparation, the growth record in the hard parts is read. It is necessary to classify the kinds of checks and relate them to the actual events causing them. Annuli can be identified by a study of homologous checks at the edge of hard parts collected at varying times during the year. Such a study permits a determination of time of completion of the annulus and also assists in identification of any checks due to other causes. As an alternative, parts may be studied from animals whose age has been determined by other methods. Special attention must be given to young animals, since they may not form an annulus during their first year, and to old animals because, frequently, they show poorly defined annuli. The process of reading is necessarily somewhat subjective, but it is fairly reliable when the reader is familiar with the life history of the group of animals with which he or she is working.

Age is usually designated as the number of annuli either in arabic or in roman numerals. Some confusion exists, because animals in their first year are designated age 0 before the first annulus is formed (usually close to 1 year after spawning). After the first annulus, the count corresponds to human birthdays,

FIGURE 6.24   An otolith of a 12-year-old Pacific halibut, *Hippoglossus stenolepis*. (Photo by International Pacific Halibut Commission, Seattle, Wash.)

Released August 27, 1957

1957 – 1958 annulus

1958 – 1959 annulus

Recovered July 18, 1959

FIGURE 6.25   A numbered razor clam, *Siliqua patula,* added two clear annuli to its shell before recapture. (Photo by H. C. Tegelberg, Washington Dept. of Fisheries, Olympia, Wash.)

for example, an animal designated as age 3 is in its fourth year. A variety of special conventions is used when additional information can be obtained. Sockeye salmon, whose life in fresh water and in salt water can be recognized separately on their scales, are designated, for example, as age 2.2 when they have two freshwater and two saltwater annuli. This species of salmon fails to form an annulus in its first year of life, so this example would be 5 years old. Another common method of designating the age of this example is 5_3; 5 shows total age, and the subscript is the number of total years in fresh water.

Much more information than age at capture can be obtained from the hard

FIGURE 6.26   A longitudinal section of a canine tooth of a 9-year-old male fur seal (*Callorhinus ursinus*) taken at sea near Yakutat, Alaska, on April 21, 1959. Scale marks in millimeters. (Photo by D. F. Riley, U.S. Natl. Marine Fisheries Service, Seattle, Wash.)

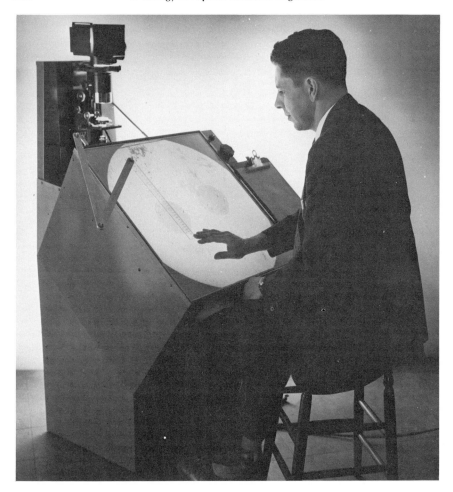

FIGURE 6.27 A special microprojector is useful for reading and measuring fish scales. (Photo courtesy of Eberbach Corp., Ann Arbor, Mich.)

parts when, as was mentioned earlier, the relationship of hard part size to body size is known. The size at all earlier "birthdays" can be estimated. Usually the easiest points to measure are along a radius to the boundaries marking the completion of each annulus and the beginning of the following year's growth (Fig. 6.28). Such measurements frequently provide extremely valuable information on the early lives of the animals.

Unfortunately, the growth of hard parts is rarely isometric. Scales and many bones do not form in the larva until the body reaches an appreciable size, and they may grow subsequently at different rates. Some early investigators assumed isometric growth and soon discovered puzzling discrepancies between

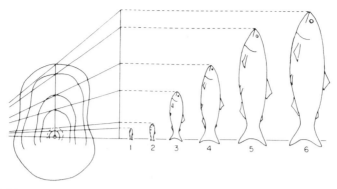

FIGURE 6.28   Diagram of approximate relationship of salmon size to scale radii. The length at capture (6) and the scale radii can be used to estimate length at the times of annulus formation (1, 3, 4, 5) and the time of migration to sea (2).

the size estimated from hard parts and the size of young animals actually caught. The most common result is the tendency for the estimated length of a young animal of specific age to decrease as the total age at capture increases. This has been called *Rosa Lee's phenomenon*, after its discoverer. It is caused by allometric growth of hard parts (Fig. 6.29) and by more subtle factors, including a marked tendency of fishing gear to select fish of certain size. Com-

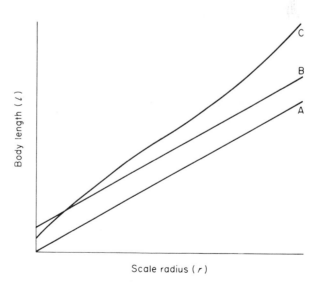

FIGURE 6.29   Types of relationship between fish scale and body length. (A) Isometric growth, $l = br$. (B) Allometric growth typical of late-appearing scale, $l = a + br$. (C) Allometric growth of type that must be fitted with curvilinear relationship.

monly, larger young fish and smaller old fish are taken by a single kind of fishing gear—a circumstance that alone can cause Rosa Lee's phenomenon.

REFERENCES

Burner, C. J., and Moore, H. L. (1962). Attempts to guide small fish with underwater sound. *U.S. Fish Wildl. Serv., Spec. Sci. Rep.—Fish.* **403**, 1–30.

Hasler, A. D. (1966). "Underwater Guideposts, Homing of Salmon." Univ. of Wisconsin Press, Madison.

Nakamura, E. L. (1968). Visual acuity of two tunas, *Katsuwonis pelamis* and *Euthynnus affinis*. *Copeia* **1**, 41–49.

Sette, O. E. (1943). Biology of the Atlantic mackerel (*Scomber scombrus*) of North America. Part 1. Early life history, including the growth, drift, and mortality, of the egg and larval populations. *U.S. Fish Wildl. Serv., Fish. Bull.* **50**, 149–237.

7 | Ecological Concepts

The three preceding chapters are about the aquatic environment, the organisms, and their biology; all are elements of aquatic ecology. But fishery scientists are concerned with more than individual living organisms and their environments. They are concerned with groups of individuals of the same species, called *populations;* with groups of populations of different species in given areas, called *communities;* and with the functioning together of communities and their environments in given areas as *ecosystems.* An individual, a population, a community, or an ecosystem can be considered as a biological unit. A *resource ecosystem* is a unit in which one element is a product of direct use to humans. Fishery scientists, who will be dealing with one or more of these kinds of units in all of their work, will be applied ecologists.

Ecology is usually considered a branch of biology, but it is also an integration of the biological with the earth sciences such as oceanography and geology. It is a unifying concept of how life exists on our planet. It may be defined as the study of the interrelationships among organisms and the interrelationships of organisms with their nonliving environments. It includes, of course, the terrestrial environment as well as the aquatic, but this chapter will focus on the aquatic.

Ecology is a new and exceedingly complex field of study, even though its concept was recognized by the Apostles in their use of the phrase "all flesh is grass." Much more recently, John Muir, the famous naturalist, eloquently described its scope in the sentence "When we try to pick out anything by itself we find it hitched to everything else in the universe." Yet the concept of an ecosystem was not applied to a simple terrestrial system until the 1930s or to aquatic systems until later. Aquatic ecology is a multidisciplinary science with no clear boundaries among the many contributing sciences. And it is, in some ways, more complex than terrestrial ecology, because few aquatic population, community, or ecosystem units have fixed physical boundaries. Both sciences and environmental units tend to be in a permanent state of turbulent flux.

Such complexities mean that our understanding of the interrelationships is far from complete, and predictions of the consequences of events may be much less reliable than desired. Efforts to model completely some relatively simple aquatic ecosystems have choked computers. Predictions of events based on ecological study (and perhaps required by law—see Chapters 13 and 14) must, therefore, include good judgment in addition to careful compilation of facts.

This chapter will omit much of the biology that is included in aquatic ecology by some authors and introduce ecological concepts related to the organic productivity of the waters. It is a prelude to the description of a major part of fishery science, the study of exploited populations, in the next chapter.

7.1 SINGLE-SPECIES POPULATIONS

The reader may have noted earlier in this book that the factors of the environment have been discussed with respect to their effect on living organisms, and the animals have been discussed with respect to their reaction to factors in the environment. All of this discussion has been an introduction to the central fishery problems of producing food and recreation from living aquatic resources as parts of ecological systems. The next step toward considering the complexities of the systems is to discuss the characteristics of single-species populations. Such populations never exist in nature because no organism exists alone; an organism, even when it is apparently isolated, may have parasites, and every organism is dependent on another organism, either living or dead, for food. Nevertheless, much of the theory about ecology has developed from studies of single-species populations, and in fishery practice, many of the concepts that have been developed from single-species populations are applied directly and usefully to the study of stocks being exploited, under the assumption that each stock lives in a constant environment.

7.1.1 Limiting Factors

Every population of a species of organisms is limited in its distribution and abundance by its relative ability either to utilize or to tolerate the factors in its environment and by its relative ability to disperse. Each population lives in a *habitat*, and it also has a *role* among the other populations in its community. Its habitat and role together form its *niche*. The niche includes the relations of the population to food, to predators, to competitors, and to the physical environment at all stages in its life. The niche of young individuals may be entirely different from that of old individuals, and it may shift from one environment to another. For example, the niche of the sockeye salmon includes its role in a gravel stream bottom for about 7 months, a lake for 1 or 2 years, coastal ocean waters for about 3 months, the surface layer of the high seas for 2 or 3 years, and a stream for the final month of its life.

As organisms have evolved to tolerate the multitude of environments on earth, each species has found a niche of its own, not completely occupied by any other species, although partially occupied by a multitude of other species. Neither in nature nor under experimental conditions do two species occupy exactly the same niche for long. Even when two similar species live together in the same place, they will differ in some aspect, such as food, tolerance to some environmental factor, or breeding requirements, and one will tend to replace the other. This singularity of one species in one niche is known as the *Volterra–Gause* or *exclusion principle*.

Each species has a set of ideal conditions under which it thrives, but it can tolerate more or less change from the ideal conditions. The degree of tolerance to a factor is expressed by the prefixes *eury-*, meaning more tolerant, and *steno-*, less tolerant. For example, a eurythermal animal tolerates a relatively large range of temperature; a stenohaline animal tolerates a relatively small range of salinity.

Each species has its own set of limiting factors. An aquatic species is limited by temperature, salinity, depth, light, dissolved oxygen, and probably by any of many other factors. Many of the elements present in minute quantities, poisons excreted by certain algae, and interactions with virus or bacteria have been shown to be limiting. Anything in the environment that affects the physiology of an organism may be at some level a limiting factor for that organism.

The limiting factors operate according to their strength or level in complex ways. Most physical and chemical factors have maximum and minimum limits. Some factors may operate only at certain times; others operate continuously. Usually, one factor will be more critical than others and will be the controlling factor. Among required chemical factors, the controlling factor will be the one available in the smallest fraction of the amount needed. This is called *Liebig's law of the minimum* and has been extended in an abstract sense to physical and biological factors.

The reproductive period is commonly the time of least tolerance. The conditions must be tolerable for spawning animals, eggs, larvae, and young, which frequently have different needs. For example, the Pacific oyster requires a minimum temperature for successful spawning but will grow and fatten perfectly well at much lower temperatures. This limiting factor is overcome in oyster culture by collecting the young in spawning areas and transferring them to growing areas.

The determination of tolerance levels is important in understanding the distribution of organisms, and it is receiving increasing attention as humans change the environment. With many chemical and physical factors, tolerance is a function of both level and time; the greater the change from the ideal level, the less the length of time the change will be tolerated. Potential limiting factors are tested at gradients of concentrations for determination of the time required for manifestation of discomfort, change in behavior, change in body chemistry, or, more commonly, death of a fraction of the organisms. The easiest criterion to use is the time of death for half of the animals, but this level is obviously different from the tolerable level; nevertheless, it is a useful index.

Even though one factor may be critical in a certain situation, often, factors operate together to create *synergistic* effects. For example, an animal contending with low oxygen is almost certainly less able to tolerate high temperature, low salinity, or shortage of food. An excellent example of the operation of three factors on lobsters is given by McLeese (1956) in Table 7.1. Lobsters that prefer cool, well-oxygenated seawater tolerate the highest temperature under conditions of high oxygen and high salinity, the least oxygen under conditions of high salinity and low temperature, and the lowest salinity at medium temperature and high oxygen.

Testing the tolerance for chemical and physical factors is rendered difficult, not only by the synergistic effects, but also by the ability of animals to become *acclimatized* and to be more tolerant afterward (Fig. 7.1). (The lobsters tested for Table 7.1 were fully acclimatized, or else they would have been far less tolerant.) The mechanisms for acclimatization are little understood but are common to many animals. Aquatic animals may acclimate to higher pressure, extreme temperature, low oxygen, and even to certain poisons.

Another kind of long-term adjustment occurs in populations where those individuals that are better able than others to tolerate certain factors survive and transmit their characteristics to their offspring—the process of *natural selection*. Such *adaptations* occur increasingly as people change the environment and create new niches. Adaptations that have been demonstrated include

TABLE 7.1   Lethal Salinities for Lobsters Subjected to Various Combinations of Temperature and Oxygen[a]

Oxygen (mg/liter)	Salinity (‰)						
	5°C	9°C	13°C	17°C	21°C	25°C	29°C
1.0	18.0	21.0	20.4	22.0	30.0	—	—
2.0	13.3	12.6	12.0	11.2	12.0	15.0	—
3.0	11.6	10.4	9.6	9.0	9.3	11.2	30.0
4.0	11.4	9.7	9.4	9.0	9.3	11.2	20.4
5.0	11.0	9.5	8.8	8.8	9.3	11.2	17.8
6.0	9.8	8.8	8.4	8.4	9.3	11.2	16.4

[a]When the lobsters were subjected *gradually* to each of the combinations indicated, their average mortality rate was 50% in 48 hr.

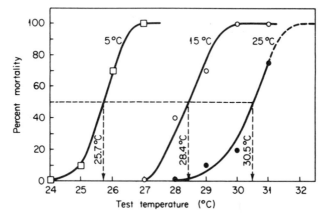

FIGURE 7.1   Mortality rates at 48 hr for lobsters acclimated at 5, 15, and 25°C, each at 30‰ salinity and 6.4 mg O_2/liter. Fifty percent lethal temperatures are shown on each curve.

resistance to disease, improved ability of predators to find and consume food, improved ability of prey to avoid and resist predators, tolerance of chemicals, and avoidance of fishing gear. Some species adapted extensively to chemical and biological stress in five to eight generations in experiments.

After each population of each species of an organism has acclimated to the maximum amount of a factor and has adapted as a result of natural selection, it is still limited in distribution and abundance by some levels of critical factors. Its distribution will tend to be like that diagrammed in Fig. 7.2, with a region where abundance is maximal, a region where occurrence is regular but not abundant at all times, and a region where appearance is occasional and temporary. In the aquatic environment, such a distribution is always three dimensional. The boundaries are complex and are usually related to different factors in different places. For example, the distribution of a coastal marine bottom fish might be limited by deep water on the ocean side, high temperature at the south end, wave motion on the land side, absence of food above the bottom, low salinity in an estuary, and a current that carries its eggs out to sea on the north side.

7.1.2 Patterns of Distribution

Any population has a characteristic structure or arrangement of its individuals, known as its *pattern of distribution*. The pattern of distribution results from the entire behavioral response of individuals to the factors of the environment. Patterns vary greatly in scale, from a few times the length of the individual to thousands of kilometers. They are not as orderly as the distribution of cells in an organism but are orderly enough to be classifiable (Hutchinson, 1953).

One of the kinds of patterns is called *vectorial*; it has already been discussed at some length but not named. This is the distribution of individuals in response to gradients in the chemical and physical factors of the environment

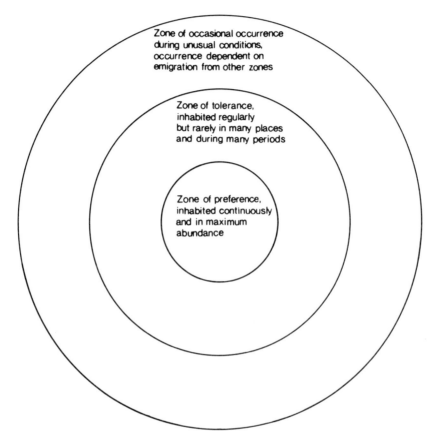

Zone of occasional occurrence
during unusual conditions,
occurrence dependent on
emigration from other zones

Zone of tolerance,
inhabited regularly
but rarely in many places
and during many periods

Zone of preference,
inhabited continuously
and in maximum
abundance

FIGURE 7.2   Diagram of distribution of organisms. [After Andrewartha and Birch (1954). Copyright 1954 by the University of Chicago Press.]

such as temperature, light, pressure, salinity, current, and bottom type. With most such factors, individuals occur in numbers according to the level of the factor. For example, the vectorial distribution in Fig. 7.3 is in response to salinity.

A second kind of pattern, the *reproductive*, is evidenced for a time by the species that care for their young. Some species of fish and many of the aquatic mammals form family groups for a few days, and some form them for many months. Another kind of distribution associated with reproduction is the pattern of distribution of the young after dispersal by the currents. Many marine species of crustaceans, mollusks, and fish have such pelagic young, and their distribution is the result of spawning time, spawning place, drifting time, and the set of the ocean currents.

Aggregations of a species for spawning, feeding, or other *social* purposes is another kind of pattern of special importance to the fisheries. This pattern has already been discussed under the topic of schooling (see Section 6.4).

TABLE 7.2 Hypothetical Series of Catches from Overdispersed, Random, and Contagious Distributions

Sample number	Overdispersed (nearly uniform)	Random	Contagious
1	5	4	0
2	5	7	0
3	4	5	0
4	5	6	45
5	6	8	0
6	5	2	0
7	5	5	5
8	4	6	0
9	6	3	0
10	5	4	0
ΣX	50	50	50
$\bar{X}$	5	5	5

The small-scale dispersion of a species in a uniform part of its environment will tend to be a *random,* or stochastic, pattern. A random distribution is a result of chance, but frequently, social factors result in a distribution of clumps of animals that may be random in size and form random distributions themselves. A distribution of clumps of individuals is called a *contagious* distribution, or *underdispersion.* The opposite distribution, *overdispersion* (Table 7.2), is rarer than the random distribution but might occur in samples from a school of uniformly spaced animals. (The reader should note that some ecologists have used *superdispersion* to mean underdispersion and *infradispersion* to mean overdispersion.)

The description of an animal distribution usually depends on the fact that a series of random samples of randomly distributed animals will contain a frequency distribution of animals per sample that conforms mathematically to the *Poisson* distribution. This distribution is the expansion

$$\frac{1}{e^m}, \frac{m}{e^m}, \frac{m^2}{2!e^m}, \frac{m^3}{3!e^m}, \frac{m^4}{4!e^m} \cdots,$$

in which e is the base of natural logarithms and m is the mean frequency of occurrence. When a comparison of a frequency distribution with the expected Poisson leads to the conclusion that contagion exists, other mathematical distributions, such as the negative binomial, may be appropriate.

The fifth and last type of pattern to be mentioned here is the *coactive;* it results from competition between closely related species. It follows from the exclusion principle (see Section 7.1.1) that only one such species can occupy a single uniform environment, but it happens in nature that few environments are uniform. Almost always, two or more competing species can exist together

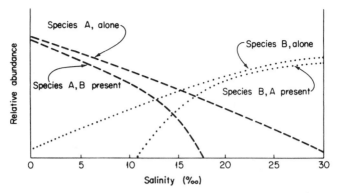

FIGURE 7.3 Diagram of the vectorial distribution of each of two species alone in a salinity gradient and their coactive distribution when existing together.

because each can find its niche in a part of the environment that it can use better than the others. The two or more species will tend to reach an equilibrium distribution that depends on the species involved and on the gradients in the physical features of the environment.

The resulting distribution is in part vectorial and in part coactive. This distribution is typical of the distributions of many organisms. To illustrate, let us assume that a single species A exists in a salinity gradient that restricts but does not prohibit the occurrence of the species at the high-salinity end, and we find a vectorial distribution of species A as shown in Fig. 7.3. If we added a competing species B that favors high salinity, we would expect that it will predominate at the high-salinity end and depress the abundance of the species that favors low salinity.

7.1.3 Dynamics

The study of a population as a living unit is usually called *population dynamics,* because interest is focused on changes in the population associated with birth and death rather than on its static composition. There are, however, numerous similarities and contrasts with a living individual (Table 7.3). A population is a unit in a space with a size, usually expressed as a density or abundance. It maintains itself in continuous life as a rule; birth and death of populations occur only rarely. It has, however, a *natality* (birthrate) and a mortality (death rate). It has a physical structure, expressed as its pattern of distribution (discussed in Section 7.1.2), and it has an age structure. A population grows at times in ways remarkably similar to those of the growth of an individual, but it may decrease in size also. Most important to the fishery scientist, it has a capacity for production.

Density   The size of a population is expressed by the total number or total weight. The latter, commonly called the *biomass,* may be expressed as either the liveweight or the dry weight. Neither number nor weight is usually measur-

TABLE 7.3 **Comparative Characteristics of Individuals and Populations**

Individual	Population
Size	Density, abundance
Morphology	Pattern of distribution, age structure
Growth in length or weight	Growth in number or biomass
Birth	Rate of natality
Death	Rate of mortality

able directly, therefore relative number and weight, or indexes, are used; these are called the *density*, or *abundance*. The common unit in fisheries is the *catch per unit of effort* by a specified kind of fishing or sampling gear. When the gear either covers a known area or strains a known volume of water, the catch per unit of effort may be transformed to a catch per unit of area or volume.

Such relative measures of density are useful when indexes of the same type are being compared. One of the central problems of population dynamics is converting the relative measure into absolute numbers or weight.

Natality is the rate at which new individuals are added to a population by reproduction. It is a broader term than birthrate, because it includes young that result from eggs, spores, or seeds, as well as those that are born. It is less than the rate of fecundity of egg-, spore-, or seed-producing organisms, because individuals are not usually counted until after hatching or germination. Natality of aquatic resource populations is rarely measurable, however, because the new individuals are so tiny, so dispersed, or dying so rapidly. Instead, the addition of young animals of a size catchable by fishing gear is measured and called the *recruitment*.

Mortality   The number of individuals in a population decreases by death, and because every organism usually produces a surplus of young, the rate of mortality at early life stages is usually the primary factor that regulates the number in a population.

The direct causes of natural death in aquatic populations are varied and not well known. Dead aquatic animals are seen rarely, and dying animals even more rarely. Even in fish catches from populations with known natural mortality rates ranging from 20 to 30% per year, the proportion of dying to healthy individuals seems much less than the expected 1 or 2/1000/day.

In fresh waters of temperate zones, dead fish are observed most commonly in spring, because at this time fish seem less resistant to disease and dead bodies decompose less rapidly than at other seasons of the year. In summer and fall fish still die, but those that die are consumed rapidly by scavengers. Small animals are commonly eaten by predators, but larger animals are eaten more rarely. Epidemics of either disease or parasites are common, even a constant menace, among cultivated populations of fish or mollusks, but they appear to be rare in natural populations. Occasionally, windrows of stinking animals may be visible on the beaches from natural catastrophes, such as shifts in ocean currents, oxygen depletion under ice, or biological catastrophes, such as blooms of poisonous plankton (red tide); but generally, mortalities among aquatic animals, even in large populations, are unnoticed and uncounted.

A consequence of our inability to observe death directly is our inability to estimate the rate of mortality directly. Almost always, it is determined from the number of survivors at various times from a group of individuals that started life at the same time. Such an estimate is usually attended by considerable difficulty (see Section 8.8) and can rarely be made at close intervals, so, in practice, an estimate of mortality rate spans a considerable period of time. Despite the certainty of death, its measurement is somewhat uncertain.

Regardless of circumstances, no population increases indefinitely; each is limited in size by some mechanism that either decreases the rate of natality or increases the chance of death for the individual as the population increases. Such a mortality is said to be *density dependent,* or *compensatory.* A number of somewhat interrelated mechanisms that regulate the mortality in this way have been identified.

Intraspecific competition for the essentials of life is the first and probably the most important limiting mechanism. As organisms increase in number, their food, space to move in, opportunities to get rid of waste, ability to avoid predators, and opportunities to transmit disease change in ways that tend to increase the mortality rate.

A second mechanism is an increase in environmental heterogeneity. As a population increases in number, it tends to occupy a larger area, and this area usually includes zones that are less than favorable for the population (Fig. 7.2). Such zones may be uninhabitable during times of unusual conditions, and the organisms in them must migrate or die.

A third mechanism is an increase in predator (including parasite) populations as a consequence of a greater number of prey and a greater ease of finding them. Such a buildup of a predator population tends to lag behind the increase in the prey population so that when the predation becomes limiting to the prey population, the predator population is still increasing. The consequence is likely to be a major decline in the prey population.

Another mechanism, the genetic adaptation of populations, operates to enhance the survival of each succeeding generation under stress. When the population is a prey species, it tends to become more adept at evading the predator. When it is either a predator or a parasite species, it tends to become more adept at catching the prey. When it is subject to a limiting factor in the environment, it tends to overcome it to some extent. The result in a varying environment is a dominance, at least temporarily, of the more quickly adaptable populations.

The effects of all of these mechanisms change regularly with season, irregularly with other climatic and biological factors, and according to the age of individuals. Predators are more active at certain times than at others and shift their preference from one food to another frequently. Usually food supplies vary during the seasons and probably affect directly the well-being of the feeders. Among the stages of animals, egg and larvae are probably most in danger (note the 10% per day, or higher, mortality rate of young mackerel in 1 year, see Section 6.6.9); young adults are probably subject to the least chance of death, and older adults are subject to a greater chance of death. The consequence of all of these variable factors is a large variability in number in the population.

7.1.4 Age Structure

The interplay of natality and mortality in a population results in a set of proportions of the different age groups. Age may be measured in any unit of time, but for fishery resource populations, a year is almost always the appropriate unit, especially when the individuals are born during a single season of the year.

Much of the usefulness of the age structure depends on its relationship to the survival rate of a *cohort,* or *age group* (a group of animals of the same age). In a stable population, the proportion of each age group at a given time is equal to the proportion of each corresponding age during a cohort's life. Because it is far quicker to obtain estimates of the proportion of each age group in a population of a given time than to wait out a cohort's life, this relationship is frequently used. From it are constructed life tables of the estimated survivorship of cohorts, and from these may be computed estimates at each age of numbers dying, as well as mortality rates and life expectancies.

Despite the ease of age determination of most fishery resource animals, complete age structure has been determined for few populations. The reasons lie in the separate distribution of eggs, larvae, juveniles, and adults for most animals and in the selectivity of all fishing or sampling gear for animals of certain sizes. For salmon, survival rates during life in the gravel, life in the lakes, and life in the sea have been estimated with reasonable accuracy, but even in the case of salmon it is necessary to assemble an approximate composite (Fig. 7.4). As the figure shows, in salmon the survival rate during the egg and larval stages is lower than during the other stages; among animals having larger numbers of eggs it is lower still. The survival rate is nearly uniform among immature and adult salmon; among many species of fish this trend in survival rate is typical.

Although complete age structure is nearly impossible to determine, *age compositions* of the catches of sampling or fishing gear are easy to determine and are of much use in the analysis of fishery resources. An age composition of halibut, a long-lived species, from a catch in a virtually unexploited population is an example (Fig. 7.5). The figure shows the typical feature of absence, or a lower proportion of the younger age groups than of the older; the younger age groups are not fully available to the fishing gear either because they are distributed differently from the older animals or because they cannot be caught as efficiently as the older animals.

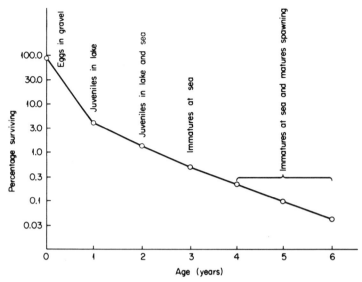

FIGURE 7.4   Estimated approximate age structure of sockeye salmon from Bristol Bay in September of each year after spawning and before death of mature individuals.

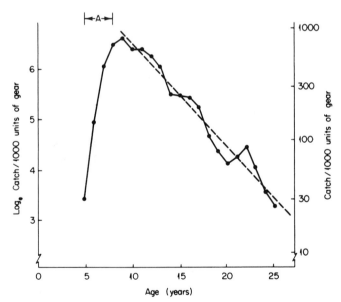

FIGURE 7.5   Age composition of catch of halibut from an unfished area in the Bering Sea, June, 1930. (A) shows the ages of reduced availability to the longline gear. The dashed line indicates a constant instantaneous annual mortality rate of 0.235 (see Section 8.8.3). [Data from H. A. Dunlop, F. H. Bell, R. J. Myhre, W. H. Hardman, and G. M. Southward (1964). Investigation, utilization and regulation of the halibut in southeastern Bering Sea. *Int. Pac. Halibut Comm. Rep.* 35, 1–72.]

7.1.5 Population Growth

When a species is introduced into a new environment it grows in number and actual weight at a rate that resembles the sigmoid (S-shaped) growth of an individual organism (Fig. 6.21). At first, the introduced organisms must acclimatize and complete their reproductive cycle. Then, if there are few environmental restraints, their natality will greatly exceed their mortality, and the number in the population will increase exponentially for a time. Later, as food, space, disease, or some other factor restrains their numbers, the rate of increase will slow, and the population will approach a size at which average natality and average mortality are equal. This population size will not be a limit in the mathematical sense, like the limit on the size of the individual, but a point of balance with more or less fluctuation around it.

Such a growth pattern has been demonstrated experimentally with many short-lived organisms. Experiments (Silliman and Gutsell, 1958) in which populations were started in aquaria with five pairs of guppies each and kept under constant conditions with a constant amount of food clearly show this pattern of growth (Fig. 7.6). For a few weeks, the total weight of guppies in an aquarium accelerated, then it slowed, and finally it reached equilibrium at about 30 g. Subsequently, the total weight fluctuated about ±5 g. Since there were no other species of fish in the aquaria, this situation was an example of intraspecific competition.

Similar patterns of population growth occur in nature when animals are introduced into a new environment. Examples of such patterns among fish are the growths of populations from the striped bass (Fig. 7.7) and the shad introduced into Pacific coastal waters from the Atlantic coast. Other examples are the growths of populations from pests that have been introduced accidentally. The American oyster drill, *Urosalpinx* (a snail called "tingle" in England), is the most serious predator on English oysters, and the American slipper limpet, *Crepidula*, is the most important competitor of oysters. The former was introduced about 1910 and the latter about 1880; both have caused much greater problems for oyster growers than similar native animals.

Exponential population growth in nature may occur seasonally. Examples are the blooms of plankton in the spring in temperate zones. A typical sequence

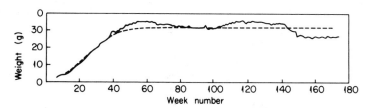

FIGURE 7.6 Growth in biomass through 175 weeks of a population started from five pairs of guppies in an aquarium. The dashed line is a fitted logistic curve. (After Silliman and Gutsell, 1958.)

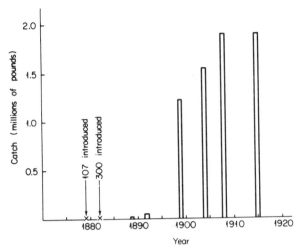

FIGURE 7.7   Catches of introduced striped bass in California in various years between 1888 and 1915. After 1915 catches were lower, but fishery was restrained by law. [After J. A. Craig (1930). An analysis of the catch statistics of the striped bass (*Roccus lineatus*) fishery of California. *Div. Fish. Game Calif., Fish. Bull.* **24**, 1–41.]

of events is the warming and enrichment of the water, the accelerated population growth of a few species of phytoplankton, and the accelerated population growth of a few species of zooplankton. With both kinds of plankton, the overwintering forms are released from their environmental restraints and stimulated to reproduce at near maximum rates until restraints become effective again.

7.2 COMMUNITIES AND ECOSYSTEMS

Any population of a species outside a laboratory or a farm lives as a part of a community of species on which it depends and which depends on it. It may be a *producer* (a plant population) that carries on photosynthesis, a *consumer* (an animal population) that feeds on plants or animals, or a *decomposer* (a bacterium or fungus population) that breaks down the wastes and dead bodies into nutrient materials again. It may depend on other organisms for support or protection, or it may provide such services. It will have parasites, and it may be a parasite population. If it is an animal population, it will be a predator as well as a prey population.

Any community of organisms has a structure consisting of the physical distribution of its members and an age distribution of each of its populations. Thus, the structure of the community is the sum of the structures of its component populations.

More important than the structure is the functioning of the community in its environment. The community and its environment comprise an ecosystem in which energy and materials are circulating and being transformed. The ecosystem will be some kind of a unit, identifiable from its geography and organisms and more or less separated from other units. It will never be a completely closed system, however, because energy and materials will continue to flow in and out. Even the ecosystem of the earth receives and loses energy and receives a small amount of materials.

7.2.1 Controlling Factors

The controlling factors of ecosystems, the geological situation, the climate, and the available organisms (Fig. 7.8), are independent of the system. They are complicated sets of factors that are largely independent of each other, and many of them vary in daily or seasonal rhythms or haphazardly around mean levels. Initially, these factors determine the kinds of organisms that can survive; eventually, they limit the activity and structure of the ecosystem. They are commonly modified within the ecosystem to change their effects. For example, light may

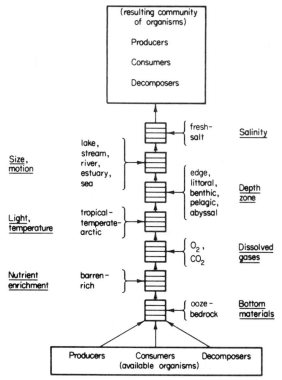

FIGURE 7.8   Independent factors that control aquatic ecosystems.

be intercepted by plants and cut off from the organisms below, or the composition of the bottom may be changed by burrowing animals.

The ecosystem is not a haphazard arrangement but is an ordered structure, as a result of evolution, that tends to be self-regulating. The patterns of distribution and age structures of organisms tend toward an equilibrium state as long as the inputs and outputs from the system remain the same. Should these change as a result of such factors as climatic change and human intervention, the equilibrium levels will shift and trigger a new evolutionary trend toward a set of organisms that will fit better in the changed system.

7.2.2 Examples of Fishery Ecosystems

Ecosystems can be classified according to the dominant biological or physical elements of the system. Many terrestrial systems are identified by the dominant biological complex, such as tundra or tropical rain forest. But most aquatic systems are identified by the dominant physical factors; examples are streams, lakes, continental shelves, or tropical seas.

Fishery ecosystems are those that support a fishery activity, either to capture wild stocks or to farm domesticated stocks. All are subject to human intervention, which ranges from removal of a fraction of a wild species in an ecosystem to nearly total control of the ecosystem of a domesticated species. Almost every one will be unlike another. The major kinds of systems are listed with some typical characteristics of each.

1. Oligotrophic lakes—Cold, clear water, usually deep, high altitude or latitude, low level of productivity, frequently dominated by salmonid fishes.

2. Eutrophic lakes—Warm, usually shallow and turbid, temperate to tropical areas, high level of productivity with abundant plant growth, frequently dominated by perciform fishes.

3. Impoundments—Variable biological elements depending on many factors, such as access by migratory animals, age of impoundment, extent of adaptation by organisms, depth and temperature, inflow of water and sediments, timing and extent of water release schedules.

4. Cold-water streams—Higher-altitude, clear-water streams in forested areas, gravel or rubble bottom with gradient above 0.2%, frequently dominated by salmonid fishes.

5. Warm-water rivers—Lower-altitude, turbid-water streams in agricultural areas, muddy or sandy bottom with gradient less than 0.2%, frequently dominated by perciform, cyprinid, or other warm-water fishes.

6. Estuaries—The most stressful environment for organisms, salinity variations from fresh to oceanic, reversing tidal circulation with strong currents, muddy bottoms and turbid water, very large accumulations of emergent plants and organic material, intensively used by humans for many purposes, usually polluted, the location of the major molluscan fisheries, essential spawning or nursery grounds for many coastal crustaceans and fishes.

7. Continental shelves—The ocean bottom to a depth of 200 m, the major fish-producing areas of the world, cod and flounder grounds in arctic and temperate areas, croaker grounds in subtropical and tropical areas.

8. Continental slopes—The ocean bottom between 200 and 500 m, important rockfish or hake grounds in some locations.

9. Oceanic surface waters—Depths to about 50 m, the domain of the pelagic species, intensely productive in regions enriched by upwelling, biological deserts in some midocean areas, immense herring and anchovy fisheries in enriched regions near shore, tuna fisheries in enriched regions farther out.

10. Oceanic near-surface waters—Depths from about 50 to 500 m, the domain of some very abundant potential resources in enriched areas, productivity limited by low dissolved oxygen in some locations.

11. Deep sea—Below 500 m, no photosynthesis, no living plants but animals found to maximum ocean depths of about 10,000 m, most animals red or black in color, many with light-producing organs, not extensively explored but little chance of fishery resources.

12. Tropical reefs—Domain of the corals, the most complex aquatic ecosystems, many species of fishes, some with extraordinary coloration.

13. Warm-water farm ponds—Depths to about 3 m, the major fish cultural sites of the world, controlled ecosystems producing at least part of the fishes' food, operated in association with agriculture.

14. Raceways, fish pens, and cages—Fish enclosures in which the animals are completely segregated by species and size and fed a complete diet, animal concentrations up to about 40% by weight of the water enclosed.

7.2.3 Interspecies Relationships

Each species in a community influences the lives of others and may be influenced by them in turn. The interrelationships may be mutually beneficial, as in the case of bacteria in the gut of a cellulose-eating animal; beneficial to one partner and of no great concern to the other, such as a barnacle on a whale; beneficial to one partner and harmful to the other, such as predator and its prey; or harmful to both partners, such as competition between two species for the same space or food. These relationships shape the evolution of both the species and the ecosystems, as organisms adapt or perhaps become extinct. Predation and competition are of special interest to the resource scientist.

Predation   In the popular concept, a predator is a rapacious animal, such as a tiger shark; in a more general sense, it is any consumer of other organisms. It may be a herbivore, eating plants, or a carnivore, eating other animals. This broad definition includes the parasites, because they live at the expense of their hosts.

When a newly introduced predator population starts to consume a prey population that has been in equilibrium with its competitors and other predators, the first consequence is an increase in the mortality rate of the prey population. Removal of some prey will leave more space and food for the remaining prey, and their chances of survival and growth will improve (Fig. 7.9). If the predator population should take still more prey, the prey population would decrease further, and the compensatory survival and growth would increase more, but never enough to restore the original biomass of the prey. The predator population can make increasing use of the prey until the maximum equilibrium (sustainable) yield of prey is reached. Should prey in excess of this level of yield be taken, the excess could be only temporary, and the consequence would be a reduced prey yield, to the detriment of the predator.

When people begin to harvest a living wild resource, such as a stock of fish, their harvest is additional predation. They can take advantage of the resiliency of the target population through its opportunities to increase in survival and growth, but in their own interest, they should not reduce the number in the target population below the level of maximum sustainable yield. They can increase their benefits in the process of harvest by taking individuals whose removal will least affect the potential yield, such as old, slowly growing, poorly reproducing individuals, and surplus males. Further, if they can control the natural balance of predators and prey, they may be able to increase the potential yield to higher than the natural level.

On the other hand, when people seek to control a pest by either capture or poison, they will need to overcome the resiliency of the pest (prey) population and reduce its number well below the level of maximum sustainable yield. A large effort may be necessary, although the task will be easier if they can select individuals whose removal will reduce the resiliency and productivity of the pest population, such as young females during their reproductive period and larvae at a critical stage.

Competition among Species   Two or more species may interfere with each other by needing the same things at the same time and place. Usually the degree of competition is related directly to the closeness of the phylogenetic relationship among the competitors, but even diverse organisms compete with each other to some extent (e.g., a seaweed and a fish cannot occupy the same space at the same time). The consequence of competition is a change in natality or mortality rates that will result in fewer numbers of at least one of the competing populations.

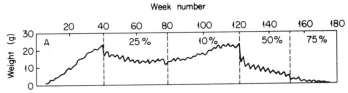

FIGURE 7.9   Biomass of a population of guppies in an aquarium that was started with five pairs and subjected to artificial predation. After week No. 40, animals were removed at the triweekly rates shown in the panels. (After Silliman and Gutsell, 1958.)

The principal mode of a competition of different species is for space and food, but many other factors may be involved. Some species may degrade the environment for others through their egesta or excreta. Some may escape predators that will then seek others. Some may harbor disease organisms that are harmless to themselves but virulent to others. Competition may occur only during adult life, or only during larval life, or between larvae and adults.

The results of competition are difficult to observe directly in nature because of the multiplicity of competing species, the changeability of environment, and the rarity of opportunity for observation of the beginning or progress of the competitive situation. Consequently, our understanding of the principles has been obtained largely from laboratory models of simplified situations. These have shown that if two extensively competing species were placed in a simple, uniform environment, one would eliminate the other, but the survivor would not always be the same one. As one species would become more abundant, it would gain competitive advantage. On the other hand, if the environment were not uniform or were changing, the two species would coexist, at least for a much longer time. Each of the species would find some advantage over the other at some place or time; they would compete less. Neither species would be as numerous as it would be if it were isolated, but the two together would be more numerous than a single species in the same environment. These findings are supported by observation of natural environments. Generally, in nature more species occur in the complicated environments than in the simple ones, that is, there are more niches to be filled, but the maximum yield of resource animals comes from the simple environments that have few species.

Predators may change the competition among prey populations. As one prey species becomes rare, the predators may shift their exploitation to a more abundant, more available species. In this case, the competition between prey species will diminish and both may survive, whereas in the case where there is no predator species, competition between species might lead to elimination of one. This principle is supported also by the observation that in nature the coexistence of many closely related species seems to be enhanced by predators.

7.2.4 Cycles of Abundance

The size and structure of populations and communities are regulated, not only by the physical factors of the ecosystem, such as temperature, substrata, and depth, but also by the interplay of intraspecific competition, interspecific competition, and predation. Either the effects of the physical factors or the interplay of biological and chemical factors may lead to cycles of abundance in the populations. Abundance cycles may follow passively recurring external influences or perpetuate themselves under certain combinations of internal factors.

We shall not be concerned here with the annual cycles of the abundance of short-lived organisms that follow the seasonal changes in climate. These may be related to the changes in temperature, precipitation, or light.

The long-term changes can influence ecosystems in ways that will change greatly the balance among populations, even if they are evidenced by trend changes of only 1 or 2°C in mean annual temperature or a similarly small change in another factor. The changes may involve a major change in numbers, or in the case of some mobile marine populations, a change in location.

Self-perpetuating cycles of abundance in certain populations occur over long periods of time in some communities. The mechanisms that sustain the oscillations are the relationship of a consumer population to its food supply, in some cases at least, and the time lags in the response of a population to changes in its limiting factors. Any community in a constant environment will tend toward an equilibrium of numbers in which consumers are keeping their food supplies near the level of maximum sustainable productivity. As the consumer population increases in number, the food supply per individual decreases, and vice versa; thus, the population of consumers is regulated by its food supply.

The population of consumers does not respond instantly, however, to changes in its food supply. Any increase in number must be caused by either an increase in natality or a decrease in mortality, usually of the younger organisms. The increase in biomass requires time for young to grow and perhaps time for additional generations to mature and reproduce.

Another kind of time lag occurs when limiting factors are tightening. An increase in mortality will probably affect mostly the younger organisms and leave a population of adults. When the limiting factor is a food supply, the adults may exist for a time in semistarvation.

The consequence of these lags is a tendency for predator populations to either increase or decrease some time after the changes in the prey populations have taken place. When the predator and prey live in an uncomplicated ecosystem, the changes in populations may be cyclic. In laboratory experiments and in a few natural populations (e.g., those of lynx and rabbit in Canada), oscillations of the kind shown diagrammatically in Fig. 7.10 have been sustained for a time. Such cycles tend to annihilate themselves when the predator population is very efficient and the prey population is very vulnerable. They tend toward a steady state when the predator population is inefficient and the prey population is hard to capture. As the complications increase, however, the systems tend to have greater stability and less oscillation. Predators can turn to alternate prey, and prey can find refuges for protection from predators.

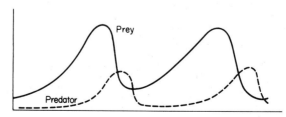

FIGURE 7.10 Diagram of a stable cycle in numbers of predator and prey.

Some natural cycles in aquatic animal populations are amazingly persistent and not adequately explained. For example, the quadrennial cycles of the several runs of Fraser River (British Columbia) sockeye salmon (e.g., to the Adams River, tributary of the Fraser, Fig. 7.11) have endured for many decades. Their persistence has been attributed in part to predation by trout during the time that the young sockeye salmon live in the lakes. It was observed that, when juvenile sockeye salmon were scarce, the trout ate a large proportion of the salmon but fared poorly themselves because they could not find enough other prey. When juvenile sockeye salmon were abundant, the trout ate a smaller proportion of the salmon but fattened as they satisfied themselves. Apparently, the trout population was limited to the number that could survive the years of scarce salmon (Roos, 1991).

7.2.5 Succession of Communities

In any new ecosystem with stable physical inputs, the community of organisms will change in composition and complexity with time. The first organisms to dominate the system will modify the environment in ways that may be detrimental to themselves and favorable to others. Organic material, both living and nonliving, will accumulate, and the number of species will increase, at least for a time. The trends will not be haphazard, but will be directional toward a climax in which the composition and functioning of the community will be relatively stable.

Such a succession of changes is analogous in ways to the growth of individu-

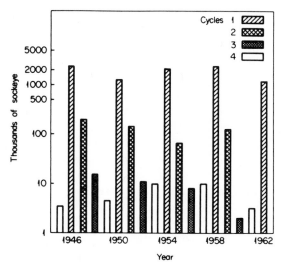

FIGURE 7.11   Annual parental populations of Adams River sockeye salmon. [After F. J. Ward and P. A. Larkin (1964). Cycle dominance in Adams River sockeye salmon. *Int. Pac. Salmon Fish. Comm. Progr. Rep.* **11**, 1–116.]

als and populations. The succession of stages follows a predictable pattern. The young structure changes rapidly; the old one changes slowly. The size (i.e., biomass of organic material) increases toward a limit. It is a biological process controlled by the community within the overall restraints of the physical environment.

The best-understood examples of succession are those in terrestrial ecosystems dominated by higher plants. The sequence of changes in a temperate climate from grassland through shrubs to pine and hardwood forests, with the associated succession of animals, has been studied extensively. Each successive dominant organism changes the environment to its own disadvantage and to the advantage of its successor. Such a sequence of changes may require more than 100 years before it reaches the relatively stable status of the mature hardwood forest.

Less well understood is the succession of aquatic populations, especially those away from the edges of waters where higher plants do not occur. In these, the one-celled plants and the animals do not modify their environment in the same obvious ways that shrubs and trees do. Nevertheless, succession does occur in reasonably predictable ways, organic matter does accumulate, and an ecosystem with constant controlling factors does tend toward stability.

When the controlling factors of the ecosystems change because of either natural catastrophe or human intervention, the patterns of succession may be reversed. The usual effect is a change in the ecosystem toward the pattern of a younger, less stable community. Often enough, when people intervene, the patterns may become essentially unpredictable.

Most wild aquatic resource populations are found in a climax state and do not provide evidence of succession, but those in which the species have changed provide examples deserving close study as possible indications of changes to be expected commonly. One of the best-documented changes in the succession of fisheries and fish species has been in the Great Lakes of North America (Smith, 1968). These fisheries have been subject not only to heavy fishing, but also to introductions of new fish species, pollution, and diversion of water. The same pattern of change was repeated in all of the lakes. It was seen first in the lowest lake, Lake Ontario, then upstream in Lakes Erie and Huron, and finally in Lakes Michigan and Superior. Gross reduction occurred in favorite fish populations such as lake trout, lake herring, and whitefish, and led to the collapse of fisheries for them. The yield of native species changed rapidly as alewives and sea lampreys invaded the lakes and as carp and smelt, introduced species, became established. The changes in the Lake Michigan fisheries, where the dominant species was the alewife, were typical (Fig. 7.12) in the late 1960s. The alewife population reduced sharply the populations of other plankton-eating species. Its abundance and the nuisances caused by its spring mortalities led to the introduction of still other species, namely, coho salmon, chinook salmon, and steelhead trout, as climax predators. These fed on the alewife population and flourished in the 1970s.

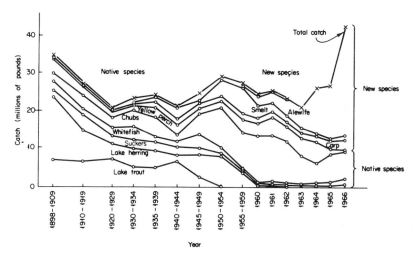

FIGURE 7.12 Catches of major species in the commercial fishery of Lake Michigan (1898–1966). (Data from Smith, 1968.)

7.3 TROPHIC RELATIONSHIPS

A better understanding of ecosystems is being sought by describing and quantifying the flow of material and energy through them. The flow includes the conversion of inorganic salts, CO_2, and water into organic material during photosynthesis, the subsequent consumption of this material by herbivores, carnivores, and detrivores, and the eventual breakdown of the organic materials into the inorganic salts, CO_2, and water. Most of the flow involves feeding habits, hence the designation *trophic*, which denotes nutritive processes.

The flow of material is *cyclic*, a temporary use of the earth's resources by living things with eventual restitution. The flow of energy is *consumptive*. The initial input from the sun sustains the photosynthesis, which in turn sustains the rest of the material cycle.

The plants that use the inorganic materials are considered to be the producers, the animals that eat plants and other animals are the consumers, and the bacteria and fungi that eventually reduce organic material to inorganic are the decomposers. In another classification, the plants are called *autotrophs* (self-feeders), and the animals are called *heterotrophs* (other feeders). The consumers, or heterotrophs, include, of course, the herbivores (plant feeders), the carnivores (animal feeders), as well as the detrivores (shredders and consumers of plant and animal debris).

The producer and consumer stages may be designated also as *trophic levels*; the plants are of the first level; the herbivores, the second; the carnivores that eat herbivores, the third; and other carnivores, the fourth or a higher level.

7.3.1 Food Webs and Material Cycles

When a plant of trophic level one is consumed by an animal of trophic level two and it in turn is consumed by an animal of trophic level three, the organisms form a *food chain*. An example of a simple, yet a major marine food chain is the diatom–euphausid–baleen whale chain of the southern oceans. Such a clear-cut chain rarely occurs, however, because many animals function in more than one level. Many tiny aquatic herbivores consume some bacteria and fungi in addition to unicellular plants. Many carnivores that feed normally on herbivores may occasionally eat other carnivores or plants. Most larger animals change food and trophic levels as they grow or migrate. The common pattern of material flow upward through several trophic levels or organisms is a *food web* more complex than the one diagrammed in Fig. 7.13.

But the material used by all organisms is eventually returned to the environment. This reverse flow includes O_2 from plants and CO_2 from animals, surplus reproductive products such as seed cases or egg shells, leaves from plants, dissolved materials lost during animal feeding, excretions and feces from animals, and finally their bodies. Such materials on land form the organic layer of the soil, where they are first shredded and consumed by detrivores and finally completely decomposed by bacteria and fungi. Such materials, after sinking in the aquatic environment, form detritus, or ooze layers on the bottom, which also support detrivores, bacteria, and fungi. During suspension in the water, dissolved or fine materials may support producers as well as detrivores such as microcrustacea, in addition to bacteria and fungi.

Thus, the material completes a cycle (Fig. 7.14). But such cycles are never step-by-step, as might be expected after starting with the concept of trophic levels. The return flow through the detrivores and decomposers is always greater from each trophic level than the flow to the next higher level.

The material cycles in the aquatic environment are not independent biolog-

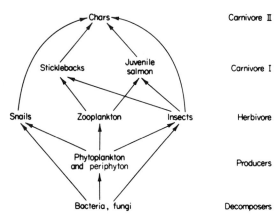

FIGURE 7.13   Diagram of a simple food web in an Alaskan lake.

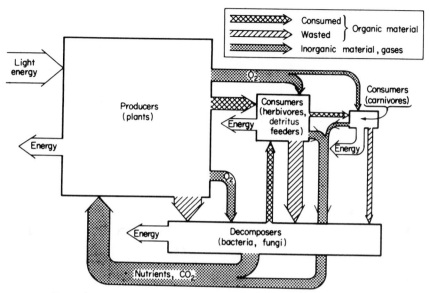

FIGURE 7.14   Diagram of the cycle of material flow in an ecosystem.

ical cycles, but require a variety of geological cycles for their closure or augmentation. The dissolved materials go with the currents, the suspended living or nonliving materials go with the currents for a time before settling out, and the larger animals may live or migrate independently of the currents. Thus, the upstream and upper water layers of lakes or oceans lose materials, and the downstream and bottoms gain them. Materials in rivers are replenished by weathering of the rocks and erosion of the ecosystems of the land. Materials that fall to the bottom in lakes and in the ocean are replenished by seasonal overturns, turbulence, and upwelling (see Section 4.7.1). Such cycles are called, appropriately, *bio-geo-chemical cycles.*

7.3.2 Energy Flow

One way of generalizing ecosystems is to study them in terms of energy flow. The assumption is made that energy is the common unit by which organic materials can be quantified as they are produced, consumed, separated into usable and unusable parts, incorporated into bodies, excreted, or otherwise changed. The energy is expressed in terms of thousands of calories, or kilocalories. Each kilocalorie represents roughly 2 g of wet flesh, or 0.25 g of dry flesh.

In the ecosystem, the sun supplies energy through photosynthesis to producers, which store organic materials, principally carbohydrates, proteins, and fats. When the producers are consumed, part of this organic matter is reconstructed to provide flesh for the next trophic level, part is oxidized to provide energy for living, and the balance is wasted as heat, unusable material, or surplus reproductive products. No energy is lost or gained in the material cycle; it is merely converted from one form to another. But the transformation of energy into useful work is never effected completely. Energy "leaks" from the cycle as heat or is "trapped" in unusable ways during the infinite variety of transformations that occur.

7.3.3 Production, Biomass, and Yield

Production of ecosystems is the step-by-step creation of living material in the trophic process, starting with photosynthesis. *Primary production* is the first step—creation of new organic material by plants using energy from the sun. Each additional step in the trophic process involves further production, which is described variously as *secondary, tertiary,* or *terminal*. The accumulated material in any trophic level, in any community, or in a population of any species at any given time has a biomass. The product harvested by humans from any of the steps is the *yield*.

Each of these concepts may be expressed in terms of the wet weight of material, the dry weight of material, the energy in kilocalories, or the weight of a component such as carbon. In practice, the yield of fishery systems is expressed in wet weight.

Primary production includes the fundamental photosynthetic process of converting CO_2 and water to an organic molecule and O_2. Measurement of the amount of primary production at a point in lake or ocean waters can be done (other methods are also used) by isolating identical samples of the water containing phytoplankton in two bottles, one lightproof and the other clear. The pair of bottles can then be exposed to conditions at the point collected for some hours, during which photosynthesis will proceed in the clear bottle and normal respiration of plants and animals will proceed in the dark bottle. The difference in the O_2 between the two bottles can be measured to estimate the amount of carbon incorporated in organic material. If water samples are taken representatively and exposed with due care, the grams of carbon produced in a body of water can be estimated.

The productivity of ocean surface waters has been found to vary about a hundredfold, from about 0.1 g of carbon per square meter per day in barren tropical waters to about 10 g of carbon per square meter per day in rich upwelling areas. (One gram of carbon equals about 10 g wet weight of flesh.)

Measurement of production at higher trophic levels is much more complex, but approximations have been made because of interest in the *trophic efficiency* of aquatic ecosystems. This may be defined for a single step as the yield of a population to its predator divided by the amount of food supplied to the population. Measurement, usually in energy units, has been attempted in two-level laboratory communities and by determination of the biomass of two or more levels in simple natural communities.

The resulting estimates have usually fallen in the range of 6 to 15% per trophic step. A widely accepted value of 10% has been used in many approximations of ecosystem efficiency. Thus, the biomass (or the *standing crop* of fishery scientists) may vary a thousandfold in a four-level trophic system with three steps. Such differences in numbers and mass of animals and plants were recognized long before the ecosystem concept, when they were called a *pyramid* of numbers or mass (Fig. 7.15). In a more modern and precise sense, it is a pyramid of energy.

The biomass of detrivores and other decomposers is not considered a trophic level, because it utilizes the 90% of material and energy that escapes from each production level. This quantity is so large in many aquatic ecosystems that it plays a major role. It is intimately associated with detritus, the dead organic matter, that may exist in large amounts due to terrestrial debris, such as leaves in fresh water, adsorption and settling of suspended organic material in estuaries, or very slow decomposition of materials such as wood over years or even centuries. In such situations, the amount of energy in the detritus, organic muds, detrivores, and decomposers may greatly exceed the energy in the food web.

When water depth is less than photosynthetic depth, the detrivores and decomposers may be mixed with producers and herbivores, as on stream bottoms where the periphyton (attached algae) cover the bottom. A similar situation occurs in estuaries, where rooted water plants grow on top of organic muds that are "mined" by many invertebrates. But in deep lakes and oceans some detrivores and decomposers function in the water column, where they utilize the organic particles during settling, and other kinds of detrivores and decomposers continue the process on the deep bottoms. The total decomposition in such waters is usually relatively complete if judged by the scarcity of organic muds deposited in the deep ocean.

7.4 EVOLUTIONARY ADAPTATION

Human harvest of selected parts of wild populations and the permanent modification of their environment are accelerating the evolution of those populations. They are adapting to new circumstances through the process of natural selection.

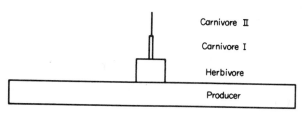

FIGURE 7.15   Diagram of the relative amounts of materials or energy in four trophic levels of an ecosystem. This succession of changes is called a *pyramid of mass,* or *pyramid of energy.*

Adaptation is any change in structure, function, or behavior of an organism that results from natural selection and by which the organism becomes better able to survive and reproduce in its environment. It has been a constant process in living populations that produce an excess of young, of which only those that are best able to compete survive to eventually reproduce. These are populations in which mortality is density dependent and competition among populations is high.

The result in a natural ecosystem is a community with populations that more or less maintain their abundance indefinitely. Each population maintains its status relative to competitors and its place in the food web as a consumer of something and a prey for something.

But when competition changes, other kinds of individuals may survive. No harvest such as the annual capture of half of the codfish over 40 cm in length from a fishing bank can proceed without affecting the natural selection. The surplus of young codfish will not completely restore the large cod population, so the competitors of the large cod will fare better and their predators will fare less well. Even the young cod will not be as abundant, so their competitors and predators will be affected. If the harvest is steady, a new equilibrium state will eventually be reached, with the new generations of organisms having a slightly different genetic inheritance than those before the harvest. They will have adapted in some aspect of structure, function, or behavior.

The rate of adaptation is a function of the length of the reproductive cycle in each organism and can be counted in generations. It may be surprisingly rapid, because it has been possible in laboratory studies to cause major behavioral change, such as an insect's ability to evade predators, in three to five generations. And salmon populations, which have an inherited ability to spawn in their birthplace, will wander occasionally and use a new spawning area intensively (perhaps one made accessible by a retreating glacier) in about the same number of generations.

A characteristic of the selection process is that it is unique in each environmental unit. In fresh water, the units will include each lower-order stream, each pond, and each river system. In estuaries, each estuary will be a unit because few organisms can migrate through the sea to other units. But in the ocean, the boundaries are less recognizable or definite, and most units are much larger. A few organisms, such as some tuna, can make regular transoceanic migrations and encounter many different domains of less migratory species.

A consequence of the occurrence of a species in different environmental units is that the range of variation in the species is always greater than it is in any population. An example among fish is the chinook salmon of the North Pacific. Each race inherits its spawning habits, which include the route of migration from the feeding ground in the sea to the river, the location in a river, and the timing of the entire process to provide a suitable environment for eggs and young. The species ascends rivers around the North Pacific from southern California to the Amur River in Asia, including several in the Bering Sea. It can be found entering a river somewhere almost any month of a year, and it may spawn from just above tidewater to as much as 1000 km upstream. Each race will spawn in its own place, in its own river, and in its own period of perhaps three weeks. Such a multiplicity of genetic units means that a controlled harvest must be exercised on each unit!

A reduction of competitive pressures occurs in a new habitat, as in a reservoir. When it is flooded, the fish inhabitants (assuming no stocking) will be the riverine fish, some of which may flourish temporarily in the new impoundment. Almost all young may survive to adulthood for a few years, as they utilize the enriched waters from flooding of the land and enjoy the scarcity of competitors or predators. But eventually, competition for food and space will be restored, and the populations will more closely resemble those from natural lakes with similar environments.

Some benefits of natural selection have been rediscovered during efforts to improve survival of hatchery-reared fish in the wild. The fish from hatchery brood stock are inevitably selected because they survive and grow well in the hatchery. But these characteristics may not aid them in finding natural food or evading predators, so domesticated strains may be crossed with wild strains in efforts to find stockable fish that are both easy to rear and survive well after release.

7.5 ECOLOGY AND FISHERY SCIENCE

Fishery scientists have done pioneering work in the application of many simpler ecological concepts to their understanding of fish populations. Even before the introduction of the ecosystem concept, aquatic animals were included in studies of zoogeography. Later came many studies of the effect of environmental factors on the physiology of fish—especially the resource animals. Next came studies of population dynamics, and now fishery regulation is strongly based on ecological principles.

Part of the reason lies with the different structure of aquatic ecosystems as compared to terrestrial ones. The aquatic producers of the water column in the deep lakes and the oceans are the phytoplankton. They are short-lived, with reproductive cycles recurring in hours or days, and they are grazed in their entirety by tiny animals, mostly microcrustacea, which have short reproductive cycles of days or weeks. The terrestrial producers, in contrast, are predominantly rooted plants, which reproduce in annual to decade-long cycles, and almost none of them are eaten in their entirety by one kind of animal. The grazing animals, which range from insects to mammals such as deer, each eat only parts of the plants, which usually survive throughout the year if not over decades.

Another major difference among resource animals is in the trophic level. The aquatic resources are almost all carnivores, even the filter feeders such as herring, which feed mostly on microcrustacea. The terrestrial resources, or their domesticated counterparts, are almost all herbivores.

These differences have consequences in the application of ecosystem principles to resource animals in the two situations. The aquatic food chains usually consist of several levels of carnivores, and all levels, including the producer, tend to be limited by the food supply. The terrestrial food chains usually consist predominantly of producer and herbivore levels, in part because humans have reduced the populations of carnivores. And the herbivore level is seldom limited by food supply; rather, it is controlled by predators, either natural or human. Should it be limited by food supply, as happens occasionally with deer populations, the consequences are long-term disasters, because the producer level takes years to recover from overgrazing.

It should be clear to the fishery student, after considering the concepts described in this chapter, that ecology is very complex. All living resource sciences, including fisheries, make extensive study of the impact of environmental factors on the resource organisms, but fishery science requires additional emphasis on food chain relationships. Even the major activity in fish farming requires management of ecosystems in fish ponds.

The practical application of ecosystem analysis to wild fishery populations will be described in the next chapter.

REFERENCES

Andrewartha, H. G., and Birch, L. C. (1954). "The Distribution and Abundance of Animals." Univ. of Chicago Press, Chicago.

Hutchinson, G. E. (1953). The concept of pattern in ecology. *Proc. Acad. Nat. Sci. Philadelphia* **105**, 1–12.

McLeese, D. W. (1956). Effects of temperature, salinity and oxygen on the survival of the American lobster. *J. Fish. Res. Board Can.* **13**, 247–272.

Roos, J. F. (1991). "Restoring Fraser River Salmon. A History of the International Pacific Salmon Fisheries Commission." Pacific Salmon Commission, Vancouver, Canada.

Silliman, R. P., and Gutsell, J. S. (1958). Experimental exploitation of fish populations. *U.S. Fish. Wildl. Serv., Fish. Bull.* **133**, 213–252.

Smith, S. H. (1968). Species succession and fishery exploitation in the Great Lakes. *J. Fish. Res. Board Can.* **25**, 667–693.

8 | Analysis of Exploited Populations

Most fishermen, in any part of the world, know that fishing is not as good as it used to be. They remember with more or less accuracy that, during early years of developing fisheries, their catches were of greater abundance, consisted of larger fish, and had a high proportion of fish of the more valuable species. They have seen increases in the number of fishermen and total catches, but not in the catch per person. If they are fishing in readily accessible fresh waters, the innovative and industrious fishermen may be blamed for the decline. If they are participating in large oceanic fisheries, the development of superior new vessels with electronic fish detection equipment and efficient new nets may be blamed. The result in either case has been despair for the resource unless something is done by government agencies.

Out of this despair have come basic questions about the use of the natural stocks of fish. What are the causes of changes in the abundance of fish stocks? Are they natural fluctuations that are beyond human control? If so, can we forecast their abundance to take maximum advantage of periods of high abundance, and protect populations during periods of scarcity? Was the decline caused by fishing? If so, what sustainable yield will be expected from a designated amount and kind of fishing? Or, as is more probable, were the changes caused by a combination of fishing and environmental factors? If so, what is the best fishing strategy? And how can the environment be controlled to enhance the fishing or to preserve a threatened species?

Investigation of such problems is a major activity of many fishery agencies that is usually called *stock assessment*. This chapter will summarize some of the basic methods, but for a detailed elaboration of the complex techniques, Hilborn and Walters (1992) should be consulted.

8.1 APPLICATION OF ECOLOGICAL CONCEPTS: A CHAPTER SUMMARY

Answers to the foregoing questions were urgently needed during the last part of the nineteenth century as the freshwater fisheries of Europe and North America seriously declined. This led to the adoption of more efficient steam trawlers, which began to exploit the fisheries of northwest Europe. Fishery scientists became concerned about the seemingly inevitable trend in such fisheries as the catches reached a peak and then declined as the amount of fishing continued to increase. Eventually the problem became clear—each population of fish in a stable environment should have an optimal catch rate. If that rate is exceeded, the sustainable catch can be increased only by decreasing the amount of fishing.

The ecological principles can be illustrated by a simple example. Consider an isolated, unfished pond containing fish of one species that reproduce and maintain themselves. Then assume also that the isolation and lack of fishing have existed for many years, there are no serious epidemic diseases, and that the population is stable in abundance and age composition. This situation is the simple one of a population in a niche in which it is limited in biomass by food, space, and other factors. Each individual must compete with others for food, space, and mates.

Now assume that people start to fish on the population with gear that can catch all fish over 20 cm long with equal efficiency. The result is an increase in the rate of mortality of the older fish and a replenishment of them only by growth of the fish that are less than 20 cm long. The expected result, therefore, is a reduction in abundance of the whole population and a reduction in the average size of the fish. This will leave more space and food for each of the remaining fish, and allow an increase in the rates of growth and survival. These increases will probably not entirely compensate for the reduction in abundance from fishing; so, as we increase the amount of fishing, we must expect a continuing decline in abundance and average size. If we removed all fish before any became mature, then we would eliminate the spawners and eventually exterminate the population. If we understand the complex and changing factors determining the size of the population, manage the fishing wisely, and leave enough spawners to supply young, then we will perpetuate the resource. Further, we should expect that the surviving fish or other animals will evolve and permit a sustained maximum catch that will perpetuate the population at a smaller size.

Thus, environment clearly limits the size of the population. Space is obviously a limiting factor among others, such as food supply, disease organisms, cannibalism, and predation. Each causes mortality that increases with density; hence, growth of the population is density dependent and slows as density increases. Growth of individuals will be limited by food supply and will decrease with an increase in density; hence, it too is density dependent. Such factors usually affect the young members of fish populations more than the older ones, and fishery scientists frequently assume that density-dependent mortality occurs primarily during the egg, larval, and juvenile stages.

These changes in a population are consistent with ecological principles and commonly occur in nature, but they are obscured frequently by the practices of fishermen who always seek profitable catches. They fish for the most valuable combinations of species and sizes on the nearest ground. As the fish on these grounds become scarcer, either with the seasons or with the years, the fishermen move to more distant grounds or seek other species. Moreover, they learn to locate the fish in particular areas more easily, and they frequently improve their nets, vessels, or navigation equipment in order to catch fish more efficiently. These changes in gear or shifts in area may obscure the characteristic reduction in abundance and size of fish in a single population.

The fishery scientist is almost always faced with questions about an existing fishery. The first step is to learn about the resource, usually from the experience of the fishermen and samples of the catch. What species are caught, where, and when? Where do these species migrate? How fast do they grow? When and where do they breed? What do they eat and what eats them? If the fishermen's information and samples of their catches are not adequate, special studies may be undertaken to tag fish and track them, to explore for them with special gear, and to study their environments.

Such information is seldom definitive, and the fishery scientist must make an early decision about the parts of the resource that should be managed as units. Ideally, a unit should be a single interbreeding population, but information about fish populations must come from the fishery, and fishermen seldom fish for a single species or for a long time in a single area. The unit will therefore be a compromise between ecological principles and the operating practices of the fishery. It will be called a *stock* and may consist of different races of the same species or even of more than one species.

The next step is to learn about the experience of the fishery, especially the history of catches and amount of fishing. This provides data on the trends in the catch per unit of effort, an index of the density of the population. The past data will be collected, and efforts will be made to obtain current data of the kind desired as accurately as possible. Usually records of the sale of commercial catches can be obtained, but lacking them, surveys of fishermen can be undertaken. Catches can be examined to obtain data on size, weight, and age.

With a series of such data over several years, it becomes possible to estimate annually the relative change in the size of the stock of each *cohort* (year group) in the stock (Fig. 8.1). The additions are *recruitment*, the quantity of young

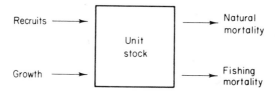

FIGURE 8.1   Diagram of the additions to and losses from a stock

animals reaching catchable size, and the increment of *growth,* the increase in size of animals after reaching catchable size. The subtractions are the catch or *fishing mortality* and the *natural mortality.*

The challenge for fishery scientists is, first, to estimate how the abundance of the stock (and its cohorts, if possible) has changed during the history of the fishery. Then they will estimate the expected yield at each level of fishing effort and forecast how the catch will change with changes in the amount of fishing effort. They may also be asked to estimate the effects of changes in the kind of fishing gear, the size of the mesh in nets, and the season of fishing with respect to the impact of each on the stock and economics of fishing.

The fishery student will discover that recruitment, growth, and mortality are rates, and their manipulation requires calculus. Much of the rest of this chapter will describe the various ways of estimating expected yields, all of which require some acquaintance with calculus. The student will discover further that the assumption of an ecosystem in a steady state may not be fulfilled because of environmental changes, and also because estimating the full complexity of changes in the ecosystem is much more difficult than determining the changes in the stock of fish.

8.2 AN EXAMPLE OF THE CATCH TREND IN A NEW FISHERY

When a new fishery develops as a consequence of the discovery of a new stock and the promotion of suitable markets, the production may increase rapidly. As profits are made, more investors are encouraged in fishing, processing, and marketing. The potential seems exciting because the fishery is harvesting an accumulated stock that has not been exploited previously. Usually the excitement is sustained by continuing discovery of new grounds, development of better gear, and improved fishing strategy.

Large new fisheries are now rare because the world's fish stocks are so well known, but the course of one unique fishery is illustrative. It is the Alaskan king crab fishery, which, prior to 1947, was virtually unknown, with no recorded production, and of course no market. The crab is large, reaching a weight of 10 kg. It lives in depths to about 200 m around the North Pacific from northern Japan to southeastern Alaska. It is a long-lived animal, requiring about 5 years to mature and living about another 10.

Fishing for the crabs started in the late 1940s, when a pioneering Alaskan fisherman with enough financial resources to withstand several years of losses began to catch a few crabs and market the frozen meat to the national restaurant trade. The fishery began in the western part of the Gulf of Alaska, principally out of Kodiak. It expanded exponentially as the stock was explored and as markets developed (Fig. 8.2). New, larger vessels were built that could fish farther from port and carry more crabs. New companies began new processing operations, and the market spread from frozen crab to include canned crab and also expanded to several foreign countries.

Production increased steadily until the mid-1960s, when the trend reversed, and by 1970 it had fallen to about one-third of the peak. The large accumulated stock of larger crabs that had been fished mostly out of Kodiak had been decimated, and the slow-growing crabs did not sustain the fishery, even though some restrictive regulations had been applied. The fishing area expanded westward to several new grounds near the Aleutian Islands and in the Bering Sea. The catch gradually increased again to a new record in 1980, only to collapse even more rapidly to about one-fifth of the second peak by 1982. In the next 10 years the catch remained between about one-eighth and one-half of the former levels.

The trends in prices were volatile. Prior to the first peak, the average price (whole weight to the fisherman) rose only from about $0.13/kg in the late 1940s to $0.22/kg during the peak year of 1966. Then, with reduced supply and established market, the price rose to $0.56/kg in 1970. At the time of the

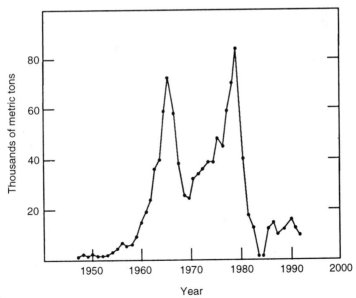

FIGURE 8.2   Trends in U.S. production of Alaskan king crab. This is an example of production trends in a new fishery on several unexploited stocks that became overfished one by one. (Data from U.S. Department of Commerce, 1945–1992.)

second peak in 1980, the price was about $2.00/kg, and with the decline in catch the next year, it reached $3.95/kg—roughly the level that has since been maintained.

The crab business changed dramatically during the 1970s and 1980s. Several less abundant and less valuable species of crabs were fished to supply the market, and the king crab catch became only a small fraction of the total. The opportunities for profits were large at times and risky at others. New vessels were paid for out of profits in as little as two years, and fishermen occasionally earned more than $1000/day during the short and arduous crab season. Such opportunities brought political pressure on regulatory agencies to avoid rational fishing regulation around sustainable yield levels.

Similar patterns of discovery, expansion, collapse, and the complexity of conservation measures have been repeated over and over in many of the world's fisheries.

8.3 STOCKS AND MIGRATIONS

Let us assume that fish of one species are being caught by a fleet operating on a fishing bank small enough so that the fish intermingle extensively during the reproductive cycle. Let us assume further that each individual has an equal chance of mating with any other individual of the same species and opposite sex on the bank, that is, the fish are a fully interbreeding group.

When the fishing on this bank is enough to affect the abundance and average size of the remaining fish, that is, the success of fishing, where else is the success of fishing affected? Would fishing for bass at one end of a 10-km-long lake affect the success of fishing at the other end? Would fishing for albacore off Japan affect fishing for albacore off California?

The geography of the migrations is critical to the answers to such questions, which depend on the extent to which each group intermingles with others. Should the intermingling be extensive, then all the intermingling groups would be affected by fishing in an area occupied by any group. Should the fishing on one group need to be controlled, then fishing would need to be controlled in all areas occupied by that group.

To ecologists, determining the extent of intermingling of such groups is a problem of determining the niche of the catchable parts of the populations. To taxonomists, the problem is one of determining the population and genetic structure of the species. To fishery scientists, this step is a critical one toward understanding the complexity of the migrations, and defining the stocks with which they must deal.

The student will find this subject bewildering because there is little agreement on the meaning of the words to define groups in the hierarchy with the rank of subspecies and below (see the discussion on nomenclature, Section

5.1). The rules of international nomenclature recognize *subspecies* as the least inclusive category, but within that category the terms *population, subpopulation, stock, race, variety, breed,* and *strain,* as well as others, are in current use. Moreover, some of these words are used for population units with legal or other nonbiological implications. For example, *stock* has been defined in some international treaties as a group of fish capable of independent exploitation and management.

Further complications arise because most species and subspecies are composed of local populations, no two of which have identical genetic characteristics. The populations are of one or more of the following categories: (1) populations that are geographically separated from each other, with slight opportunity for genetic interchange; (2) a series of gradually changing, contiguous populations, called *clines;* and (3) a set of sharply differentiated contiguous populations with zones of hybridization in between (*stepped clines*).

In this volume, therefore, the words *population* and *stock* have been used without exact definitions, so some explanations of usage are in order. *Population* is used to mean a biological unit, and *stock* is used to mean an exploited or management unit. It will be obvious that the ideal definition of *stock* is that of a single interbreeding population, but this condition is so rarely demonstrable, either because of scanty data or because of the rarity of isolated interbreeding populations, that *stock* must be more or less arbitrarily defined. It is a unit capable of independent exploitation or management and contains as much of an interbreeding unit or as few reproductively isolated units as possible. As such, it may include several species, for example, the stock of Antarctic whales or the stock of several species of salmon that intermingle in a fishery. Thus, by usage in this volume, a *stock* is a management unit defined for operational purposes, whereas a *population* is the actual biological unit, and may never be completely described.

8.3.1 Migrations

Migration is a seasonal habit of many aquatic animals and an adaptation to increased abundance in a varied environment. It enables them to use rich feeding areas, to protect eggs and larvae as may be necessary, and to avoid unfavorable temperatures, salinities, or other changes in aquatic climate. Some animals can find all of their requirements in one particular current system or in one location, but most species migrate at least short distances for feeding, spawning, or overwintering. Some may make transoceanic migrations, such as certain whales, tunas, and salmons. The value of the adaptation is illustrated by the tendency among closely related species for the migratory kinds to be most abundant. Most of the major food fishes, crustaceans, and mammals are migratory.

Migrations are accomplished either by passive drift or by active swimming. The eggs or larvae of many marine species and a few freshwater species are pelagic, and may drift for hundreds of kilometers. When they do, a migratory cycle is completed by the return migration of either juveniles or adults so spawning will recur in the same place.

Understanding the diverse migratory cycles of each stock of aquatic animals poses some major biological challenges. Usually the cycles occur according to the season, and the periods of active swimming follow periods of fattening or maturation of sexual products. The active migration must be a response, whether imprinted or inherited, to some kind of guidance information. The simpler migrations may be accomplished by random searching or by following a gradient of depth, temperature, or salinity. Some fish follow a current gradient upstream or downstream. Some, such as salmon, probably find and follow the distinctive odors of certain streams by means of highly developed olfactory systems. Those that migrate on the high seas perform remarkable feats of navigation by methods about which we can only speculate at this time. They may receive guidance from the sun, from the magnetic field of the earth, or from the current systems of the ocean. Finding out about migrations is important for the utilization of the stocks, as well as for the solution of challenging biological questions, and they will probably receive a great deal of attention in the future.

8.3.2 Intermingling

Determining either the extent of intermingling during the course of migration or the converse, the degree of reproductive separation, is a difficult task because we cannot recognize individual animals throughout their lives; we must rely instead on a variety of circumstantial evidence. The best evidence is furnished by attaching tags to animals and recapturing them later. Evidence of migration from marked animals will establish that intermingling is occurring but may not indicate the amount, because tagged fish rarely can be caught more than once and fishing is never distributed evenly over the animal's range. Other evidence may be sought from studies of larval drift; from correlations in population characteristics, such as abundance, age composition, or incidence of parasites; from structural similarities, such as shape or number of body parts; and from chemical similarities, such as antigens or resistance to disease. Any evidence of these kinds may be useful to the fishery scientist who must make decisions about the geographic limits of stocks.

Population Characteristics   Fishermen have learned the migratory habits of many juvenile and adult aquatic animals by accident. They try to operate in areas where animals congregate either to feed or to spawn, and they avoid areas where animals are scarce or absent. Sometimes they can follow migratory groups, for example, the mackerel on their northward migration off eastern North America (Fig. 8.3). Always they watch for evidence of migration in the changing relation between catches and time, current, temperature, or catches elsewhere, because their success depends largely on being at the right place at the right time.

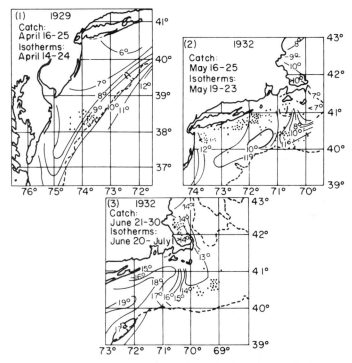

FIGURE 8.3   Distribution of mackerel catches (shown by dots) at intervals during the spring in relation to surface temperatures (°C). [Adapted from O. E. Sette (1950). Biology of the Atlantic mackerel (*Scomber scombrus*) of North America. Part II. Migrations and habits. *U.S. Fish Wildl. Serv., Fish. Bull.* 51, 249–358.]

The biologist, too, can learn much from an analysis of catches according to environmental data, including time, location, and apparent shifts in centers of abundance. When, in addition, observations on age composition, sexual condition, and food habits are available, he or she can make inferences about the age groups involved and the reasons for the migrations.

A distinguishing characteristic of certain populations is the presence of unique parasites. When these are acquired by young animals and carried throughout life, they are good marks. For example, a large proportion of the sockeye salmon from the Bristol Bay area of Alaska carry *Triaenophorus*, a cestode parasite of the muscles that does not occur in any other sockeye salmon. *Dacnitis*, a nematode of the intestine, occurs in a small proportion of the sockeye salmon of the Kamchatka area and nowhere else. The occurrence of these parasites in sockeye salmon taken at sea is a useful indication of the origin of the individuals (Fig. 8.4).

Marks and Tags   Hundreds of ingenious ways of marking aquatic animals for later identification have been devised. All have shortcomings, so the fishery

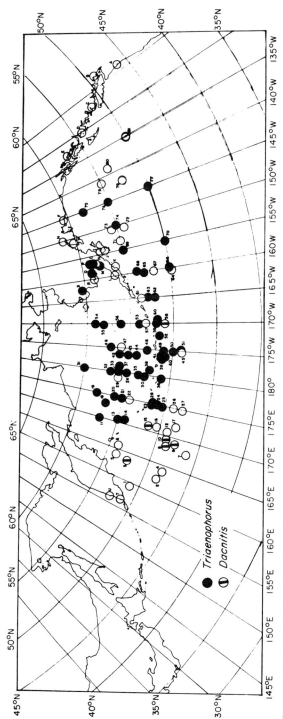

FIGURE 8.4 Distribution of sockeye salmon samples taken in 1959 and occurrence of *Triaenophorus* and *Dacnitis*. [From L. Margolis (1963). Parasites as indicators of the geographical origin of sockeye salmon, *Oncorhynchus nerka* (Walbaum), occurring in the North Pacific Ocean and adjacent seas. *Bull. Int. N. Pac. Fish. Comm.* 11, 1–156.]

scientist must be familiar with many techniques and the problems of using them. Ideally, the mark should be applicable to the animals without injury and with no subsequent effect on the rate of growth, rate of maturity, behavior, or liability to capture either by fishing gear or by predators. Marks fall into two categories: those that will identify a group, and those that will identify individuals. The former are commonly known as *marks,* and the latter are usually known as *tags.*

Fish marking includes clipping minor fins, punching holes in fins or opercle, branding with heat or cold, tattooing, and chemical marks such as dyeing with vital stains or spraying with fluorescent plastic particles. In some cases, dyes that color the flesh or substances that are deposited either in scales or in otoliths are placed in the food. Generally, marks are used on animals that are too small to carry a tag of suitable size and on batches of large numbers of animals (e.g., several hundred thousand).

Almost all marks and marking procedures risk injury to the fish in some ways, so benign practices are essential. The amputation of fins causes a slower rate of growth. Handling leads to some injuries, especially in wild fish. Further, few marks are recognizable by the public, so identification at the time of recapture must be managed by trained people. Despite these difficulties, marks have been used extensively to supply essential data on the migrations and survival of salmonid fish, especially those reared in hatcheries.

Tags that are large enough to carry a number have a great advantage in providing individual identification. External tags with instructions for return can be found and returned by the public, and thus do not necessitate an inspection program. Such tags should be permanent, noncorrosive, and nontoxic. They are usually made of metal such as stainless steel, or nickel, or of plastic (Fig. 8.5). One of those devised early that is still one of the most useful is the Peterson tag; it consists of a pair of disks joined by a metal pin passed through the body of the fish. Other tags in common use are made of spaghettilike plastic tubing that can be passed through the body and tied in a loop or attached to an anchor embedded in the muscle.

Especially useful for small, delicate fish are small strips of either metal or plastic that are placed internally, or miniature Peterson tags that are bar-coded. Some of the metal ones are magnetized or made radioactive, and are detected when the fish are passed through the field of an electronic detector. Such tags have been used for large-scale experiments on fish, such as herring, that are handled largely by processing machinery to which a detector can be added.

Tags on large animals, such as whales or large tunas, usually must be applied without removal from the water. They are delivered either from a gun or on the head of a harpoon. Usually such tags consist of an anchor that embeds in the musculature and a colorful plastic trailer that serves to call attention to the tag.

Tags on crustaceans are lost when the animals molt unless they are attached to the musculature near the isthmus where the shell splits during molt. Such tags have been devised for blue crabs and king crabs (Fig. 8.6).

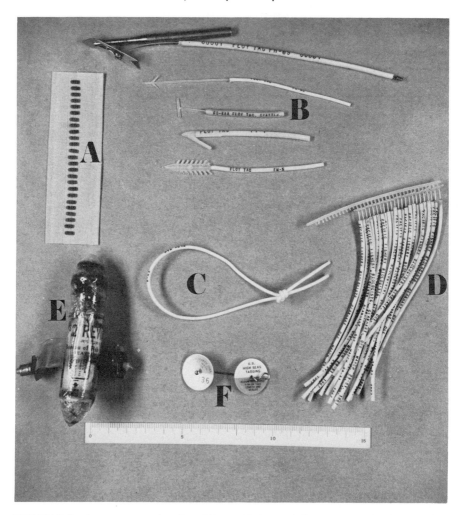

FIGURE 8.5 An assortment of modern fish tags. (A) Internal plastic tags suitable for fish about 10 cm long or for small crustaceans. (B) A number of dart tags. The largest is shown attached to the metal tip of a small harpoon that is used to insert the tag in a tuna or spearfish. All are inserted into the musculature of the back. (C) A spaghetti tag of flexible plastic that is inserted through the back of a fish near the dorsal fin. (D) A partially used "clip" of dart tags that are inserted with an automatic tag applicator. (E) A sonic tag that can be attached to the back of a fish and followed with acoustical instruments. (F) A Petersen tag of two plastic discs joined by a stainless-steel pin inserted through the back of a fish. (Photo by W. F. Royce, Univ. of Washington, Seattle, Wash.)

Shelled mollusks that cannot be tagged are marked by filing a place on the shell until it is smooth and numbering it with a special quick-drying paint. The mark must be waterproof and must last for a long time, and few paints are satisfactory. Batches of mollusks may be marked by mutilation of the shells

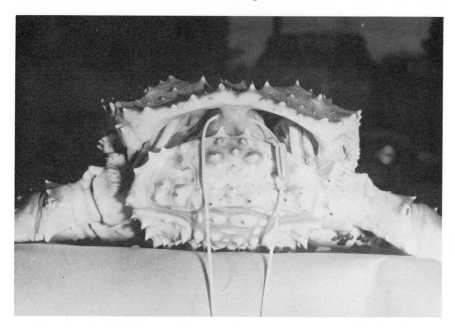

FIGURE 8.6 A spaghetti tag threaded through the musculature of a king crab such that it will be retained throughout the molt. (Photo courtesy of K. K. Chew, Univ. of Washington, Seattle, Wash.)

with either a file or an awl. Fish and larval mollusks may also be marked in batches with chemical marks.

The information obtained from tagging and marking programs is used in two general ways: to provide evidence of migration and intermingling, the topics of this section, and to provide information on growth and mortality rates (see Sections 8.8 and 8.9). For either use, it is important that the tags remain attached or that the marks remain recognizable, and that the fish not be handicapped for the duration of the study. Any loss of tags, mortality, or change in catchability of the fish must be measured. In addition, the tag or mark reporting must be adequate, and the rate of overlooking or nonreporting must be known. Various advertising and reward systems are used to encourage the return of tags.

Structural Characteristics There is a tendency toward similarity among individuals of any isolated interbreeding population because of exchange of genetic material, and a tendency toward difference from other populations of the same species because of evolution in response to different environmental conditions. The differences may not warrant designation of separate subspecies, but may suffice to separate the individuals of one population from those of other populations. Fishermen frequently recognize stocks from subtle differences in color or shape of the body. Biologists have quantified differences as *morphological* when they relate to size of body parts and *meristic* when they refer to number of body parts such as scales, fin rays, or vertebrae. Differences of color in tone or intensity are difficult to quantify, but differences in distribution of color are occasionally easy to measure.

The relative size of any body part varies according to the total size of the animal together with an allometric growth factor (or change in body shape with size) and random variability within the population. A comparison of body parts between populations is made by measuring the three kinds of variability and their differences. Sometimes a part will grow at the same rate as the whole body (isometrically); then the relative size of the part may be expressed as a ratio, or percentage, or more commonly a per mille. When, as frequently occurs, the part grows allometrically, then the ratio changes according to the size of the fish, and one determines the relationship by regression* analysis.

An example of allometric growth in the fin of a yellowfin tuna is shown in Fig. 8.7. The graph shows that this fin grows at a different rate than the body does, and an average ratio of length to body length is misleading at most lengths. This trouble was avoided in the analysis of the data on yellowfin tuna by transformation of the data to logarithms so that the data could be brought approximately to a straight-line relationship and the regression lines could be fitted to the data separately for each of the three segments (i.e., small, medium, and large). Then the size of the part was calculated from the regression for the central length in each segment.

The number of body parts, such as vertebrae, scales, fin spines or rays, gill rakers, and pyloric cecae, can be counted after careful specification of what is or is not to be included in the count. This step is especially important because the differences in counts between populations may be small and may be masked if different counters interpret fused vertebrae, branched fin rays, buried gill rakers, and so on, in different ways. Counts usually change only occasionally with the size of the fish.

After either morphometric or meristic data have been obtained for samples from two or more populations, the means may be compared. They may be compared by a statistical test of significance, that is, a computation to determine the chances of the means being drawn from a single population. Chances of one in twenty or less are usually taken to indicate that the means differ significantly and are probably from different populations. The test results depend, however, on the number of observations in the sample; the greater the number, the more likely that real differences may be expected to be found routinely when samples are large enough. Under these circumstances, the test for statistically significant differences will lead only to unimportant conclusions.

Of more worth and extensive use by taxonomists is the concept of overlap

---

*Regression is the statistical term for a functional relationship of one variable to another—in this case the size of the body part on the size of the fish. In this and the ensuing discussion the student should consult a statistical text.

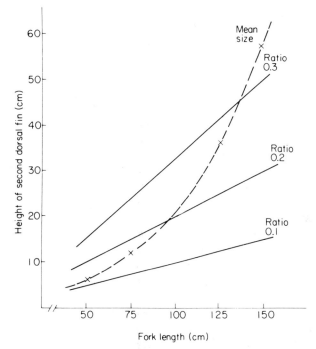

FIGURE 8.7   Growth of the second dorsal fin of yellowfin tuna, an example of allometric growth. [Data from W. F. Royce (1964). A morphometric study of yellowfin tuna *Thunnus alba-cares* (Bonnaterre). *U.S. Fish Wildl. Serv., Fish Bull.* **63**, 395–443.]

of frequency distributions (not to be confused with geographic overlap). When the counts of a character from two samples are plotted, commonly they show differences (Fig. 8.8). These are expressed frequently in terms of the mean, twice the standard error of the mean, the standard deviation, and the range. Commonly, these are plotted in a form that shows at a glance the nature of the distribution. When the black bars just meet, the overlap is 16%. In the example given, the difference is highly significant statistically, and the overlap is greater than 16%.

Similar statistics for comparison of the size of body parts may be obtained from regression analysis, which takes into account the effects of changes in total body length. Instead of an ordinary mean, the mean value is estimated from the regression for body length and is used for comparison. Instead of the standard error of the mean and the standard deviation, similar measures of the distribution around the regression line may be used.

In practice, a comparison of two samples by use of the overlap of a single character is rarely satisfactory, because organisms that differ in one character differ also in others. The number of possible characters for comparison is huge, but using more than one is statistically difficult. A great variety of sums, ratios,

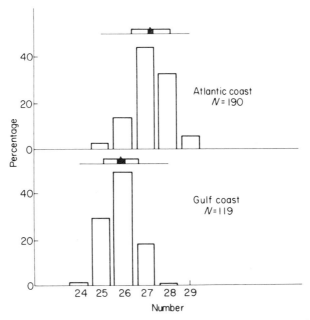

FIGURE 8.8 Percentage distribution of articulated dorsal fin rays in samples of weakfish, *Cynoscion regalis*. The diagrams show a small triangle for the mean, a black bar for twice the standard error of the mean on either size of the mean, a hollow bar for the standard deviation on either side of the mean, and base line for the range. [Graphical method from C. L. Hubbs and C. Hubbs (1953). An improved graphical analysis and comparison of samples. *Syst. Zool.* **2**, 49–56, 92.]

and products have been used. Comparisons have been attempted with coefficients of racial likeness, multiple factor analysis, and many other methods. The most satisfactory mathematical approach is based on a measure of difference between means in units of the standard deviation. It allows consideration of multiple characters by adding to the difference between means for the second and additional characters only to the extent that they are not correlated with characters previously considered (Royce, 1957).

Some tagging and morphometric studies of fish populations have shown that fish migrate from one location to another, but that fish in the two locations show morphological differences. These two findings demonstrated by the tag returns show that intermingling is occurring; the morphological differences show that intermingling is not complete. When the morphological comparison includes consideration of several characters, then the overlap may provide an estimate of the maximum amounts of intermingling that could have occurred if the differences are not related to mortality.

Biochemical Differences   Genetic differences between closely related populations may appear as biochemical differences as well as structural differences. Some differences, presumably biochemical, may be detectable as behavioral differences, as tolerance to environmental factors, or as resistance to disease. Other differences may be detectable by direct biochemical tests of the composition of the blood or other tissues.

Comparison of populations on the basis of behavior, tolerance of environmental factors, or resistance to disease must be done with living animals. For this reason, such studies are made on domesticated animals more commonly than on wild populations. Further, they are made frequently for the purpose of selective breeding to isolate and preserve characteristics that make the animals more amenable to aquacultural practices. Such comparisons have led to the discovery of races of oysters that are resistant to the Malpeque disease, trout that are resistant to furunculosis, and trout that grow better because they are less aggressive or better tolerate certain water conditions. A few wild populations have shown behavioral differences, for example, sockeye salmon fry that migrate upstream upon emergence from the gravel instead of the usual downstream movement.

Comparison of wild aquatic populations by use of biochemical methods has advanced rapidly during recent decades with the development of extrasensitive methods of separating organic compounds and the application of immunochemical methods to the problems. These methods have been applied in comparisons of a number of body tissues and seem especially promising in the case of blood, because the differences in blood types among some animals have been shown to be genetically controlled.

The serological comparison of blood depends on the presence of antigens (substances, such as proteins, that cause the production of antibodies) either in the cells or in the sera. When antigens are sought in the blood cells of an animal, these are separated from the sera and injected into a laboratory animal, for example, a rabbit, that will produce antibodies. When the antigens are sought in the sera, this is injected separately. After some time and perhaps several injections, the laboratory animals may produce antibodies. When brought together with the blood of an animal similar to the original antigen producer, these antibodies will agglutinate the cells, or form a precipitate with the serum.

The technique of comparing cell antigens is to bring together the cells and the serum containing antibodies in a test tube and measure the amount of agglutination (Fig. 8.9A). Sera containing antigens have been tested against a serum containing antibodies by being allowed to diffuse through agar on Ouchterlony plates (Fig. 8.9B). The presence and amount of precipitation are measured. Sometimes the antibodies in a serum can be partly agglutinated, and the remainder can be used for additional tests. Such tests are extremely delicate and require great care in the handling of the samples to avoid either chemical changes or bacterial contamination.

In addition to serum containing antibodies produced by laboratory animals, ordinary sera from any animal or plant may be used in the search for antigenic differences. The chances of such sera giving a reaction to one population of a

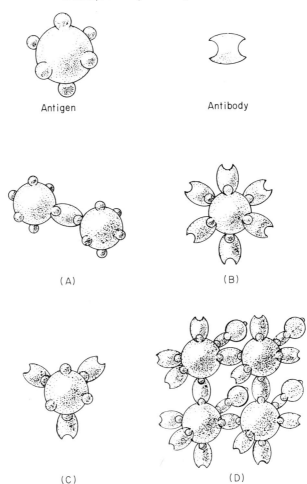

FIGURE 8.9A   Diagrammatic representation of the lattice theory of antigen–antibody reactions. (A) Soluble primary aggregate formed in excess antigen. (B) Primary aggregate formed in excess antibody. (C) Primary aggregate formed at zone of optimal proportions. (D) Insoluble lattice formed by interaction of the primary aggregates shown in (C).

species and not to another are very small, but such comparisons have become so easy with the new chemical techniques that thousands can be tried. Limited work on fish has produced some encouraging results. One trial consisted of a comparison of skipjack populations by the use of the bovine sera and a comparison of sockeye salmon populations by the use of pig sera.

The special usefulness of blood typing for recognizing interbreeding populations arises from the occurrence of certain blood types as alleles in populations. Such blood types are revealed by the presence or absence of certain antigens that are maintained in constant proportions in interbreeding populations, sub-

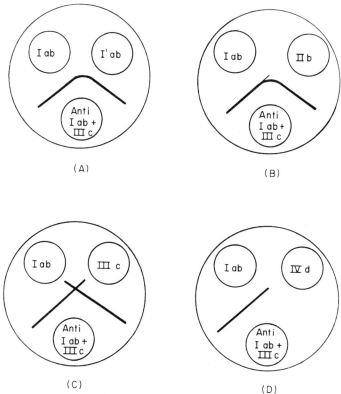

FIGURE 8.9B Types of comparative reactions in agar-plate diffusion method of precipitation analysis. (A) Reaction of identity, complete fusion of line. (B) Reaction of partial identity, "spur" formation or partial fusion of line. (C) Reaction of nonidentity, crossing of lines. (D) Lack of reaction, absence of line. [After Ridgeway *et al.* (1962).]

ject only to mutation and random variation. This phenomenon, in its general sense, is known as the *Hardy–Weinberg Law.* It permits a comparison and judgment about interbreeding of populations when samples are large enough to permit the establishment of frequencies with sufficient limits of accuracy. Some important studies of this kind have been made on sockeye salmon in the North Pacific (Ridgeway *et al.*, 1962). Several samples of the sockeye salmon's sera were compared on Ouchterlony plates with serum containing antibodies produced by a rabbit after injection with serum of sockeye salmon from Cultus Lake, British Columbia. The comparison produced 14 lines of precipitation by antigens, of which 2 were missing in 8 samples of sockeye salmon from the Okhotsk Sea, but were present in 10 of 11 samples from the Naknek River, Alaska. More extensive sampling and tests showed that 95.8% of the sockeye salmon from the American areas had neither antigen I nor antigen II, and 92.1% of the Asian sockeye salmon lacked these antigens. These comparisons were extended to samples from the high seas and used in an attempt to show area and time of intermingling.

A continuing effort is being made to identify members of populations by means of highly sensitive techniques that will detect such chemical differences. Comparison of muscle and blood proteins by electrophoretic methods and those of elemental composition by v spectrum analysis have shown promise.

8.4 COLLECTION OF BASIC DATA

After defining a stock that is being fished, the fishery scientist will seek to measure the effects of fishing on the stock. This can be done only by correlating the changes in the amount of fishing with the changes in growth, distribution, reproduction, and, especially, total mortality rate. The last is measurable indirectly either through marking experiments or changes in catch per unit of effort, age composition, or total size of the stock.

The bases for such correlations are accurate and consistent data that are obtainable only from suitable, continuing, and carefully planned statistical systems. The fishery scientist will want data on the amount of fishing according to time and unit area, and the catch in weight or number by species, time, and area. The area divisions must be chosen to agree with the known or probable stock locations, and sometimes must be designated by depth also. The time divisions should agree with season, fishing periods, fish habits, and so on. Usually days, months, and years are suitable units.

Such scientific data may be useful to the fish trade, especially when additional data are available on trends in price and value, employment, amount and value of processed products, kinds of craft and equipment used, quantities in storage, and exports and imports. When the fishery is recreational, then data on the number of people, their expenditures, their time spent, and the equipment used are useful to the trade that supplies services and equipment. In addition, many of these data provide an essential basis for government decisions about food supplies, port facilities, water usage, and taxation.

The statistical systems for collecting scientific data on the effects of fishing on the stocks will be combined, therefore, with systems used for other purposes. The fishery scientist will use as much data as possible from the general statistical system and design a subsystem to collect supplementary biological data. Usually for food fisheries, the general system will supply complete data on catch according to species, time, general location, and weight, and on the amount of fishing by time. Additional information on the exact location, time, and depth of fishing, on the catch by area, on the size and sex, and on the age composition by species, time, and area will be obtained by a special system, usually a sampling system. For recreational fisheries, any general system is likely to provide only data on catch, species, time, and general location. Such systems are relatively inaccurate and need to be checked as well as supplemented by carefully designed sampling systems.

Any fishery statistical system requires carefully defined objectives, supervision to ensure the continuity of comparable data, and cooperation of the people who supply the statistics. Usually, cooperation is achieved by making it easy to supply the data, by keeping personal data confidential, and by making the need for the data known to those concerned.

8.4.1 The General System of Fishery Statistics

The general fishery statistical system is usually the responsibility of the fishery agency rather than of a general statistical agency. The system is frequently a costly part of the fishery agency's activity. It is an activity that also deserves the close attention of the fishery scientists because the analysis of the effect of fishing depends on the reliability of the basic statistics. These are collected in ways that fit the practices of the fish trade. A number of systems are in common use.

Record of First Sale Almost all food fish are sold by fishermen rather than bartered, and almost always fishermen and/or buyers keep a record of the sale. In many countries, a third copy of this first sale is required by law to be furnished to the fishery statistical system. Normally the sales record shows the names of seller and buyer, and the amount and value of the sale by trade categories. In addition, the record may contain the name of the vessel, the kind of gear, and the area where the catch was taken. The state usually furnishes free of charge the blank forms showing the information desired (Fig. 8.10). This

		STATE OF WASHINGTON FISH RECEIVING TICKET					**No. S**	**0602**		

DEALER'S NAME   **DAHL FISH CO.**   STATION
RECEIVED FROM:
NAME OF FISHERMAN OR FIRM: ADDRESS:
BOAT NAME: PLATE NO.
PLACE CAUGHT: GEAR USED: DATE OF LANDING   19

			DEALER – DO NOT WRITE IN THIS SPACE					
	DIST.	DEALER 200	PORT	BOAT	TICKET NUMBER	GEAR	CATCH AREA	

	SPECIES AND DESCRIPTION	UNITS EFFORTS	SPECIES CODE	POUNDS	NO. OF FISH	PRICE	AMOUNT
	DOVER SOLE		205				
	ENGLISH SOLE		206				
	PETRALE SOLE		207				
	ROCK SOLE		209				
	LING COD		231				
	TRUE COD		241				
	SABLEFISH		221				
	ROCKFISH		251				
	PACIFIC OCEAN PERCH		254				
	FLOUNDER		212				

DEALER'S RECEIVER SIGNATURE_____   TOTAL
I Certify that these Fish or Shellfish were taken: (Check One) Inside Territorial Waters ☐, Outside Territorial Waters ☐, and that all other information on this ticket is true and correct
FISHERMAN'S SIGNATURE_____   **DEALER'S COPY**   60

FIGURE 8.10   A fish sales slip used by the state of Washington for gathering fishery statistics. One copy each goes to the dealer, the fisherman, and the state.

system is probably the best that has been devised to provide reasonably accurate data on the catches, but it should be routinely evaluated.

Records of Fishing   Many commercial fishermen keep personal records of their catches and fishing activities in either a vessel log or a business log. These may provide invaluable information to the fishery scientist on location and amount of catches and amount of fishing by area. Special log books showing the information desired are supplied free of charge to fishermen by some fishery research organizations. The information is copied by the research organizations, and the log book is returned to the fisherman.

A personal log for the season is required in some recreational fisheries. This may be a punch card in which the fisherman notes the capture of a fish by punching a hole and adds the data and location. This log is turned in at the end of the season.

Interviews with Fishermen   When large numbers of fishermen land their catches in one location, it may be possible for an interviewer to talk to either a sample or all of the fishermen as they land, in order to obtain details of their catches, gear, and location of the fishing for each trip. In addition the interviewer may be able to perform *market sampling,* that is, to examine, measure, and weigh catches, and to collect scales and other data. Such a system may provide a valuable supplement to the records of catches obtained from sales records or log books.

In recreational fisheries where anglers are scattered in boats or along streams, they may be visited in the field and their catches can be examined and counted. This procedure is called a creel census. Frequently it will provide the best possible data on recreational catches, but the records must be adjusted to allow for the catches of anglers who are not interviewed, and the catches made by anglers after they have been interviewed.

In recreational areas where access can be controlled, fishermen may be required to show their catches when leaving.

Periodic Canvass   In places where daily records are not obtainable, the only alternative may be an occasional canvass. Village leaders may be asked to estimate the season catches by the village. Fish buyers may be asked to estimate their total purchases for the season. A sample of fishermen may be asked to estimate their annual catches. Such a method may produce statistics with large errors, but it may be the best possible.

Related Statistics   A broad variety of statistics can be related to fish catches by some kind of ratio. Freight records or export records can be related to total catches by an estimate of the fraction shipped. Boat rental records of a recreational fishery can be combined with records of catch per unit for estimation of totals. Village census data can be combined with estimates of catch per family. Either fishermen or vessel-licensing records can be combined with estimates of the catch per person or per vessel. Counts of vessels made from the air can be combined with estimates of the catch per unit. Such data may be the best means of estimating trends in catches, or it may provide useful checks on other methods.

8.4.2 The Subsystem of Biological Statistics

Usually the general system of catch statistics will not provide sufficient information on species, area of catch, size composition, or sex composition for fishery scientists to determine the effect of fishing. They will use the general system as a statistical base and obtain the additional data from catches by a research vessel or from samples of the catch by the fishery.

Sampling presents special difficulties for fishery scientists. When the samples are to be used as the basis for valid statistical inferences about the (statistical) populations from which they have been drawn, samples must be representative, that is, they must have been obtained by a method that ensures that the characteristics of all possible samples drawn by the method will bear a known relation to the corresponding characteristics of the population being sampled.* When the samples meet this requirement, the individuals in them must have been chosen by some type of *random selection,* possibly by *random sampling* or a modification of it, such as stratified random sampling.

Random sampling in the statistical sense is an exact concept and does not mean haphazard selection of samples. It means that every member of the population being sampled has an equal chance of appearing in the sample. Some statisticians argue that a random sample cannot be drawn consciously, but that some mechanical randomization (such as dice or random number tables) must be used. They point out that if samples were just grabbed, even if every individual had an equal chance of being in the sample, the individuals in the grab samples would tend to resemble each other more than individuals in truly random samples; consequently, the estimates of variability in the grab samples would overestimate the significance of the differences between the means of the grab samples and underestimate the variability in the population.

If the samples are to have the characteristics of a sampled population, the latter must be specified with care. Usually a sample may be considered to have been drawn from any of several populations. For example, biologists sample fish in a bin on the deck of a boat. They might consider their sample to be from the population in that bin, from the catch of the net that filled that bin among other bins, from the day's or the month's catch of the vessel, from the catches of the fleet, from the populations of the fish on the bank, or from the population of fish along the whole coast. They must specify the *sampled population* with its units and probably will do so with a *target population* in mind, to which they would like their conclusions to apply. A common circumstance is to specify the sampled population as the catch of a fleet from a fishing ground during a period in order to gain knowledge of the target population, consisting of the stock of fish on those fishing grounds.

Sampling programs should be designed for achievement of the best balance between accuracy of results and cost. Accuracy of results depends in general on two factors: *variance* of the estimated mean and *bias,* or systematic error.

---

*The student who is not familiar with sampling theory should consult a modern statistical text.

Variance of the mean is caused by random errors and is reducible by procurement of larger samples, by more efficient sampling of strata, or by sampling in stages. Bias may be much more serious, however, because rarely can one examine whole aquatic populations to provide checks on sampling procedures, and agreement among repeated samples is no guarantee of lack of bias. The problems in fishery sampling differ in character, depending on whether the information sought concerns the structure of the population or its abundance.

Sampling to determine structure (i.e., species, sex, length, or age composition of catches) is a relatively straightforward problem of stratifying and randomizing the units to be sampled. Bias may develop when all market categories, types of gear, kinds of vessels, fishing grounds (within the area occupied by the stock), and landing ports are included in the sample. It is desirable to consider each possible unit as a stratum to be sampled at random. A more subtle but still serious bias may arise from selection of the fish for measuring. Commonly, the larger fish are on top of containers because they either tend to "surface" when loaded or are placed there to attract buyers. Hence, it is desirable to measure the fish in entire containers or in portions from top to bottom. Another source of bias is related to time and handling after capture. Fish in rigor mortis are significantly shorter than they are when not in rigor and have been pressed beneath other fish in the hold. Oysters may become shorter after handling because of breakage of the thin edges of the shell. Fortunately the biases can be detected fairly easily by small comparative studies and can either be avoided or measured.

Different and even less tractable problems arise in determining a standard measure of the fishing effort. The measure of effort is affected not only by the variability in space and time of the fish stocks themselves, but also by the variability in habits and efficiencies of the fishing unit that exploits them. Difficulties arise not so much because of the difficulties of obtaining samples that form a large fraction of the landings, but because of the nonrandom behavior and changes in efficiencies of fishing nets. Fishermen tend to concentrate where they think fish are abundant; hence, usually the average catch per unit will be above the average expected from fishing the area occupied by the whole stock (Fig. 8.11). They may lose efficiency by concentrating so much that they interfere with each other, by being unable to find concentrations and spending a large amount of time scouting, or by missing a seasonal run of fish as a result of a price dispute or weather factors. On the other hand, they may gain in efficiency by learning better how to find and catch fish through experience and by using better navigational aids or fish-finding devices. An even more important and subtle source of variability is the steady improvement in vessels and gear. Commercial fishermen try to obtain vessels that are larger and more powerful and that can handle bigger fishing gear, reduce running time, and lose less time from repairs in port or weather at sea. Such changes have brought increases in efficiency of fishing units of 50% or more within a few years.

Accounting for such variations and trends requires an intimate knowledge of

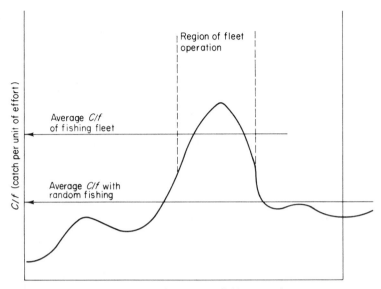

Distance along a line across a fishing ground

FIGURE 8.11 Diagram of the distribution of a catch per unit of effort across a fishing ground.

both fish and fishermen. The uneven distribution of effort can be accounted for by division of the area and time into small units and suitable weighting of each. Changes in efficiency of fishing units can be accounted for by a special determination of the changing ratios of the catch per unit of effort among different gear and vessels.

8.5 AVAILABILITY AND GEAR SELECTIVITY

After determining the structure of the catch from a stock by a sampling program, the biologist estimates the structure of the populations comprising the stock. The catch may be a large fraction of the stock and it may be sampled with care, yet its structure will be certain to differ from the structure of the populations because of differences in availability of the animals and the selectivity of the gear used.

Availability is defined generally as the fraction of the stock that lives in areas where it is susceptible to (i.e., may encounter) a given fishing gear during a given season. Availability varies commonly with species, sex, and size of animals, and each of these factors may vary with time or location. This variability arises primarily from different feeding and spawning migrations and different behavior. For example, the rosefish of the northwestern Atlantic can be caught by ordinary trawls in the day but not at night, because they move up in the water; the large individuals among the halibut of the northeastern Pacific prefer rough bottoms and the small individuals prefer smooth bottoms; the 1- and 2-year-old members of the fur seals of the Bering Sea remain at sea during the breeding season, when the 3- and 4-year-old males are available for capture on the shore; and in many species, size groups tend to segregate in pursuit of food. When the stock includes more than one species, the availability will differ because no two species can occupy exactly the same niche (Fig. 8.12).

The availability to different kinds of fishing gear is certain to differ also, for reasons other than the selectivity of the gear (discussed next). Each combination of gear and vessel works better in certain places than in others; bottom trawl nets require relatively smooth bottom, line gear can operate in canyons and other rocky areas, purse seines can fish from the surface to a considerable depth, and so on. For example, the introduction of purse-seine gear in the Lofoten cod fishery produced catches of larger fish than those taken by either gill nets or longline gear (Fig. 8.13). The purse seines also produced occasional catches of up to 90% of a single sex.

Even though individuals in a stock encounter the fishing gear, they may not be caught because the meshes or hooks are selective. Generally the meshes of trawls and seines will retain fish whose girth is greater than approximately the circumferences of the opening, but the meshes of a gill net will hold only those fish whose girth and shape permit them to wedge in the meshes; smaller fish wriggle through and larger fish twist out. Hooks tend to retain only fish of a limited range of sizes, because large fish are able to break free and small fish are unable to grasp the bait. The meshes and slats of pots used to catch crustaceans allow small individuals to escape.

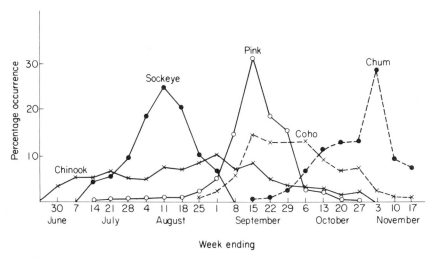

FIGURE 8.12 Seasonal occurrence of salmon in Fraser River gillnet catches. [From G. A. Rounsefell and G. B. Kelez (1938). The salmon and salmon fisheries of Swiftsure Bank, Puget Sound, and the Fraser River. *U.S. Fish. Wildl. Serv., Bur. Fish.* **49**, 693–823.]

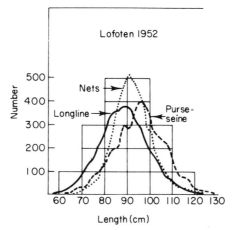

FIGURE 8.13   Length composition of Norwegian cod taken with different fishing gear. [After G. A. Rollefson (1956). Introduction to problems and methods of sampling fish populations. *Rapp. P. V. Reun. Cons. Int. Explor. Mer.* 140, 5–6.]

The relation between size of mesh and size of fish retained depends not only on the relation of mesh circumferences to fish girth, but also to a lesser degree on the size of the net twine, stretch of the fiber, and hanging of the net (i.e., the way in which the net spreads out while fishing). With some kinds of nets and species the behavior is also important, because in parts of trawls and seines, a large mesh with small twine will guide fish to a bag where the actual filtering occurs. Thus, commonly, the selectivity of nets must be determined by comparative fishing experiments. In these experiments, nets of different construction are fished side by side, and the operation is repeated as many times as are necessary to minimize the variance between the means of successive sets. One can determine the selection of trawl nets also by covering the bag with a finer mesh or by rigging a special trouserlike bag with different mesh in each leg (Fig. 8.14).

The catch by species and sex from such comparative fishing is measured, and the results are expressed as a length selection curve, or ogive. The percentages of each length group are computed and plotted (Fig. 8.15) and, when necessary, are adjusted to satisfy the assumption that the fishing powers of the nets are equal. Selection curves are compared normally at the points at which 50% of the fish are retained.

8.6 RECRUITMENT

Recruitment, R, is defined as the number of fish of a single age group entering the exploitable phase of a stock in a given period by growth of the

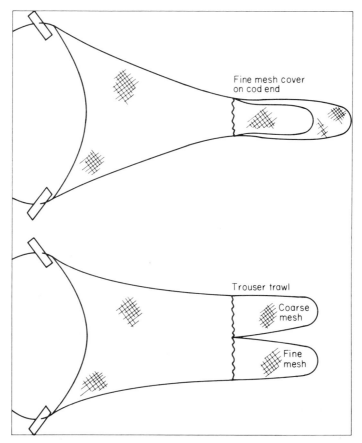

FIGURE 8.14   Diagram of otter trawl nets rigged for mesh selectivity experiments. The mesh in the forward part guides the fish to the cod end, where the filtering occurs.

smaller younger individuals. It is also defined as the number of fish from a single year group arriving in an area during a given period where fishing is in progress, even though the fish may be so small that chance of capture is negligible (e.g., at the time that larval stages descend to the bottom). The latter is a more definite event than the former, but generally the former is more easily measurable. The term should be used with the former meaning and with the added advantage that, in many stocks of fish, the individuals mature at about the same time that they become exploitable; the size of the exploitable stocks is probably an index of the number of spawners.

Recruitment is of special interest, first, because of its relation to subsequent yield of the year group. It is the first obtainable index of the abundance of the year group and is therefore a useful base for prediction. In the special instance of stocks that produce only an occasional *dominant year class* (Fig. 8.16), the appearance of abundant young is a sign of good fishing to come.

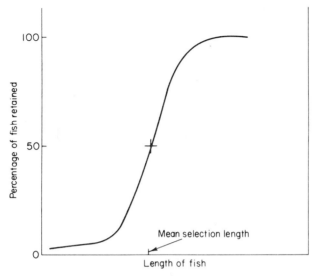

FIGURE 8.15 Hypothetical selection curve for a trawl net. The selection is never precise and the lower end of the curve may not reach zero because some small fish become mixed with the bulk of the catch.

More important to understanding the dynamics of the stock is the relationship between the number of spawners and the ensuing recruitment. This relationship is one of the two key ones to be determined in the analysis of a fish stock; the other is the relationship between the amount of fishing and the catch for a given recruitment.

The general relationship between the number of spawners and the ensuing recruitment conforms to three factors derived from the concepts of the single-species populations (see Chapter 7). First, when there are no spawners, there is no recruitment exclusive of immigration, which should be accounted for separately. Second, all populations except those headed for extinction have adapted to their niche, have a capacity for growth, and have a resiliency that enables them to recover from adversity. Third, populations are limited in number by natural factors that increase the mortality rate as the number of animals increases. These factors and the ensuing mortality are called *compensatory*. Other factors, which are density dependent, become less effective as density increases and are called *depensatory*.

The compensatory factors affecting fish populations include the following (from Ricker, 1954, pp. 562, 563).

> 1. Prevention of breeding by some members of large populations because all breeding sites are occupied. Note that territorial behaviour may restrict the number of sites to a number less than what is physically possible.
> 2. Limitation of good breeding areas, so that with denser populations more eggs and young are exposed to extremes of environmental conditions, or to predators.

3. Competition for living space among larvae or fry, so that some individuals must live in exposed situations. This too is often aggravated by territoriality—that is, the preemption of a certain amount of space by an individual, sometimes more than is needed to supply necessary food.

4. Death from starvation or indirectly from debility due to insufficient food, among the younger stages of large broods, because of severe competition for food.

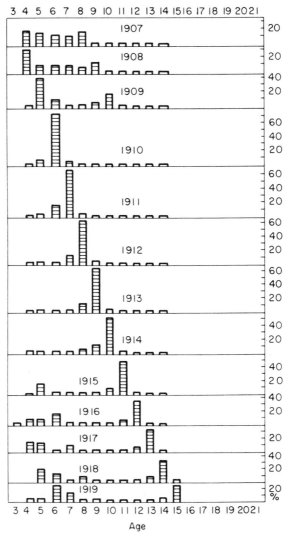

FIGURE 8.16 Percentage age composition of samples of the catch of spring herring along the Norwegian coast. Note the predominance of the 1904 year class. [From J. Hjort (1926). Fluctuations in the year classes of important food fishes. *J. Cons., Cons. Perm. Int. Explor. Mer.* **1**, 5–38.]

5. Greater losses from predation among large broods because of slower growth caused by greater competition for food. It can be taken as a general rule that the smaller an animal is, the more vulnerable it is to predators, and hence any slowing up of growth makes for greater predation losses. Since abundant year-classes of fishes have often been found to consist of smaller-than-average individuals, this may well be a very common compensatory mechanism among fishes.

6. Cannibalism: destruction of eggs or young by older individuals of the same species. This can operate in the same manner as predation by other species, but it has the additional feature that when eggs or fry are abundant the adults which produced them tend to be abundant also, so that percentage destruction of the (initially) denser broods of young automatically goes up—provided the predation situation approaches the type in which kills are made at a constant fraction of random encounters.

7. Larger broods may be more affected by macroscopic parasites or microorganisms, because of more frequent opportunity for the parasites to find hosts and complete their life cycle.

8. In limited aquatic environments there may be a "conditioning" of the medium by accumulation of waste materials that have a depressing effect on reproduction, increasingly as population size increases.

Floods, droughts, extreme temperatures, and other environmental changes may be noncompensatory, but even these nonbiological factors may be more devastating to large populations than to small ones (and hence compensatory) by killing the individuals that cannot find shelter.

Compensatory mortality factors appear to affect most of the younger stages of aquatic animals—the eggs, larvae, and juveniles—which are produced in great abundance by many aquatic animals. Such factors affect the mature individuals also, but probably much adult mortality is noncompensatory. In many stocks, the mortality due to fishing is primarily among adult individuals.

When we represent the relationship of recruitment to the number of spawners, it is convenient, and frequently accurate, to use the size of the stock as an index of the number of spawners. We expect a curve showing an average relationship within the limits shown in Fig. 8.17. All such curves must go through the origin and cross the line at which recruitment just replaces the parent stock, or the "45-degree line." If they remained entirely above the 45-degree line, they would be increasing without limit; and if they remained entirely below it, they would be headed for extinction. At higher levels of parent stock, the recruitment may either approach an upper limit or decline to nearly zero. When the curve tends toward an upper limit, the population size tends toward stability; whereas when the curve descends sharply toward a zero recruitment, the population size tends to oscillate.

Stock recruitment curves tend to be obscured by variability in the relationship. Great variability and lack of demonstrable relationship seem to be associated with species having young in large numbers and with very high mortality rates. These are the species that tend to produce occasional dominant year classes. For some such species (Fig. 8.18), there appears to be little reason for

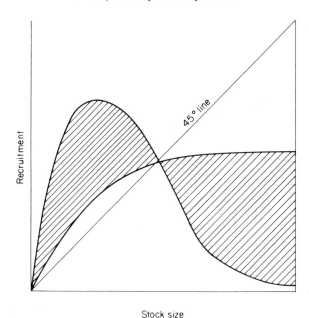

FIGURE 8.17   A diagram of the bounds of average recruitment curves. The shaded areas indicate the extent of variability, which presents changing management problems.

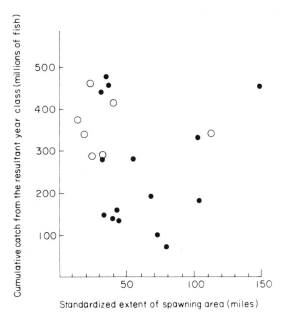

FIGURE 8.18   Recruitment data for British Columbia herring. The dots represent the west coast of Vancouver Island, 1937–1952. The circles represent the lower east coast of Vancouver Island, 1947–1953. [From N. Hanamura (1961). On the present status of the herring stocks of Canada and Southeastern Alaska. *Bull. Int. N. Pac. Fish. Comm.* **4,** 67.]

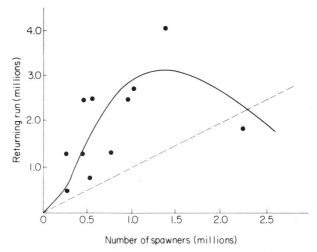

FIGURE 8.19 A stock—recruitment curve for sockeye salmon in the Wood River, Alaska. Stock equals number of spawners and recruitment equals number returning from the sea. The dots represent spawner-recruit data and the line represents recruits equal to the number of spawners. [Data from F. J. Ossiander, ed. (1967). Bristol Bay red salmon forecast of run for 1967. *Alaska Dep. Fish Game Inform. Leafl.* **105**, 1–51.]

concern about a scarcity of spawners, because a few of them seem as likely to produce a dominant year class as a multitude. When the relationship is evident, even when there is variability about the relationship (Fig. 8.19), it may provide a basis for superior management strategy (see Section 13.4).

8.7 STOCK SIZE

The *stock size* is either the number N of individuals in or the weight P of a stock at a given time. Both measures are usually necessary, so it is customary to use information on size composition and length—weight relationship for conversion of one to the other. The terms *abundance* and *density* are used by some authors to mean the total number of individuals, but these terms are assigned more properly either to number per unit of area or to unit of fishing effort. The term *biomass* may be applied to the weight of the stock. Some authors use *population* with approximately the meaning assigned to *stock* in this volume and hence use *population size*.

The stock size per se in terms of P is useful as a measure of the utilization of aquatic space. The total weight of one species or of all species of fish in a body of water may be reported as the *standing crop* in units of the weight per unit of area as a rough index of productivity.

The change in stock size is much more useful. It may be related to corresponding changes in the environment, in fishing, and in populations of food organisms, predators, and competitors. Further, the change in stock size, or in size of some of its parts, provides the fundamental statistic for estimation of survival and mortality rates. The determination of stock size is, therefore, an important step in the study of an exploited stock. Many scientists have devised a variety of methods for determining stock size.

8.7.1 Direct Counts

Whenever the direct count method can be used it is preferred, because it is usually cheaper and more precise. Fortunately, it is possible in a wide variety of environments. It is usually accomplished by counting the individuals in a known fraction of the area occupied by the stock and capturing simultaneously a sample in order to add information on the size and age structure. When a pond can be drained without the loss of fish, the entire stock may be available for examination. Salmon migrating up a river can be counted from observation towers (Fig. 8.20) or booths in fishways with a high degree of accuracy and sampled occasionally with a seine. The accumulation of spawning salmon in a stream can be estimated with fair accuracy by trained observers in airplanes. Sessile animals, such as clams or oysters, can be counted or examined in quadrants that are selected at random and contain a known fraction of the beach area. Clams and other invertebrates in deeper water can be sampled with a dredge (Fig. 8.21) that is known to bring up a definite area on the bottom. Fish with territories in streams or on coral reefs can be counted by scuba divers.

FIGURE 8.20 Migrating salmon are counted easily in the Kvichak River, Alaska, from a tower placed near the path of migration. Surface ripples are reduced by a small boom. (Photo by W. F. Royce.)

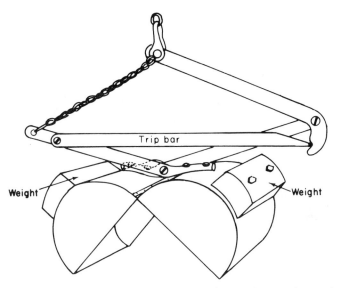

FIGURE 8.21 A Petersen-type dredge is used to sample animals on sandy or soft bottoms.

The catch of any mobile fishing net can be used for direct estimation of stock size when the selectivity of the net and the area covered are known. Both of these matters are difficult to determine, so the selection of sampling area is standardized, a constant efficiency is attempted, and the catch per unit of effort, C/f, is used as an index (see Section 8.7.4).

The adaptation of sonar gear for fish detection has stimulated numerous attempts to use the instruments to count fish (Fig. 4.12). As with fishing nets, the selectivity of the echoes and the area covered must be determined. With the addition of a counter or integrator to the electronic equipment, the instrument will yield a total count from a record of the echoes, but may still be most reliable as an index of change. Such sonic equipment must be used in conjunction with fishing nets, however, for reliable identification of the species of animals causing echoes.

8.7.2 Correlated Populations

Sometimes breeding populations of aquatic species can be estimated from the production of eggs or the numbers of nests (or redds for salmon). Estimation from eggs has been attempted for a few marine species that have pelagic eggs. The information required is the number of eggs per female or per unit of weight of females, the sex ratio of the mature animals, and the total number of eggs laid in the sea during the spawning season. The former is relatively easy to obtain, but the latter is extremely difficult to procure because of the sampling problems. The entire geographic range must be sampled with plankton nets during the complete period of spawning. The period may be much longer than the time needed for hatching and larval development of a single batch of eggs. The eggs and larvae are subject to high rates of mortality; therefore, the rates of development and mortality must also be known and adjustments must be made. Despite these difficulties, the numbers of eggs have provided important clues to the presence of large fish populations that are not being exploited (e.g., anchovies off the coast of California).

A rough estimate of the numbers of eggs laid by spawners has been developed for the stocks of herring along the coasts of British Columbia and Alaska (Fig. 8.18). The eggs are laid in the intertidal zone, where they adhere to the bottom or to vegetation. Visual estimates are made of the length of beach covered (in miles) and the density of spawn. These can be converted very roughly to an estimate of the total number of eggs by a count of the eggs in a subsample of the beach and an estimate of the number of females from the data on fecundity.

The nests of certain fish, such as salmonids and basses, may be counted, and the count can be used as an estimate of the number of spawning pairs. Counts of Pacific salmon nests (redds) are made at times to provide information on the utilization of stream areas for spawning. Usually such counts are made in limited and easily accessible areas and are considered to be an index, because sampling of all spawning areas and during all periods may be difficult.

8.7.3 Marked Members

Marked or tagged fish may provide information on stock size besides providing information on migrations and intermingling of stocks (see Section 8.3.2). The use of this method is indicated especially for small, discrete freshwater stocks that support recreational fisheries, and for which complete catch statistics are difficult to obtain. The basic objective is to establish a population of marked fish that will be subject to the same probability of recapture as the unmarked population. Thus, the fish to be marked are taken at random from catches of the gear in use and distributed at random among the unmarked population.

When a sample of fish, m, is marked and released into a stock with a number N and then another catch C is made, the estimated N is

$$\hat{N} = mC/r, \tag{1}$$

in which r is the number of marked fish recaptured. But usually the fish to be marked must be caught and released over a period of time, during which some of the marked fish may be recaptured. These data can be used in a *multiple census* (also called a *Schnable type estimate*, after its originator), for which the simplest estimate is

$$\hat{N} = \Sigma \ (m_t c_t)/\Sigma r_t, \tag{2}$$

in which c_t is the total sample on day t, m_t represents the total marked fish at large on day t, and r_t represents the recaptures on day t. An example of the computations is given in Table 8.1.

TABLE 8.1   **A Multiple Census of Crappies from the North Half of Foot's Pond, Indiana**[a]

5-day period	$\Sigma m_t c_t$	Σr_t	$\hat{N}$
1	2850	1	2850
2	5710	3	1900
3	8410	4	2100
4	12,470	5	2490
5	21,660	9	2410
6	36,540	15	2440
7	36,540	15	2440
8	45,980	17	2700
9	61,060	19	3210
10	67,900	22	3090

[a]Recomputed from data used by Ricker (1975). Recaptures were made in traps used to catch the fish for marking.

All of this is deceptively simple because of the difficulty of establishing a marked population of sufficient size to yield an adequate number of recaptures and subject to the same probability of recapture as the unmarked population. The pitfalls are numerous and include the following.

1. The formulae given here are positively biased approximations when numbers of recaptures are small. They have been modified by several people to provide more precise estimates.

2. The marked population may decrease in size during the experiment because of deaths caused by marking or loss of marks.

3. The unmarked population may vary in size during the experiment because of recruitment.

4. The rate of exploitation on the marked and unmarked populations may differ because the marks make the fish more liable to capture (some tags catch on nets) or less liable to capture (some marked fish are injured or disturbed and behave differently).

5. Marks may not be recognized or reported by fishermen.

6. The marked fish do not distribute themselves at random in the stock, and random sampling of the stock is not possible. Some fish may have territories near fixed gear, and they may be recaptured repeatedly. The stock may range over bottoms and depths that cannot be fished. The range of migration of the stock may not be known accurately.

Despite these difficulties, the method is important and may be the only one possible under many circumstances. Investigators must use it with special efforts during the design of the program and its analysis to compensate for any known problems.

8.7.4 Catch per Unit of Effort

Most stocks of fish in the sea and in large lakes do not behave in ways that permit direct or correlated counts, and they are either too large, too remote, or too variable to permit estimation of stock size by means of marked members. For such stocks, the catch per unit of effort, C/f, is a useful index of the size. Even without a measure of the relationship of the index to the actual N, frequently one can assume reasonably that each unit of a fishing operation, say one day's fishing by a trawler, captures a fraction k of the stock being fished. It follows that f units capture a fraction fk of the stock and

$$C = fkN, \ C/f = kN, \tag{3}$$

or, in words, that k is an unknown constant.

Much needs to be done to ensure the assumption that k is constant, if management is to be trusted. The fishing effort is usually expressed as a unit of time fished by standard units of gear and vessel, such as angler hours, size of nets, trawling hours times horsepower, or "soaking" time of fixed nets. Account is taken of nonfishing time spent in travel and lost to bad weather. The effect of a change of gear by one vessel can be estimated by comparing the vessel's catches with those of other vessels not making the changes. Any cyclical changes in behavior of fish with days or seasons are estimated and adjustments made. Any long-term trends, either in gear efficiency or in skill of fishermen as a group, or in migrations of the fish, are used to adjust the measure of effort.

8.8 SURVIVAL AND MORTALITY

Survival, S, and mortality, $1 - S$, its counterpart, are determined by a comparison of the numbers of a cohort or a group of cohorts alive at successive ages. The cohort(s) may be composed either of marked animals or of a year class(es) of animals in a stock as determined by age analysis and sampling of the catches.

A comparison of this section and the preceding one on stock size will reveal that the basic data for determination of stock size and survival rate of a cohort are similar, if not identical, in many cases. It is obvious from formula (1) that m can be considered a cohort, and

$$r/m = 1 - S, \tag{4}$$

in which $1 - S$ is the fraction caught and S is the estimated fraction surviving if we assume no natural mortality. The similarity is so close that many authors have considered determination of stock size and of mortality rate as a single topic. It is desirable, however, to consider these methods separately, because most methods of determining stock size are useful for either stocks or fractions of stocks of unknown age composition, whereas the methods of determining mortality rate for changing cohort size are useful usually for stocks with a known age structure.

8.8.1 Fractions and Rates

It is essential to change from the fractions S and $1 - S$ to instantaneous rates for studies of mortalities. Let us assume that the rate of decrease is proportional to N in a cohort, then the rate of decrease at age t is

$$dN/dt = -ZN, \tag{5}$$

in which Z is the instantaneous mortality coefficient.* This is related to the fraction S,

$$S = N_1/N_0 = e^{-Z_t}, \tag{6}$$

and if one should need to compute the number surviving at some time t after t_0,

$$N_t = N_o e^{-Z_{t-t_0}}. \tag{7}$$

Usually computations of mortality involve both natural mortality M and fishing mortality F. If these occur simultaneously and constantly during period t, then

$$F + M = Z. \tag{8}$$

Sometimes it is useful to divide $1 - S$ into the fractions dying from fishing, E, and from natural causes, D. These are related to F and M as

$$E = (1 - S)F/Z \tag{9}$$

and

$$D = (1 - S)M/Z. \tag{10}$$

---

*The concept of instantaneous rates troubles many students, who may be helped by an example. Let us assume that a population of fish at the beginning of a year is $N_0 = 1000$ and is subject to an annual expectation of death of 0.8. Survival, of course, is 0.2. It follows that the mortality during a year is

$$0.8 \cdot 1000 = 800, N_t = 200.$$

Now suppose that the year is divided in half, and the annual expectation of death also is divided in half, so that in the first half-year

$$0.4 \cdot 1000 = 400, N_1 = 600.$$

In the second half of the year the same mortality rate is applied to the remaining fish,

$$0.4 \cdot 600 = 240, N_1 = 360,$$

and total mortality is $400 + 240 = 640$ instead of 800.

If the year is divided into four parts, similar computations indicate that $N_1 = 410$; if it is divided into ten parts, $N_1 = 431$; if it is divided into a very large number of parts, $N_1 = 449$. The latter is the result of an instantaneous rate, $Z = 0.8$, divided n times (with n very large) and applied to n fractions of the period. In this example, the fraction dying in one year is $(1 - S) = 0.551$. Z is always larger than the $1 - S$ and may be greater than 1.

As a simple example of such rates, let us consider again the pond with no spawning area that is stocked with 1000 catchable-sized trout and is found to contain 600 when drained a year later. The catch resulting from 50 days of fishing effort during the year was 100 ($E = -0.1$), and the presumed deaths from natural causes were 300 ($D = 0.3$). Thus,

$$S = 600/1000 = 0.6$$

and from (6)

$$S = e^{-Zt}.$$

Changing S to natural logarithms and substituting, we have

$$\log_e 0.6 = -Z = -0.511.$$

The instantaneous total mortality rate, A, is equal to the natural logarithm (with the sign changed) of the fraction surviving, that is, the complement of the fraction dying. Note that this is an annual unit of the instantaneous rate. For a period of 12 months ($t - t_0 = 12$), the value of Z for the month would be $0.511/12 = 0.043$.

The total instantaneous mortality rate, Z, may be divided into M and F if we assume that these occurred in a parallel way, by the proportion of E and D:

$$F/E = M/D = Z(1 - S). \tag{11}$$

In this example,

$$F = 0.1/0.4(0.511) = 0.128$$

and

$$M = 0.3/0.4(0.511) = 0.383.$$

8.8.2 Effort as an Index of Fishing Mortality

To assume that the catch per unit of effort (C/f) is an index of stock size N is equivalent to assuming that the total amount of effort is an index of fishing mortality F:

$$f = k'F. \tag{12}$$

The relative constancy of k' can be ensured by the same precautions used to ensure the constancy of k (see Section 8.7.4).

8.8.3 Catch Curves

Now let us suppose that we reflood the pond and leave therein the 600 fish, undamaged by any handling. Let us suppose further that fishing effort (and fishing mortality) and natural mortality continue at the same rate, that 360 fish are in the pond when it is drained at the end of the second year and 216 fish are in it at the end of the third, and that the catches in the intervening periods are 60 and 36. Let us recall that C/f is an index of N and is useful for estimating survival, thus

$$\frac{C_1/f}{C_0/f} = \frac{N_1}{N_0} = S \qquad (13)$$

$$\frac{60/f}{100/f} = \frac{36/f}{60/f} = \frac{600}{1000} = \frac{360}{600} = 0.6.$$

The series of catches per unit of effort, C_i/f_i, plotted against years forms a *catch curve* that is comparable to the *survivorship curve* of demographers. On a logarithmic scale, such a catch curve (Fig. 8.22) is a straight line with slope equal to Z; in this case $Z = 0.511$.

The hypothetical population of trout in the pond used in this example was either a cohort or a set of cohorts, depending on whether all of the trout were of the same age; it makes no difference in the example. There was no spawning area, and hence no recruitment. Catch was counted directly, and natural deaths were estimated indirectly by a difference.

Such a hypothetical situation is similar to a population of marked individuals subject to a fishery. When we assume that natural mortality and loss or nonreporting of marks occur at constant rates, then the catches per unit of effort or marked individuals provide a catch curve and an estimate of total instantaneous mortality, Z. The hypothetical situation is also similar to a cohort that is identified by determination of the age structure of a stock in

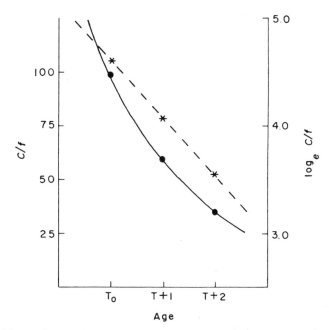

FIGURE 8.22 Arithmetic (dots) and logarithmic (asterisks) catch curves from hypothetical cohort experiencing a constant mortality rate of $Z = 0.511$.

successive years. When the cohort is uniformly available to the fishing gear at different ages, then the catch curves per unit of effort of the cohort in successive years provide a catch curve. When both recruitment and mortality can be assumed to be constant, then the age structure at any time provides a catch curve. Also, when age structures of catches from constant fishing for a series of years are available and can be averaged so that the effects of varying recruitment and mortality are smoothed, then the average age composition will provide a catch curve.

With all of these assumptions about recruitment, natural mortality, and steady states, it is surprising that so many catch curves provide useful information on total mortality. Many do resemble those in Fig. 8.23, with an ascending left-hand limb until recruitment is complete and then a nearly straight right-hand limb. The occurrence of such straight right-hand limbs is regarded as evidence that recruitment and natural mortality are sufficiently steady to give confidence in the method. Furthermore, when catch curves do not have a straight right-hand limb, there is reason to suspect that recruitment or catchability varies, or that the population is not in equilibrium.

8.8.4 Natural and Fishing Mortality

A contrast to the relative ease and accuracy with which the total mortality rate Z can be determined is the relative difficulty of estimating the division of the total mortality rate into fishing mortality rate F and natural rate M when

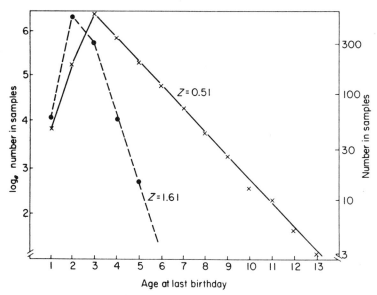

FIGURE 8.23 Catch curves for the Pacific pilchard. The numbers of individuals in the samples were assumed to be proportional to the age structure of the catch. Crosses indicate 1925–1933; dots indicate 1937–1942. (From Silliman, 1943.)

both are occurring. Natural mortality is equal to total mortality during periods when there is no fishing, but when fishing is occurring the simultaneous natural mortality must be estimated mathematically.

The natural mortality rate of catchable-sized fish in stocks that have never been fished is related to the maximum age of the fish to be expected. For example, if fish were of recruitment size at age 3 and sampling indicated that there were only 1% as many fish at age 13 as there were at age 3, then the average natural mortality rate $M = \log_e 100 - \log_e 1/10 = 0.46$. If only 1% remained after 20 years, then average $M = 0.23$. (Note the estimated mortality rate of 0.235 for an unfished halibut population attaining about 25 years of age as shown in Fig. 7.5.)

When such special circumstances do not permit direct estimates of natural mortality, the method derives from the relationship between total mortality rate Z and total fishing effort f when both are varying. An estimate of M for the California stock of pilchard was obtained by Silliman (1943) from catch curves for two separate periods, within each of which the recruitment R, stock size N, and effort X were relatively stable, but between which the effort X and the total mortality rate Z were greatly different. The catch curves (Fig. 8.23) showed that $Z = 0.51$ for the period 1925–1933, and $Z = 1.61$ for the period 1937–1942, when X was four times as great as for the first period. Assuming that natural mortality M was equal in the two periods and that fishing mortality F was in the same proportion as total effort X (availability was equal in the two periods), one may write two simple simultaneous equations:

$$F_1 + M = 0.51,$$
$$F_2 + M = 1.61,$$

and

$$F_2 = 4F_1.$$

The solution (rounded off) is

$$F_1 = 0.37,$$
$$F_2 = 1.46,$$

and

$$M = 0.15.$$

A graphical solution of the same problem (Fig. 8.24) illustrates the principle involved—extrapolating mathematically to zero fishing effort.

If fishing effort X and total mortality should both vary continuously, then their relationship would provide a similar possibility of estimating M. An illustration in the form of a simple linear regression is shown for Georges Bank haddock in Fig. 8.25. In this example, $M = 0.0$ although this value is not very reliable, because the variability in the data is relatively large. In addition, a

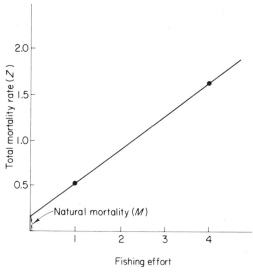

FIGURE 8.24   A graphical method of estimating the natural mortality rate for the Pacific pilchard.

mathematical refinement is desirable to overcome the slight error of using the average of a year's data to estimate the abundance at the midpoint of the year.

Regardless of the difficulties in determining M when fishing is occurring, most estimates are lower than the M that would be expected on the basis of maximum age of the animals in unfished populations. M is frequently estimated to be only about 0.1 or 0.2 under conditions of heavy to moderate fishing on stocks that fall short of reaching the 25- to 50-year ages that would occur in unfished populations subject to these natural mortalities. It is clear that the reduction in stock size that accompanies fishing diminishes the natural mortality rate—a compensatory change from reduced environmental pressures on the survivors.

8.9 GROWTH

Growth of aquatic animals is a highly irregular process (see Section 6.7); it varies with age, sex, season, climate, reproductive cycle, and population size. It is especially irregular during larval and postlarval periods, as rapid changes take place in body form. A complete description of growth in mathematical terms has not yet been formulated, although many attempts have been made.

One regularity is present, however, in the growth pattern of many aquatic animals after the juvenile states; their growth, in either length or weight, slows and approaches an asymptote (Fig. 8.26).

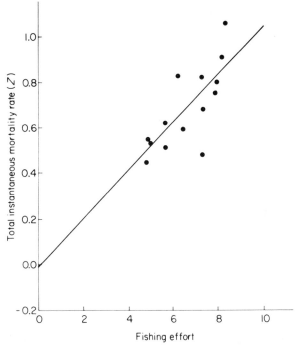

FIGURE 8.25 Extrapolation of relation of total mortality rate to fishing effort by linear regression analysis for estimation of natural mortality rate. [Data for Georges Bank haddock from 1933–1945 from H. A. Schuck (1949). Relationship of catch to changes in population size of New England haddock. *Biometrics* **5**, 213–231; and W. F. Royce and H. A. Schuck (1954). Studies of Georges Bank haddock. Part II. Prediction of the catch. *U.S. Fish Wildl. Serv. Bull.* **56**, 1–6.]

8.10 YIELD MODELS

Let us now assume that we have identified a unit stock subject to fishing and have estimated some or all of the factors that have been described earlier as influencing its size. We will then develop an appropriate model from which we can predict yield or other characteristics of the stock that may be a guide for action.

Fishery scientists have pioneered the application of mathematical models to fish stocks. Such models all require statistical methods for management of the input data, and some require calculus for analysis. Some have been elaborated into computer simulations of changes to be expected in a stock under varying circumstances. All require mathematical and statistical changes to be expected in a stock under varying circumstances. All require mathematical and statistical concepts that cannot be used in an introductory fishery course, so the following discussion will be largely descriptive of how the models are used. The student

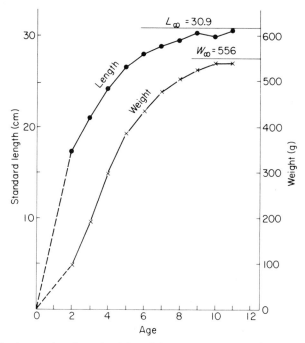

FIGURE 8.26   Average lengths and weights of ciscoes, ranging in age from 0 to 12 years, from Vermilion Lake, Minnesota. (Data used by Ricker, 1975.)

will find discussion of derivation and computational techniques, as well as examples of application, in Ricker (1975), Gulland (1988), and numerous other sources.

It is essential to emphasize at the outset that all fishery models are approximations. A complete ecological system is impossible to measure, let alone model in detail. Simplifying assumptions are made, which usually include the notions that stock growth is density dependent, that measurements of stock characteristics are reasonably accurate, that a level of stock abundance that will produce a desired sustainable yield can be maintained, and that all genetic components of the stock (different species or races) are similarly affected by fishing.

The density dependence of stock growth is a troublesome concept that is best illustrated with examples. No natural population can increase in size indefinitely, and one that reaches a steady state without human intervention can be assumed to have reached the upper part of the sigmoid curve of population growth (Section 7.1.5). This limit can be the result of any one or a combination of environmental factors, such as space, temperature, salinity, or food (Section 8.6). The effects of the factors causing stock growth and stock decline are equal. After fishing takes place, the effects of the natural factors causing stock decline are lessened because the stock is smaller. Fishing is said to be taking *surplus production* (Fig. 7.9).

But no steady state persists for long, and every change will trigger a series of population changes that will ripple through the ecosystem. The changes may include several trophic levels, each of which may require a time equal to a few generations of the organisms involved to again reach equilibrium at the top of the sigmoid curve. And some stocks may never reach equilibrium. An example might be a stock in a northern lake subject to occasional winterkill (suffocation under a persistent ice cover), which would be survived by only a few adults. This fish stock might never have enough juveniles to utilize the space and food available during the growing season, and adults might never be abundant enough to permit harvest. Prediction of the impact of such changes must be largely empirical and might be based on observed stock changes after a series of winterkills.

8.10.1 The Surplus Production Model

The most generally applicable model is the relation of stock size to fishing mortality rate. Stock size is usually indexed by the catch per unit of effort; fishing mortality rate is calculated by carefully refined measures of the total effort. Application of the model requires a long series of data, and no attempt is made to identify separately the rates of recruitment, growth, and natural mortality. Any changes in these factors caused indirectly by fishing are assumed to have reached equilibrium.

The model (sometimes called *logistic*) derives from the prevailing tendency for population growth to follow a sigmoid curve (Section 7.1.5) in which the rate of natural increase is at a maximum at the point of inflection and decreases gradually to zero either at zero population size or at the limiting population size. In the assumed equilibrium state, the catch is equal to the rate of natural increase, and the stock size is a function of fishing effort decreasing from its natural maximum with no fishing to zero at some very large effort (Fig. 8.27).

When fishing effort is related to equilibrium yield there is, of course, no yield at zero effort and no yield if the fishery has caught all of the stock. The relationship is an inverted parabola with a maximum equilibrium yield at half of the maximum natural stock size (Fig. 8.28).

Analysis of the yield–effort relationship in a number of fisheries suggests that, commonly, the maximum equilibrium yield is obtained at a stock size somewhat less than half the maximum, perhaps a third, and that the function is more complicated. Nonetheless, the data available are commonly so variable that the simple function provides a useful model.

8.10.2 The Recruitment Model

In many fish populations, the recruitment of young to the next generation may be largely independent of parent population size (Figs. 8.17 and 8.18). This is consistent with the fact that most fish species produce very large num-

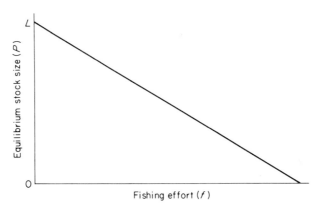

FIGURE 8.27   Relationship of stock size to fishing effort.

bers of eggs, even millions per female per season in a few, and a high survival rate from only a few spawners can replenish the stock.

But in a few populations, the number of recruits is reasonably predictable from the number of spawners. The model is especially useful for populations of Pacific salmon, in which recruitment occurs mostly just before spawning and the relationship of number of spawners to number of future recruits has been sufficiently consistent (Fig. 8.19). The maximum equilibrium yield is estimated to be the number recruited less the number of spawners at the point where the recruitment curve is most distant from the replacement line.

Application of this model to salmon requires the assumption that the density-dependent mortality occurs principally in the freshwater spawning and nursery areas, not on the ocean feeding grounds. In many situations, the assumption is supported by knowledge of limited areas of good quality for spawning and nursery.

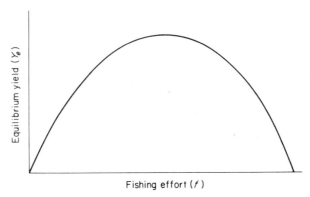

FIGURE 8.28   Relationship of yield to fishing effort.

8.10.3 Yield-per-Recruit Models

After an individual animal reaches recruitment size (i.e., it is available to the fishery) it will continue to grow and have a chance of death due either to fishing or to natural causes. Its annual growth increment in weight will increase for a time and then decline to nearly zero at its maximum size (Fig. 8.26). Its chance of natural death each year in an equilibrium population will be approximately constant (Fig. 7.5). Of course, its chance of capture will be related to the amount of fishing effort and the selectivity of the fishing gear.

The problem is to estimate growth, natural mortality, and fishing mortality rates and to incorporate these in a model that provides an estimate of the yield to be expected from various amounts of fishing effort. The fishery must provide data on the age and size composition of catches, the rate of growth by age and size, and the catch per unit of effort by age. Such data are obtained by the subsystem of biological statistics and, fortunately, age determination is possible for most aquatic animals. Given these data over a period of years, it is possible to estimate how each cohort (year group) has changed in size year by year, and how the changes have been related to the kind and amount of fishing.

If a stock in equilibrium with constant recruitment can be assumed, then the fate of a cohort over a period of years is equivalent to the fate of all cohorts in a single year. The effect of changes in the rate of fishing mortality can be predicted in order to determine the sustainable yields. In addition, the effect of changing recruitment size (e.g., by mesh regulation) on the total yield can be predicted. A further extension of the models may consider the effect of changes in both fishing mortality and recruitment size to produce what has been called a *eumetric fishing curve*.

If recruitment varies erratically (Fig. 8.16), a model can be tailored to estimate the fate of each year class as it passes through a fishery. Such models are especially useful for prediction of desirable catch rates year by year, because large year classes can be harvested more completely and small year classes can be given more protection. Further recruitment can sometimes be predicted before the cohort appears in the fishery, and the desirable catch can be estimated earlier. An example in oceanic stocks with drifting eggs and larvae is the use of a known relationship between wind direction during the postspawning period and subsequent recruitment of the cohort (Section 6.6.9).

8.10.4 Trophic and Other Models

The results during the past five decades of applying the single-species models discussed in the foregoing have ranged from striking successes to utter failures. As should be expected, the successes have occurred when the assumptions were reasonably well met. The successes, which have been somewhat fortuitous, have usually involved fish stocks with identifiable populations of long-lived species at high trophic levels, subject to steady fishing, and with steady recruitment. An example is the North Pacific halibut. The failures have occurred when major changes in abundance have not been forecast, and they have usually involved short-lived species at intermediate trophic levels that have been strongly influenced by density-independent factors. Examples are stocks of anchovies and sardines that have collapsed. Other failures (which will not be considered here) have occurred when excess fishing capacity has been impossible to reduce, or when the objectives of regulation have been changed under political pressure.

The shortcomings of the single-species models are being approached on both theoretical and pragmatic levels. On the theoretical level, more complex models are emerging that incorporate both abiotic and biotic factors, especially trophic and competitive relationships among species. On the pragmatic level, forecasts are being gradually improved using biotic and abiotic correlations that are measurable and useful, and that respond directly to questions being asked. Many of these efforts to improve the models are descriptive or conceptual, rather than theoretical and analytical. Some of the principal directions of this research are discussed in the following.

Trophic models have been alluring because most of the enormous productivity of the producer level is lost before the harvest of the upper trophic levels. The energy stored from one level to the next is roughly 10%, but the factor is highly variable for reasons that are not known. Steele (1974) reported on a major theoretical effort to simulate a plankton ecosystem. He considered the physical parameters of vertical mixing, light, and temperature; the chemical parameters of nutrient uptake, nutrient mixing, and nutrient regeneration; and the biological parameters of zooplankton grazing rate, assimilation, respiration rate, growth of zooplankton, herbivore mortality, and food chain efficiency. These parameters were used in a computer model and varied in an attempt to simulate some actual changes that had been observed in the North Sea. The main value of the model was identification of the shortcomings of existing theory as a guide for future work.

Another complex model (Patten, 1975) included all trophic levels in a shallow cove of 4.6 hectares off a reservoir. This computerized model predicted effects of perturbations in the cove, including temperature, phosphorus enrichment, and an introduced predator. A conclusion was that the predator caused far-reaching changes in the system.

A complex competitive relationship between large oceanic stocks at intermediate trophic levels was described by Skud (1982). He reviewed the changes in abundance of herring and mackerel in the northwest Atlantic and observed that environmental factors were positively correlated with changes in abundance of the dominant species, while being negatively correlated with changes in the subordinate species, regardless of which species was dominant. He also noted similar correlations with changes in dominance of sardine and anchovy stocks.

The implication is clear that single-species models are inadequate with such stocks, because both dominant and subordinate species are subject to the density-dependent status of the dominant species. Further fishery management implications are that dominance tends to be relatively stable over many years, and heavy fishing on the dominant stock or major environmental change may result in a reversal of dominance.

Much simpler models that use physical and chemical factors have improved predictions of fish production in ecosystems with clear-cut physical boundaries, such as ponds and lakes. Ryder (1965), who needed a method of estimating potential fish production from lakes with little fishing, used a simple ratio of total dissolved solids to mean depth, called a *morpho-edaphic index*. This ratio had a statistically significant relationship to fish production in a group of lakes with similar climate and substantial fishing intensity, and the regression produced useful forecasts of potential fish production. The method has been extended to include other variables, such as regional climate (Schlesinger and Regier, 1982), and provides a way of explaining more than 80% of the variability in the maximum sustainable yield of lakes.

A more sophisticated approach in ponds and lakes is to describe the major kinds of fish communities (Tonn and Magnuson, 1983) and relate the type of fish community to several easily measured environmental factors. Work on one group of lakes enabled the accurate prediction of the species assemblages to be found in another group of lakes in a similar climatic area.

Reservoirs, when new and if they are subject to major changes in water withdrawal, have populations that change rapidly. A steady-state assumption is absurd, but the sequences of changes may be predictable. O'Heeron and Ellis (1975) developed a computer model of the time series in observed relationships among many physical and biological variables in reservoirs.

Conceptual models based on a simple predator–prey relationship are used in extensive aquaculture. The practice is to use a prey species of fish (trophic level 3) that feeds largely on invertebrates (trophic level 2) and in turn supports a desired predator population (trophic level 4). This model has been used widely in North America, where thousands of farm ponds are managed by controlled stocking and harvest of bluegill sunfish and largemouth bass.

Another conceptual model in aquaculture uses several species of omnivores (Trophic level 2 or 3) that maximize the use of the production level in fertilized ponds. It has been applied to intensive aquaculture in the tropics, where several species of carps in a properly managed pond can produce as much as 50 times the level of production expected from similar natural waters (see Section 11.2.1).

REFERENCES

Gulland, J. A., ed. (1988). "Fish Population Dynamics. The Implications for Management," 2nd ed. Wiley, New York.

Hilborn, R., and Walters, C. J. (1992). "Quantitative Fisheries Stock Assessment." Chapman & Hall, New York/London. (Contains extensive bibliography.)

O'Heeron, M. K., Jr., and Ellis, D. B. (1975). A comprehensive time series model for studying the effects of reservoir management on fish populations. *Trans. Amer. Fish. Soc.* **104**, 591–595.

Patten, B. C. (1975). A reservoir cove ecosystem model. *Trans. Amer. Fish. Soc.* **104**, 596–619.

Ricker, W. E. (1954). Stock and recruitment. *J. Fish. Res. Board Can.* **11**, 559–623.

Ricker, W. E. (1975). Computations and interpolation of biological statistics of fish populations. *Bull. Fish. Res. Board Can.* **191**, 1–382.

Ridgeway, G. J., Klontz, G. W., and Matsumoto, C. (1962). Intraspecific differences in serum antigens of red salmon demonstrated by immunochemical methods. *Bull. Int. N. Pac. Fish. Comm.* **8**, 1–13.

Royce, W. F. (1957). Statistical comparison of morphological data. *In* "Contributions to the Study of Subpopulations of Fishes" (J. C. Marr, ed.). *U.S. Fish Wildl. Serv., Spec. Sci. Rep.—Fish.* **208**, 7–28.

Ryder, R. A. (1965). A method for estimating the potential fish production of north temperate lakes. *Trans. Amer. Fish. Soc.* **94**, 214–218.

Schlesinger, D. A., and Regier, H. A. (1982). Climatic and morphoedaphic indices of fish yields from natural lakes. *Trans. Amer. Fish. Soc.* **111**, 141–150.

Silliman, R. P. (1943). Studies on the Pacific pilchard or sardine (*Sardinops caerulea*). 5. A method of computing mortalities and replacements. *U.S. Dept. Inter., Spec. Sci. Rep.* **24**, 1–10.

Skud, B. E. (1982). Dominance in fishes: The relation between environment and abundance. *Science* **216**, 144–149.

Steele, J. H. (1974). "The Structure of Maine Ecosystems." Harvard University Press, Cambridge, Mass.

Tonn, W. M., and Magnuson, J. J. (1983). Community analysis in fishery management: An application with northern Wisconsin lakes. *Trans. Amer. Fish. Soc.* **112**, 368–377.

9 | Aquacultural Sciences

Aquaculture, like agriculture, is an ancient art that has been advanced by scientific practices, principally during the last 100–150 years. In aquaculture, however, there have not been changes as revolutionary as those that have occurred in agriculture through the application of scientific findings to management of soils, selective breeding, nutrition, and control of disease, which have allowed a 10-fold increase in the productivity of labor. Nevertheless, with increasing prices for fishery products, which are to be expected as a consequence of the leveling off of production from the capture fisheries, the application of science to aquaculture has been increasing rapidly.

The natural aquatic producing systems support populations of many species, a few of which may be useful to people. A small portion of each population of the valuable species can be removed at a sustained rate without a failure of the producing system. But the world's production from the traditional capture fisheries has approached the sustainable limit of about 100 million MT annually. Perhaps we can better appreciate this by considering that if people still depended on hunting and gathering from the land, human population would probably be only a tiny fraction of what it has become as a result of agriculture.

Aquaculture, like agriculture, is basically a human circumvention of the factors that limit natural populations. This usually starts with special protection for young animals or seeds of plants, which are always produced in excess. Then the young are nurtured and protected from predators and diseases, which usually requires an enclosure—a raceway, pen, cage, or pond—for mobile animals. When they are enclosed they cannot forage naturally, so food must be

267

delivered or the enclosure must be managed to produce food. Any such concentration of organisms enhances the opportunities for disease transmission, so special efforts are needed to prevent disease by provision of good food, good sanitation, good environment, and perhaps application of drugs as may be appropriate. But the organisms that have evolved to cope with a natural environment are not ideal for their sheltered conditions, so they are selectively bred to fit the growing conditions that can be provided. They may also be selectively bred for more marketable characteristics. Such selective breeding usually involves complete control of the reproductive process and, when possible, the usual annual reproductive cycle is modified to obtain a continuous output of marketable organisms.

Such a modified animal or plant is regarded as domesticated. Practically all agricultural production is based on such domesticated organisms, which only remotely resemble their wild origins, and nearly 2000 aquacultural organisms are extensively domesticated genetically. They include several species of trouts and salmons, the common carp, and many species of ornamental fish. Some other aquacultural organisms come from wild parents or have not been selectively bred because they happen to be amenable to culture. Among these are mollusks such as oysters and clams, milkfish in the southwestern Pacific, and most of the crustaceans that are cultivated—especially the shrimps.

Culture of any aquatic organism, animal or plant, is probably possible. In fact, private and public aquaria maintain between 2000 and 3000 species, and most of these are capable of reproducing to continue the species indefinitely in captivity. But large-scale production of fish for food must compete in the markets with fish from the capture fisheries and other protein foods. Production systems must be planned, therefore, around species that have been proved to be manageable under the conditions that can be provided.

Thus, aquaculture poses challenges to many scientific and engineering disciplines. The choice of plants and animals may challenge the geneticist. The confinement, transport, or protection of organisms may challenge engineers from several specialties. The care of the young will challenge the microbiologist and the ecologist. The nutrition will challenge the biochemist and the physiologist. The control of disease will challenge the pathologist and epidemiologist. Such diverse disciplines cannot be addressed in this brief chapter, but their contribution to the common goal of increasing organic production from the waters will be explained.

9.1 CONTROL OF ENVIRONMENT

Application of any breeding, nutritional, or disease control requires some degree of physical control. In agriculture, control may range from minimal herding of semiwild stock to intensive culture in which each individual animal or plant is bred, fed, given protection, and harvested under controlled conditions. Likewise, aquacultural control ranges from minor enhancement of environment to intensive culture.

The physical control required by cultivated aquatic biota depends on whether they are free during most of their lives, like fish, or sessile during most of their lives, like oysters or rooted plants. Fish require an enclosure, which is usually an isolated pond, whereas sessile animals or plants do not require a separate body of water or even barriers in the water. With either kind of organism, the major objectives of control are to ensure a desirable physical environment, to arrange the optimal density of the cultivated organism, and to exclude all competitors or predators.

Legal control is as important as physical control with any kind of organism. Commercial aquaculture can be practiced only when the people who culture the animals or plants can also control their harvest and sale. The aquaculturist cannot possibly operate like the fisherman, who seeks wild fish and owns them only after they have been caught.

9.1.1 Water Quality

Aquacultural organisms spend their lives in the water, which must be tolerable for them at all times and ideal for them as much as possible. Its quality will be a major factor in the success or failure of the fish-farming operation. The quality needed will vary with the species, with its life stage as egg, larva, juvenile, adult, and spawner, with its daily and seasonal cycles of activity, and with its acclimatization (Section 7.1.1).

Temperature is an important factor because it is directly related to development and growth—up to a point. Domesticated rainbow trout, for example, tolerate temperatures from just above freezing to about 25°C, but the optimal range of temperature is 10 to 16°C for subsistence and 10 to 13°C for spawning. Lower temperatures result in slow growth, or even in no growth at 1 to 3°C. Higher temperatures also result in decreased growth, as well as increased susceptibility to disease. The best combination of growth and good health occurs at about 15°C.

Other species, of course, have different ranges of tolerance and preference. Many warm-water species tolerate temperatures to 35°C, and prefer the range of 20 to 30°C during their growing seasons.

Dissolved oxygen (Section 4.5.1) is a critical factor, because its rates of consumption and replenishment must always be in balance. In raceways and flowing-water ponds, the water supply is preferably saturated with oxygen at the intake, and the inflow must be sufficient to supply the animals at all times. In stillwater ponds, where the inflow is used only to make up for losses due to evaporation or seepage, the oxygen must be supplied by photosynthesis or solution from the air. Such ponds are usually managed as complete ecosystems, and all of the plants and animals in them consume oxygen in the absence of sunshine. There is, therefore, a constant danger of lack of oxygen during nights following overcast days, when the pond is pushed toward maximum production.

Most carnivorous animals that live in streams or near the surface of lakes or the ocean require oxygen at levels of 4 mg/liter or higher. Some omnivorous animals such as carps, which are commonly grown in stillwater ponds, can tolerate levels of 1 mg/liter.

Many other qualities of water must be considered. Hard waters with calcium carbonate and small quantities of magnesium, manganese, iron, and zinc are usually desirable. Cadmium, copper, lead, chlorine, and ammonia can be harmful above very low concentrations. Most pesticides are extremely toxic. And a water supply with fish in it is a constant source of disease.

Fish farms that raise carnivorous fish on complete diets will also discharge water polluted by the fish and food wastes, and perhaps by chemicals used to treat the fish or by disease organisms present on the farm. Their effluent may require treatment comparable to treatment of urban sewage.

Effective management of the water is vital for the success of the farm. The design may require elaborate engineering, and the quality may require chemical analysis during operations.

9.1.2 Fish Ponds

A pond needs barriers against fish migrations at inlet and outlet, and an arrangement to allow complete drainage that provides control over the kinds and numbers of organisms in the pond. The animals in the pond, including fish disease organisms, can be reduced or eliminated either by drying or by chemical sterilization. The pond can then be filled with water and stocked with the numbers and kinds of animals desired. Usually the water is admitted without sterilization. It brings with it various innocuous beneficial organisms, such as phytoplankton and zooplankton, as well as a variety of pests, such as insects. Terrestrial pests that eat fish, such as birds or mammals, can be screened out as necessary. The result, which is attained in many well-managed ponds, is a population of a single species of fish under control with respect to reproduction, diet, diseases, and pests.

Such a regime is used for intensively managed ponds to which all food is supplied, for example, trout ponds. When the water is supplied from springs in which no fish live, diseases are largely eliminated. When the temperature is near the maximum preferred by trout year round, it allows a maximum rate of growth when the food is adequate and suitable.

On the other hand, when the pond is managed to produce all or a large part of the fish's food, as in the case of carp or milkfish ponds, the control is much more complicated. The needs of fish during larval, growing, and breeding stages must be met and, in addition, the fish food organisms must be controlled so that food of the needed size is provided in the correct amounts at the proper time. Providing such food may involve additions of fertilizers to make available needed elements, control of the pond bottom to provide suitable spawning areas, and control of rooted vegetation so that the proper amount and kind are furnished. The objective is an ecological system that produces all or a large part of its organic material through photosynthesis and transforms the synthesized material into flesh of a desired species as efficiently as possible.

The pond need not be managed for a single species or even for aquatic animals alone. In many ponds of Southeast Asia that produce much of the animal food naturally, two or more species that occupy different ecological niches can be grown together to achieve a greater total production. These can be different fish or fish and shrimp. Further, ponds can be managed for simultaneous or alternate crops of fish and rice.

9.1.3 Enclosures Other Than Ponds

No other enclosure for aquatic animals can be as satisfactory as a pond, but various screens, barriers, and cages have limited usefulness. Net enclosures can hold juvenile and adult animals where they can feed on natural food brought in by the currents as well as on artificial foods. Such enclosures are also useful for holding live animals (e.g., lobsters) for market. A major disadvantage of any aquatic net enclosure is the speed with which it becomes fouled, especially in salt water, when the openings are small enough to retain larval fish. For this reason, other kinds of barriers are eagerly sought, such as sound barriers, electrical fields, or curtains of bubbles, but most such barriers have proven to be of limited usefulness. One promising barrier to limit the movement of oyster predators on the bottom is a line of sand treated with an insoluble poison.

9.1.4 Beds and Racks for Sessile Organisms

Some larval mollusks and leafy algae attach themselves to shells or other objects. The objects, called "collectors," are arranged to suit the needs of the organisms. Some oyster, mussel, or algal collectors are strung on wires and hung from racks in order to raise the organisms off the bottom and out of reach of many of their pests. Such a system also allows a large number of organisms to grow per unit area. For example, mussels may produce as much as 300,000 kg/ha/year when suspended from racks in a good growing area.

Mollusks that do not attach themselves when in the larval stage, such as clams, are commonly grown on a suitable bottom, called a "bed." Oysters can be grown in this manner also, either attached to the collectors or separated from them. Animals in a bed can be given limited protection from pests; they can be easily recaptured either for transfer to a bed with better growing conditions or for market.

9.1.5 Partial Control of Open Waters

Almost every body of water is capable of producing fish, whether it is being stored, transported, or used for other purposes, such as transportation or hydroelectric power. Much effort and ingenuity by biologists and engineers is directed toward maximizing the production of fish for food or recreation from waters that are being used for other purposes. Usually such waters are public, but are capable of being placed under some of the environmental or population controls exercised in private aquaculture. Such a program of management in open waters may be called *extensive aquaculture,* in contrast to intensive aquaculture, in which the animals or plants are confined and controlled completely.

When the body of water has been modified without regard to fish production, the controls that are economically feasible may be very limited. When the water is polluted, the pollutants must be treated or diverted. When the stream is laden with silt from eroding land, the forestry or agricultural practices must be changed. When a dam has been built without a fish transport system, one might be needed. Making the necessary changes in existing practices or structures may be very expensive. It is much more efficient to plan to include fish production among the water uses before structures are built or management plans are fixed.

In ponds or small lakes, the populations of organisms can be controlled to a substantial extent. Unwanted animals can be removed entirely by poisoning, or partly by selective netting or spot poisoning. Unwanted plants can be cut or poisoned. Large populations of stunted fish can be reduced by destroying spawn, either by changing water levels or by poisoning nests. Desirable species can be introduced from hatchery stocks.

Some control can also be exercised over the water quality. Temperature below dams can be changed by varying the depth from which the water is taken from the lake above. Dissolved oxygen concentrations can be increased by reducing organic content or by draining off the layer of water with low oxygen content above a dam or by destratifying the water. The latter can be accomplished by pumping compressed air to the bottom and releasing it through small holes over a large area. Fertility can be controlled by restraining or adding sewage or artificial fertilizers. Shelter for small fish can be provided by leaving uncleared areas as land is flooded, or by adding brush piles and rocks.

Streams, too, can be managed for enhancement of their productivity. Migratory species, especially salmon and eels, must have easy and safe passage. Natural obstructions can be removed, dams can be built with fish ways or other transport devices, and turbine or canal intakes can be screened. Small streams with relatively stable flow can be provided with dams and deflectors that will maintain pools and sheltered areas. When the streams are full of silt, they may benefit from the management of the farms or forests above in ways that will reduce soil erosion.

Even the conditions for animals in the sea can be managed to a limited extent. Shelter can be provided by piles of rocks or wrecked autos, and areas of rubbish dumps can be managed in ways that will attract fish.

9.2 AMENABLE SPECIES

Thousands of species of aquatic animals and plants can be used by people, but relatively few are produced by aquaculture. Several hundred brightly colored species are reared for ornamental purposes; only a few dozen species are reared for food. The principal food species are the carps, especially the true carp and the silver carp; the trouts, especially the rainbow trout; the catfishes, especially the channel catfish; the milkfish of Southeast Asia; the shrimps of eastern Asia; and several species of oysters. Other groups show promise of rapidly increasing production: tilapia, eels, salmon, mollusks other than oysters, freshwater shrimps, and crayfish. Aquatic plants are grown extensively in eastern Asia.

In theory, almost any animal or plant can be grown in captivity. The environment can be modified to accommodate the needs of the species, and the species can be acclimatized or adapted to the environment. But modifying the environment may be very costly, and modifying the species may take many generations. Selecting an appropriate species will greatly economize and shorten the process.

The primary requirement of any food species is that it support a profitable operation. It should reproduce easily in captivity and supply an abundance of young. Its young should be hardy and easy to feed. It must be economical to feed, either because it eats cheap plant materials or because it economically converts expensive protein food into flesh. It must also tolerate crowding in a limited space. It must be either adapted or adaptable to the water available, whether it is fresh or salty, warm or cold, or polluted or clean, because the water cannot readily or economically be changed. The search for amenable species or varieties is, therefore, fundamentally a search for animals to fit available aquatic environments that are not being fully utilized. The possibilities include marshes, irrigation systems, agricultural lands and waters, and waters receiving either organic waste or heat. A possibility is use of drainage from irrigated lands that is too salty for reuse on the land, if it does not contain pesticides.

Aquaculture that supports angling for recreation is less restricted by cost factors and can give much greater weight either to the fighting ability or to the appearance of the fish than to the amount of protein produced for food. New species can be introduced into new environments, in which they can be expected soon to maintain themselves, or in which they must be maintained by continued stocking. Exotic species, such as trout in tropical mountain streams, can be raised so that a great variety of recreational opportunities can be provided. Combinations of species can be placed in isolated water such as ponds or reservoirs so that a natural food chain or varied kinds of angling can be furnished.

Aquaculture for the production of ornamental fish may not be sharply limited by cost factors, because the special challenge of many of the tiny colorful species is to raise and keep them in captivity. Their value is determined in part by their scarcity and the difficulty of raising them. The search for colorful species with unusual social habits has stimulated extensive ichthyological exploration of little-known tropical waters and attempts to culture any colorful species that will live in small aquaria.

9.3 SELECTIVE BREEDING

Farmers discovered long ago that plants and animals tended to resemble their parents and, hence, they chose superior individuals as brood stock. Archeological studies have yielded evidence of gradual improvement in domesticated species that was almost certainly due to selection.

Carp, which have been cultivated for 3000 or 4000 years in eastern and southern Asia, and at least 600 years in Europe, have been selected for such features as small head, high body, thick back, resistance to disease, suitability for various waters, and number of scales. Four fairly distinct morphological varieties exist: the original fully scaled carp, the mirror carp with a few large bright scales, the line carp with rows of scales only near the lateral line and dorsal fin, and the leather carp, which is almost scaleless.

The development of trout breeding and culture during the middle of the nineteenth century was accompanied by much selection of brood stock, accidental and purposeful. Selection was accidental because the brood stock had to survive the rigors of confinement. It tended to become resistant to epidemic disease and tolerant of artificial feeds and feeding practices, but poor at finding food and avoiding predators in the wild.

Controlled and scientific breeding programs in trout cultural stations have sought to develop strains that will resist disease, convert food to flesh efficiently, mature early, reproduce at various seasons, produce large numbers of eggs, and look attractive. Recently, at public trout hatcheries that rear trout for release in natural waters, some attempts have been made to select for good survival in the wild. Such an ability is, however, difficult to accomplish in a hatchery. Ideally those that do survive in the wild should be used as breeders, but obtaining their progeny is rarely possible.

The process of selective breeding involves choosing superior parents for either crossbreeding or inbreeding. The relative merits of each have been subject to much argument. Crossbreeding tends to produce young that, on the average, have a higher growth rate than the young produced through inbreeding. It has been discovered, however, that deleterious characteristics occur occasionally through mutations that tend to be recessive. These characteristics can be eliminated from the population through inbreeding followed by rigorous selection to eliminate the individuals with undesirable characteristics. After several generations, "purebred" lines can be produced, which may then be crossed so that "hybrid vigor," or special combinations of desirable characteristics, can be obtained.

Few fish-breeding programs have developed closely inbred lines. The closest approach to a purebred stock may be a stock of rainbow trout at the University of Washington in Seattle that was inbred and selected by Professor L. R. Donaldson for more than 40 years. These inbred trout tolerated poor water conditions at the hatchery, grew rapidly on a rich diet, matured early, and reproduced effectively.

Fully controlled selective breeding of bivalve mollusks had not been possible until the recent development of methods of inducing spawning and caring for the larvae (see Section 6.6.9). There has been, however, widespread transfer of stocks with desirable characteristics, such as shipment of Japanese spat to western North America, where the spawning of stocks has been limited and uncertain. Also to be noted is the identification and transfer of disease-resistant strains of Atlantic oysters in eastern North America after epidemics decimated certain populations.

9.3.1 Hybridization

A cross of two closely related species is frequently possible and may produce offspring of desirable characteristics. Such crosses occur fairly frequently in fresh water, especially among members of the Cyprinidae, the minnow family. The progeny are usually sterile and occasionally all of one sex. Sterile or monosexual fish may be especially valuable for fish culture in cases of excessive breeding. *Tilapia* breeds so effectively at a length of only about 8 cm that it tends to overpopulate the ponds. Consequently the discovery that a cross of an African male *Tilapia* with a Malayan female would produce only male offspring had immediate promise.

Crosses among the trouts, chars, and salmons have been tried, and numerous hybrids have been produced, a few of which have desirable characteristics. One successful cross in North America was that of two chars, male brook trout and female lake trout. Their progeny, called *splake*, are fertile. They have been distributed widely in eastern Canada and the United States to provide recreational fishing.

9.3.2 Control of Spawning

Selective breeding of aquatic animals and distribution of either eggs or larvae are much easier when humans can intervene in the natural spawning process in ways that allow positive selection of the parents, collection and easy isolation of the young, and procurement of a supply of eggs whenever desired. Such intervention may be the only way in which mating can be accomplished between animal varieties that tend to breed at different times or that will not mate naturally.

One of the great advantages of trout and salmon for aquaculture arises from the ease with which their normal spawning can be circumvented. The maturing fish are held in a pond in which they can be caught easily for determination of their reproductive condition. When they are "ripe," a few days before they would spawn naturally, the eggs and sperm can be pressed (stripped) from the bodies of the females and males into a pan in which the eggs are fertilized (Fig. 9.1). The fertilized eggs can withstand handling and shipping for a day or two before incubation in running water.

Additional control of rainbow trout has been possible through selective breeding for a time of spawning. The wild stocks usually spawn in the spring.

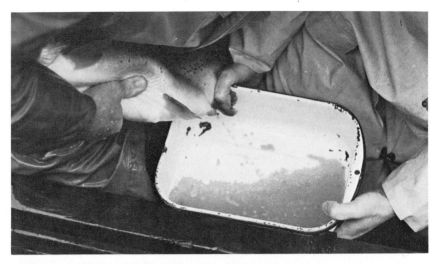

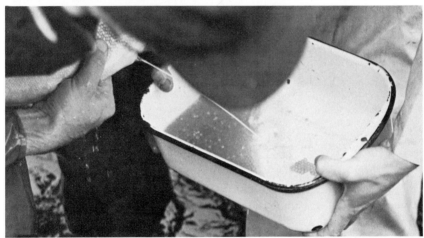

FIGURE 9.1   Stripping eggs from a large female rainbow trout (top) and stripping sperm from a male (bottom). (Photos by J. O. Sneddon, Univ. of Washington, Seattle, Wash.)

Small groups of trout have been induced to spawn earlier by holding them in warmer water, by lengthening their winter days artificially, and by feeding for more rapid growth. A few stocks could be stripped in the autumn, and the process of selection augmented. Now stocks are available that can be stripped in almost any month of the year, and eggs can be supplied almost continuously for hatchery operations.

Usually common carp have been allowed to mate naturally. They have been placed in special spawning ponds, from which the larvae (fry) are collected after spawning and hatching The larvae are transferred to nursery ponds or may even be shipped some distance. The control of the process has formerly been limited to placing selected breeders in proper spawning ponds in which the fry can be easily captured.

A greater degree of control of reproduction has recently been obtained by injecting extracts of fish pituitary glands or other hormones into newly mature carp breeders. After such injections, the breeders will spawn within a few hours under environmental conditions that might otherwise prevent spawning. Such practices have begun to make fry available during longer seasons and have permitted crosses of varieties that would not normally mate. Similar injections have been given to fish of numerous other species so that mating can be stimulated or eggs and sperm can be stripped.

Additional control of fish breeding would be possible if sperm could be stored for use when eggs became available. There have been numerous attempts to keep sperm by placing them in various physiologically neutral solutions and by quick freezing. Some success is apparent, and the establishment of sperm banks as a means of expediting breeding is likely.

The control of molluscan spawning has been especially difficult and important, because mollusks usually spawn during a brief period in summer and obtaining "sets" of larvae of the common commercial species is too uncertain. It has recently been discovered that spawning can be obtained any time of year by conditioning the adults and then stimulating them to spawn. The conditioning is primarily a control of temperature—either raising artificially the temperature of the water in winter to about 20°C for 3 or 4 weeks to advance maturation, or cooling the water in summer to delay maturation. Spawning may be stimulated by a sudden increase in temperature of a few degrees, and by adding gonadal material from ripe individuals of the same species to the water. The development of this technique has been essential for the further development of crossbreeding and inbreeding

9.4 NUTRITION

Almost any kind of confinement for an animal results in a lessening of the animal's ability to find the amount and kind of food it requires. It follows that keeping any kind of animal in confinement requires providing food for at least part of the animal's needs. In intensive culture of animals, the food that must be supplied is frequently the most critical and expensive item in the entire operation. Consequently, supplying the right food at the right time is a common determinant of the success of the culture operation.

Nutrition is the science of defining the interaction of an animal and its foods for the purpose of determining quantitatively the adequate food supply for different functions and stresses. It is a science that states the nutrient requirements in terms of components, such as lipids, amino acids, vitamins, and elements. It also compounds rations, develops feeding standards, and devises feeding methods. Nutrition involves the ways in which each nutrient is used by the body through the process of digestion and metabolism, whether the nutrient is expended, stored, or secreted, and whether it is wasted in the process. Nutrition involves the nutrients and body processes at all stages of life, especially the processes of growth and reproduction.

The feeding of aquatic animals in captivity remains largely an art, except for intensive feeding of fish such as trout, salmon, and channel catfish. Some of the feeding of pond fish, such as milkfish, is done indirectly through fertilization of the pond, and through its management as a producer of live natural food for the fish. Captive mollusks are fed by placing them where they can filter their natural food from the passing water. Recent trials indicate that oysters will eat some starch from corn or wheat. The techniques of breeding mollusks require a supply of living food for the larvae, which usually must be grown in special cultures.

9.4.1 Diets

The early trout and salmon culturists discovered that trout and salmon would grow on a diet of meat if their natural diet of zooplankton, insects, and fish was not available. Because meat was and still is expensive, cheaper diets were sought, both by trying cheaper kinds of meat and by substituting grain products. Livers, lungs, hearts, and some kinds of fresh fish are excellent food for young trout or salmon, but these items are too expensive for most trout farmers. Consequently a search for better and cheaper diets has been pressed, especially at the U.S. Western Fish Nutrition Laboratory at Cook, Washington, and the Cortland Hatchery at Cortland, New York. The latter has operated under a cooperative agreement among the New York Conservation Department, the U.S. Department of the Interior, and Cornell University.

Nutritionists proceed to determine essential nutrients by starting with two or more lots of standard healthy animals and feeding one lot a diet deficient in a single nutrient while simultaneously feeding another lot a standard diet. Faster growth in weight is the usual criterion of superiority, subject to the conditions that diseases do not appear and that the increase in weight is not largely deposited fat instead of muscle. Whole rations and various levels of either foods or nutrients are tested in similar feeding trials with similar performance criteria. Nutritionists may study further the utilization of nutrients by determining their wastage in digestion, places of storage in the body, and rates of egestion or excretion. They will be especially alert for evidence of diseases associated either with deficiencies in diets or with toxic effects. They will make all of these kinds of studies when the fish have reached various ages, especially when the feeding starts, when rapid growth occurs, when reproduction takes place, and under varying climatic conditions.

Fish in general have quite different nutritional requirements than other domestic animals, and the many species of fish differ among themselves. Nutrition scientists have identified many essential diet components, including numerous

TABLE 9.1 Dry Trout Feeds Developed by the U.S. Fish and Wildlife Service (Values Are Percent of Feed by Weight; MP = Minimum Protein)[a]

Ingredient	Starter diet	Fingerling diets		Production diet
Fish meal (MP 60%)	45	34	35	26
Soybean meal, dehulled seeds				
Flour (MP 50%)	15			
Meal (MP 47.5%)		10	20	25
Corn gluten meal (MP 60%)		6		
Wheat middlings, standard	9.35	19.3	13.3	17.3
Yeast, dehydrated brewer's or torula	5	5	5	5
Blood meal (MP 80%)	5			
Whey, dehydrated	5	10	10	10
Fish solubles, condensed (MP 50%)	5			
Fermentation solubles, dehydrated		8	8	8
Alfalfa meal, dehydrated		3	3	3
Soybean oil	10	4		
Fish oil			5	5
Vitamin premix, No. 30[b]	0.4	0.4	0.4	0.4
Choline chloride, 50%	0.2	0.2	0.2	0.2
Mineral mixture[c]	0.05	0.1	0.1	0.1

[a]From Piper et al. (1982).
[b]Vitamins per pound of premix: A, 750,000 IU; B, 50,000 IU; E, 40,000 IU; K, 1250 mg; ascorbic acid, 75 g; biotin, 40 mg; B_{12}, 2.5 mg; folic acid, 1000 mg; niacin, 25 g; pantothenate, 12 g; pyridoxine, 3500 mg; riboflavin, 6 g; thiamine, 4000 mg.
[c]Mineral mixture (g/lb): $ZnSO_4$; $FeSO_4 \cdot 7H_2O$, 22.5; $CuSO_4$, 1.75; $MnSO_4$, 94; KIO_3, 0.38; inert carrier, 251.37.

vitamins, amino acids, minerals, and fatty acids. After determination of the components required in the diet, the available feed materials are analyzed and a diet formula is recommended. Examples from Piper et al. (1982) are given in Tables 9.1 and 9.2.

In addition to the essential components, the diet needs to be prepared in particles that will not disintegrate before the fish eat them and are of a size suited to the size of the fish. The quantity required and frequency of feeding are also important and will vary rapidly as the fish grow or as water temperature changes.

9.4.2 Larval Feeding

A critical time in the life of most aquatic animals is when they start to feed. They must have food of the correct size and shape, and this food must supply the essential nutrients and be nontoxic. It must be abundant enough to be captured easily without an excessive expenditure of energy, but not so abundant that it chokes the feeding animals or produces excessive waste products.

Many young fish that live pelagically in fresh or marine waters start feeding

TABLE 9.2 Recommended Amounts of Vitamins in Fish Feeds
(Values Are Amounts per Pound of Feed and Include Total
Amounts from Ingredients and Vitamin Premixes[a,b]

| Vitamin | Units | Warm-water fish feeds | | Salmonid feeds |
		Supplemental diet	Complete diet	
Vitamin A	IU	1000	2500	1000
Vitamin D_3	IU	100	450	R[c]
Vitamin E[d]	IU	5	23	15
Vitamin K	mg	2.3	4.5	40
Ascorbic acid	mg	23	45	50
Biotin	mg	0	0.05	0.5
B_{12}	mg	0.005	0.01	0.01
Choline	mg	200	250	1500
Folic acid	mg	0	2.3	2.5
Inositol	mg	0	45	200
Niacin	mg	13	45	75
Pantothenic acid	mg	5	50	20
Pyridoxine	mg	5	9	5
Riboflavin	mg	3	9	10
Thiamine	mg	0	9	5

[a]From Piper et al. (1982).
[b]These amounts do not allow for processing or storage losses, but give the total vitamins contributed from all sources. Other amounts may be more appropriate under various conditions.
[c]R = required, amount not determined.
[d]Requirement is affected directly by the amount and type of unsaturated fat fed.

when they are between 3 and 8 mm in length and when they have nearly consumed the food stored in the yolks of their eggs. For example, food eaten by larval sardines, which range in length from 4 to 6 mm, is usually copepod nauplii between 25 and 125 μm in diameter. They are estimated to require an average of 3.5 nauplii/hr at a water temperature of 14°C and twice that number at 19°C.

If such larvae are to be reared in captivity, they must usually be fed living food. Carp larvae of about 5 mm in length eat plankton in their nursery pond, which is managed to produce the plankton. Other captive larvae, such as plaice, shrimp, and many ornamental fish, are commonly fed the nauplii of the brine shrimp *Artemia*. These are easily grown from *Artemia* eggs, and the eggs are commercially available in most parts of the world.

Substitutes for living food for larvae have been extensively sought but are rarely satisfactory. Apparently the best have been the ground bodies of fresh mussels or squid, but the requirements of a nearly continuous supply of nutritious particles of the correct size that will not contaminate the environment are difficult to meet. An added complication is the rapid growth of the larvae; they double in length in a few days and require different and larger foods.

Trout and salmon larvae, which are 20 to 30 mm in length, are much larger than the larvae of most other bony fishes when they start to feed. They accept and grow readily on nonliving food, such as ground meat or ground liver. They do require, however, food particles of the correct size, frequent feeding, and amounts of food equal to amounts consumed so there will be no excess to pollute the water.

Many of the larvae of mollusks start feeding at sizes of only 50–100 μm and have similarly critical food requirements. For example, larvae of the American oyster probably require naked flagellates for their first food but later are able to utilize phytoplankton with cell walls, such as *Chlorella*. The rearing of mollusks is now dependent on the culture of satisfactory living food supplies for all sizes. Such food has been supplied by growing combinations of algal species in fertilized seawater that has been treated with antibiotics for reduction of bacterial growth and with pesticides for reduction of infestations with zooplankton. Both of these chemicals must be used in quantities that will not damage the extremely sensitive molluscan larvae.

9.5 PATHOLOGY AND MEDICINE

The diseases of aquatic plants and animals are diverse and little known. They have been studied mainly from two viewpoints: first, by naturalists who have described the disease organisms found on wild hosts, and second, by pathologists who have described and tried to control the diseases of captive plants and animals. A large proportion of the latter have been concerned with the diseases of trouts and carps, which are, therefore, probably better known than the diseases of any other groups of fish. Fortunately for our understanding, the common diseases of trouts and carps are also common among many other kinds of fish.

9.5.1 Identification of Disease Organisms

Animals and plants commonly harbor many associated organisms, which may be harmless at times, occasionally pathogenic, or regularly pathogenic. It is not easy to determine the causal agent of a specific disease, and usually the identification of a causal agent depends on satisfying the postulates of Robert Koch, an early microbiologist. These were stated in essence as follows:

1. The causal agent must be associated in every case of the disease as it occurs naturally.
2. The agent must be isolated in pure culture.
3. When the host is inoculated with the isolated organisms under favorable conditions with suitable controls, the characteristic symptoms of the disease must develop.
4. The causal agent must be isolated and identified as that which was first isolated.

The pathogenicity of an organism is always subject to a combination of factors. The organism must of course be present, the host must be susceptible, the organism must emerge again to be transferred to a new host, and the organism must survive both in the host and in the environment between hosts. When an organism kills its host without emerging, it too dies. Thus the more successful disease organisms are those that do not seriously damage their hosts and that, by some means, tolerate easily the transfer between hosts.

9.5.2 Kinds of Diseases

Diseases are commonly classified according to the primary causal agent or factors, but it must be kept in mind that an organism weakened by one disease is often subject to others. Consequently, identification of the primary cause may be uncertain, or treatment may be necessary for several diseases. A few of each of the several kinds of diseases are mentioned in the following.

Environmental Aquatic animals in confinement depend to a large degree on suitable water for their good health. They tolerate considerable variations in salinity, pH, temperature, and dissolved gases, but when exposed to less than optimal factors, they become susceptible to disease. Excessively acid water (pH < 5.0) and excessively low temperatures may seriously weaken some kinds of carp. One form of disease, commonly called "popeye" in trout, is caused directly by water that is supersaturated with nitrogen. The gas comes out of solution as a bubble in the eye socket and causes the eye to protrude. Captive animals are also commonly subject to physical injury from nets or sorting equipment, as well as from predators or fighting among themselves. Salmon are occasionally subject to sunburn.

Dietary Diseases that are caused directly by deficiencies in the diet include anemia, due to a varied vitamin deficiency; goiter, due to a deficiency of iodine; blue slime disease of trout, due to a deficiency of biotin; and dietary gill disease, due to a deficiency of pantothenic acid. A disease attributed to overfeeding with poorly balanced diet is lipoid degeneration of the liver, to which rainbow trout are especially susceptible.

Viral Few viral diseases have been positively identified, but many infectious illnesses are thought to be caused by viruses. Their identification is especially difficult because of the problems of isolating and identifying them, and the prevalence of secondary infections by bacteria, which influence the course of the disease. One disease of trout that is caused by a virus is infectious pancreatic necrosis.

Fungus Fungi or water molds of the Saprolegniaceae are common in all fresh water and frequently infect fish of many species. They appear as cottonlike growths on the skin, but in many instances they are secondary invaders of lesions caused by injury or bacteria. They also appear quickly on dead bodies or eggs. Other fungi may invade the internal organs of fish, such as the air bladder and intestine. Still other fungi, especially *Dermocystidium*, are probably the causative agents of the worst-known infectious diseases of oysters. *Dermocystidium* attacks the body of the oyster; other fungi may attack the shell and the attachment of the adductor muscle.

Bacteria   Numerous bacterial diseases of fish have been identified. Many are evidenced by external lesions on gills, body, fins, or caudal peduncle. Furunculosis, which looks like a boil or ulcerlike skin lesions, is caused by the bacterium *Aeromonas*. An ulcer disease of trout is caused by *Hemophilus*, a small rod-shaped bacterium (Fig. 9.2). Bacterial gill disease, evidenced by excess irritation and necrosis of gills, is caused by several species of *Myxobacteria*. Other *Myxobacteria* cause columnaris, a disease marked by yellow open sores on the head and body. Other infectious internal diseases caused by bacteria include kidney disease and infectious dropsy, which is one of the serious diseases of carp.

Internal Parasites   One should expect to find internal parasites in any animal that is examined, whether it appears healthy or not. They occur in bewildering variety; most of them are relatively harmless, but a few cause serious trouble for their hosts. Especially prevalent and rarely harmful are the intestinal worms, such as tapeworms (Cestoda) and the thorny-headed worms (Acanthocephala). More serious when they occur in the musculature are the roundworms (Nematoda).

Most of the internal parasitic worms have a life cycle involving occurrence in a vertebrate host and transfer to an invertebrate host (Fig. 9.3). The invertebrate hosts are commonly planktonic Crustacea. Larval fish commonly feed on the planktonic Crustacea, so infection of young fish occurs very early in life. A secondary invertebrate host is frequently a snail. Occasionally a secondary vertebrate host is a fish-eating bird or another fish.

The parasites enter the host along with the food, but some have free-swimming stages, which enable them to penetrate the skin or gills. Many can also migrate within the host at some stage by penetrating the walls of the intestine or blood vessels.

Other internal parasites of fish include protozoans, such as *Myxosporidia* and *Octomitis*. The latter is a small flagellate that causes infectious disease in

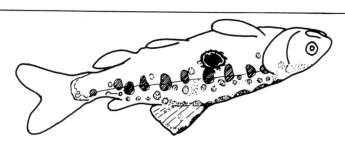

FIGURE 9.2   A small rainbow trout goes "belly up" before death from ulcer disease.

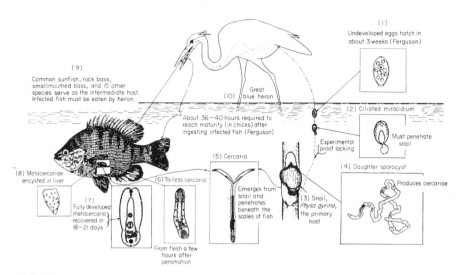

FIGURE 9.3   Diagram of the life cycle of a white liver grub, *Neodiplostomum multicellata*, that infects many species of the family Centrarchidae in North America. [After G. W. Hunter, III (1937), "Parasitism of Fishes in the Lower Hudson Area, A Biological Survey of the Lower Hudson Watershed," pp. 264–273. New York State Conserv. Dept., Albany, N.Y.]

trout and salmon. A haplosporidian, *Minchinia*, has caused drastic mortalities in oyster populations. Also serious at times are *Trypanosoma* species, which live in the blood of fish and are frequently transferred by leeches (Hirudinea).

External Parasites   External parasites, like the internal ones, also appear in bewildering variety and usually cause little harm to the host. The external parasites of fish that may cause serious disease include larval mollusks, parasitic crustaceans, parasitic worms, and protozoans. The larval mollusks are the drifting larvae of certain clams. They attach temporarily to the gills and body of fish and may be harmful if they are very abundant. The crustaceans, frequently called sea lice, can attach themselves to the skin, either with a sucking disk or with the head, which actually penetrates the flesh. These, too, are harmful only when unusually abundant. More likely to be abundant and troublesome are the external trematode worms of the genus *Gyrodactylus*, and other closely related genera. These worms, which are about 1 mm long, attach to gills and fins by means of hooks. They are likely to be especially dangerous to young carp and young trout. Unlike the internal parasites, this animal reproduces hermaphroditically without an intermediate host. The young even have developing eggs or young before birth, and one adult may carry three successive generations within it.

Other troublesome external parasites of fish are the protozoans *Costia* and *Ichthyophthirius* (Fig. 9.4). The latter is widespread among both captive and wild fish and causes serious mortality when it is abundant.

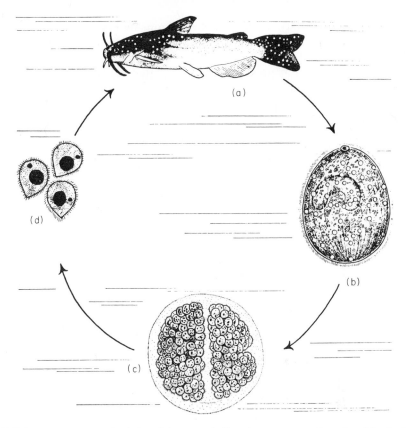

FIGURE 9.4   Life cycle of *Ichthyophthirius multifilus*. (a) Parasites in skin of catfish; (b) free-swimming parasite after leaving fish; (c) encysted parasite that has divided into a large number of young; (d) young parasites that have left the cyst and are searching for a fish host. (After Davis, 1965.)

Very different are the external parasites of oysters. Many plants and animals live on oyster shells with no effect on the oyster, but some invade the shell or body with serious consequences. The boring sponges, clams, and mudworms (Annelida) invade the shells and can seriously weaken the oyster.

9.5.3 Diagnosis and Treatment

A skilled fish farmer will always be familiar with the appearance of healthy fish and watchful of any abnormal behavior or appearance. If any familiar symptoms of illness appear, they will probably indicate a need for action, such as correction of water quality, a dietary deficiency, or some kind of stress. When these methods fail, a pathological examination is necessary. It will start with an autopsy on freshly killed fish and may involve a wide variety of histological, bacteriological, virological, and other techniques of examination.

Treatment may involve correction of environmental problems and isolation of the diseased pond. A variety of therapeutic treatments are available, a few of which are mentioned in Section 11.4.2. But the areas of fish and invertebrate pathology and disease treatment comprise the medical sciences of the fisheries—areas that are far too complex for this introductory text.

REFERENCES

Davis, H. S. (1965). "Culture and Diseases of Game Fishes." Univ. of California Press, Berkeley.

Piper, R. G., McElwain, I. B., Orme, L. E., McCraren, J. P., Fiowler, L. G., and Leonard, J. R. (1982). "Fish Hatchery Management." U.S. Dept. of Interior, Fish and Wildl. Serv., Washington, D.C.

Part III

Application of Sciences

Summary

THE CAPTURE FISHERIES

Commercial fishing is a business that supplies essential protein food and is an
important economic activity in a majority of countries, especially those on
the seacoast.

Recreational fishing is also a widespread and economically important activity
in most developed countries.

Subsistence or small-scale commercial fishing is an occupation of millions of
poor people who live near coastal and inland waters in most countries.

Production from the world's fisheries quadrupled between 1950 and 1970, and
has approached or exceeded the maximum sustainable yield for most of the
conventional wild stocks.

Fish prices have been rising more rapidly than prices of other animal protein
foods, thus stimulating more intensive fishing plus improvements in methods
of preservation and marketing, while also encouraging more social conflicts.

Fishery scientists are increasingly involved in improving fishing methods, providing general information services, and advising on investment strategy.

THE CULTURE FISHERIES

Leveling out of the production from capture fisheries is stimulating rapid increases in aquacultural production worldwide.

One major kind of aquaculture is fish farming of animals in ponds that are fertilized to produce all or a large part of their food. Many of these animals are omnivorous. Such fish farming is primarily for human food and is frequently associated with field crops, poultry, or mammalian production.

A second major kind of aquaculture is pond, raceway, or cage culture in which the animals are fed a complete diet. Many of these animals are carnivorous.

A third major kind of aquaculture is for filter-feeding mollusks in salt water.

A fourth major kind of aquaculture is for attached aquatic plants.

A fifth major kind of aquaculture is for several hundred species of nonfood animals. Many are sold for ornamental purposes; others may be grown to be sold as bait.

Fishery scientists will play an important role in aquacultural development through research on physiology, nutrition, pathology, and genetic modification of organisms on the path to domestication.

FOOD FISHERY PRODUCTS

Fishery products are exceedingly diverse because hundreds of species are caught and prepared in many ways.

About 40% of the world fish catch is used for animal food or industrial products.

The traditional methods of preserving fish for food have been drying, smoking, salting, pickling, and fermenting.

Large and increasing proportions of the catches are now canned or frozen.

A large proportion of fishery products enter international trade.

Fishery scientists will play increasing roles in quality control, product development, market intelligence, and investment strategy.

REGULATION OF FISHING

Fish in the public waters are common property resources.

Governments regulate fishing in the public waters with many objectives that include obtaining an optimal sustainable yield, allocating the fishing rights, promoting orderly fishing, preventing waste, and protecting public health.

A major and controversial problem is allocation of fishing rights among special groups as the resources fail to supply the needs of everyone who wants to fish.

Most fisheries within 200 miles of shore have been allocated to the coastal states under new international fishery laws.

The international fishery laws require scientific data for fishery regulations.

The area within 200 miles of shore, plus the inland fisheries, produces about 99% of the world's catch.

Fishery scientists in this major field of employment will face increasing challenges to solve the mixture of biological, environmental, economic, social, legal, and other problems.

AQUATIC ENVIRONMENTAL MANAGEMENT

This is a rapidly increasing field of employment for fishery scientists.

It involves fishery scientists working with water users, with many diverse sciences, and with the value judgments of many people.

Water has many uses of great social significance, many of which conflict with use by fish.

New environmental laws in many countries are requiring scientific data to support decisions that give much more protection to fish and their environment.

The impacts of environmental change on most waters are scientifically, socially, and politically complex.

Many methods of improving the environment for fish in fresh and coastal waters have been developed and are being used.

FISHERY DEVELOPMENT AND RESTORATION

Special government assistance to most kinds of fisheries is commonplace in many countries.

Further development of subsistence or small-scale commercial fisheries is a desired sociopolitical or economic objective in many countries.

Previous efforts to develop small-scale fisheries in many countries have seldom been successful because of poor administration and unsuitable technology.

Organization of extension systems for fisheries in developing countries is essential for the encouragement of development.

Success requires a broad range of scientific, technical, social, economic, political, financial, and managerial skills.

10 | The Capture Fisheries

"First we eat them, then we sell them, and then we play with them." This was a statement by Clement Tillion, an Alaskan legislator, with respect to the priorities accorded to the capture fisheries of that area. All three uses, which we call subsistence, commercial, and recreational fishing, are described in the earliest written accounts of fishing and persist today in most parts of the world, although the priority of each varies from place to place.

This chapter provides a brief general description of how, what, where, and by whom fish are caught, the overall trends in production, and the emerging role of fishery scientists in fishing. Reserved for later treatment is the role of fishery science in regulation of fishing and fishery development, which will be discussed in Chapters 13 and 15, respectively.

10.1 FISHING GEAR

Fishing must have been important for the survival of many prehistoric people, if we judge by their cave drawings and surviving artifacts. They fashioned their wood, bone, or stone tools into hooks, arrows, and spears. And they probably dug mollusks, herded fish into enclosures, or caught them with simple nets, as did the Egyptians at a later time (Fig. 10.1). Since then, people have made ever more ingenious devices that have been based on better knowl-

FIGURE 10.1 In an Egyptian village near the Nile River, about 2500 B.C., people caught and prepared fish for food while a leader became renowned for his prowess at spearfishing. [Redrawn from P. E. Newberry (1893). "Beni Hasan," Part II. Archeological Survey of Egypt. Kegan, Paul, Trench, Trubner & Co., London.]

edge of the animals or plants and on the materials that have become available. Now there are no animals or plants anywhere in the fresh waters or oceans that cannot be captured.

Most of the types of fishing gear are listed in Table 10.1. Each has evolved to its present forms through many modifications that were made to catch particular kinds of fish in given situations with greater efficiency or, in recreational fisheries, with greater pleasure. The greatest changes have occurred within the last century in the commercial gear, as mechanical power has become available for vessel propulsion and gear handling, and as artificial fibers that resist rot have been used for nets. Now most commercial catches are made with otter trawls or purse seines, but significant quantities are made with trap nets, gill nets, or hooks.

TABLE 10.1   A Classification of Fishing Gear

1. Hand diggers and collectors—rakes, tongs, or shovels for mollusks, crustaceans, or burrowing animals; knives and other tools used by divers to collect mollusks or sponges
2. Trained animals or birds
3. Spears and harpoons—hand-held spear with barb, arrow on a line, harpoon with detachable head for large fish, explosive harpoon for whales
4. Stupefying devices—poisons, explosives, electricity
5. Attracting devices—hook and line with lures or bait, broadcast bait, lights
6. Herding devices—bubble curtains, lines, splashers
7. Stationary entangling nets—gill nets with single wall of mesh, trammel nets with multiwalled mesh
8. Stationary enclosures—pots for eels, crustaceans, and octopi; brush or rock enclosures; small fyke or bag nets for river fish; large corral or trap nets for coastal fish
9. Mobile nets—lift nets (frequently used with a light), falling or cast nets, seines pulled on bottom toward a fixed point, floating purse seines, trawls pulled along bottom or in middepths
10. Towed dredges—rakelike bar fitted with a bag for oysters or scallops; conveyor type assisted by hydraulic jets for clams

Trawls are cone-shaped nets that are towed through the water (Fig. 10.2). The earliest ones were held open at the mouth by a beam and pulled on the bottom by sailing vessels, hence they were called "beam trawls." After fishing vessels were outfitted with mechanical power, it was discovered that the bottom trawls could be held open by a pair of kitelike devices, called "otter boards," and made much larger.

FIGURE 10.2   Diagram of an otter trawl.

FIGURE 10.3 A Soviet stern trawler of the Mayakovsky class. It displaced about 3700 tons, was 85 m long, carried a crew of about 110, pulled a trawl 38 m wide, and froze up to 30 tons of fish daily. Several hundred trawlers of about this size were operating on the major trawling grounds of the world in the 1980s. (Photo courtesy of the U.S. Natl. Marine Fisheries Service.)

More recently, as hydrodynamically stable otter boards have been developed, four of them have been attached to square midwater trawls. Many modifications have ensued, such as long towing warps that herd the fish toward the net, rollers on the bottom lines that permit operation on rocky bottoms, and even echo sounders or television cameras attached to the net in order to position it exactly according to depth and the fish. The largest bottom trawls now have openings about 100 m wide and are operated from large vessels (Fig. 10.3).

Seines are flat nets with floats on top and weights on the bottom. They may be operated in contact with the bottom, as from a beach, in which case they would be launched from a boat, set in a semicircle, and pulled back to the beach with the catch. The other major type, the purse seine, has larger floats that keep the top of the net on the surface (Fig. 10.4). It gets its name from the purse line in the bottom that closes, or purses, the bottom after the net has been set from a boat around a school of fish. Purse seines may be very large, as long as 1000 m and as deep as 200 m.

Nets of such a size were not practical until rot-resistant artificial fibers became available. Nets must be carried in piles on vessels and cannot be cleaned of fish slime or spread out to dry for weeks at a time, and under such conditions the use of cotton or linen nets, even when treated by preservatives, was never completely satisfactory. Now the artificial fibers are not only rot-resistant but stronger as well, and nets using their finer threads will sink more rapidly and purse more easily.

Gill nets are long, flat nets with very fine threads that entangle fish that encounter them. They have floats on one edge and weights on the other, and

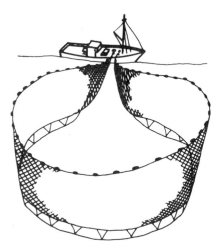

FIGURE 10.4   Diagram of a purse seine.

can be constructed to float upward from the bottom or downward from the surface according to the relative size of floats and weights. Such nets are usually constructed of monofilament nylon, which is both very strong and nearly invisible in the water. They are used extensively by small-boat fishermen in sheltered waters because they can be hauled by hand or by simple hauling machinery.

Hook and line was one of the earliest methods of fishing, and it remains the most common today. The fish are enticed to bite the hook by edible bait or by one of an endless variety of artificial attractants.

Commercial practices include holding the lines in the hand while fishing for bottom fish or attaching them to a moving boat while trolling for fish closer to the surface. Sometimes the lines are attached to rods, as in tuna and mackerel fisheries in which a school of fish is concentrated near the boat by live bait thrown overboard. Crewmen then angle with stout rods, short lines, and artificial lures in the milling school of fish.

But the most advanced commercial method is an adaptation of longline or setline fishing. Hooks on short branch lines are attached at intervals to a longline, which is divided into sections of a size that is convenient to handle (Fig. 10.5). The hooks are baited and sections are set from a moving boat, usually once each day. After "soaking" for a few hours the line is retrieved with whatever has been caught. Modern longline fishing has developed especially to catch large, widely scattered surface fish that are difficult to encircle with nets or fish on bottoms that are too rough for netting.

The longline method can be designed to capture many different kinds of fish in different environments. One fisherman may fish about 100 m of line with a few dozen hooks that can be set on the bottom in fresh water or a coastal lagoon. At the other extreme is the high-seas fishery for tuna, shark, and

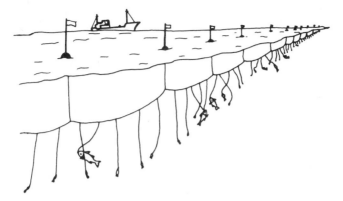

FIGURE 10.5   Diagram of a longline, one of the kinds of hook gear.

marlin, in which a vessel with about twenty crewmen may set and retrieve 50 km of line each day. Such gear is suspended from surface floats and has branch lines about every 30 m. The line is retrieved by special machinery at a speed close to the cruising speed of the vessel. Another major longline fishery is for halibut in the North Pacific. Such lines are set on the bottom between anchors fitted with buoy lines.

The most widespread use of hooks is for recreational angling because there are so many recreational fishermen. They may bait the hooks, but seem to have an inexhaustible attraction to artificial lures that they offer to the fish with different lines and rods and reels in ways that increase the fun but do not necessarily catch more fish.

As for the other kinds of fishing gear, they seem to be limited only by our knowledge of the animals and the devices that might be used to entice them, disable them, entangle them, surround them, or dig them.

10.2 STATISTICAL PERSPECTIVES ON THE CAPTURE FISHERIES

Fishing is an exceedingly diverse activity practiced by millions of people in all parts of the world. Some of the major features can be extracted from the statistical records that are now kept by most nations. In this volume, statistics for 1990 or later will be used whenever possible.

Systematic collection of statistics on world fish catches began after the formation of the Food and Agriculture Organization (FAO) of the United Nations during the 1940s. Its fishery section published its first volume for the year 1947 and attempted to compile complete statistics for one reference year prior to World War II, 1938. FAO relies on national statistics on catches, landings, and fishery commodities and it has played a major role in the standardization of the international statistical system.

The recent "Yearbooks of Fishery Statistics" appear in pairs; one volume on catches and landings and one on fishery commodities. Each of the books itemizes production from subsistence and commercial fisheries, including aquaculture. Neither includes recreational or ornamental fish catches, so these must be obtained from each country that attempts to compile them.

Similar and, in some cases, more complete statistics are kept by many nations. The annual volumes of the "Current Fishery Statistics" of the United States and the "Annual Statistical Review" of Canadian fisheries will also be used here.

10.2.1 Major Groups of Resources

One of the unique characteristics of fisheries is the diversity of resources that are used and the products that are produced. For example, in 1978 FAO listed about 870 taxonomic categories of organisms from 214 families that are harvested somewhere in the world. Most of these are, of course, fish for human consumption, but also included are 19 families of crustaceans, 24 of mollusks, 8 of mammals, 12 of other animals, 4 of aquatic plants, plus 8 others that produce nonedible products. Many other species of aquatic animals and plants are also caught and sold for recreational or ornamental purposes, but there is no worldwide record of these.

The nominal catches for the year 1992 are given in Table 10.2. These are the weights as if the animals and plants were fresh and whole, which they usually are when landed for their first sale. The actual edible proportion varies greatly, of course, according to the species, size, and condition, but some approximate conversions are as follows: 7% for oysters, 50% for shrimp, 25 to 40% for most fish, and perhaps 60% for a few especially muscular fish such as tuna.

It may be noted that the leading group of pelagic species—herrings, sardines, anchovies, etc.—and the leading group of bottom fish—cods, hakes, haddocks, etc.—together made up more than one-third of the world's production, whereas about one-fourth was made up of the hundreds of miscellaneous species. In addition to the animals, the aquatic plant harvest totaled 3,317,000 MT.

United States fisheries production in 1992 (Table 10.3) totaled 9,367,000 MT valued at $3,678,000,000. (The tonnage differs from the FAO total for the United States given later because the FAO includes shell weight of mollusks and fish produced by aquaculture.) The five major species groups comprised about two-thirds of both total value and total tonnage of U.S. landings.

Canadian production in 1992 totaled 1,324,000 MT of which 46,000 came from fresh water. The total was valued at C$1,471,000,000 to the fishermen and $1,805,000,000 after processing (Table 10.4).

10.2.2 Major Products

The total aquatic animal production reported by FAO is divided roughly into about 70% for human food and about 30% for animal food or industrial

TABLE 10.2 Nominal World Catches of Fish, Crustaceans, and Mollusks in 1992[a]

Group	10³ MT	Percentage
Inland waters	15,579	15.9+[b]
Marine areas	82,534	84.1
Total world	98,113	100.0
Herrings, sardines, anchovies, etc.	20,369	28.1+
Cods, hakes, haddocks, etc.	10,542	12.3
Jacks, mullets, sauries, etc.	10,486	10.7
Carps, barbels, and other cyprinids	5852	5.7
Redfishes, basses, congers, etc.	7037	7.2+
Tunas, bonitos, billfishes, etc.	4365	4.5+
Mackerels, snooks, cutlass fish, etc.	3309	3.4
Shrimps, prawns, etc.	2912	3.0+
Squids, cuttlefishes, octopi, etc.	2758	2.8+
Clams, cockles, arkshells, etc.	1447	1.5+
Oysters	1696	1.7
Salmons, trouts, smelts, etc.[c]	1607	1.6
Tilapias and other cichlids	940	1.0+
All others[d]	24,792	27.7+

[a]From FAO (1994). "Yearbook of Fishery Statistics," Vol. 74. FAO, Rome.
[b]Indicates percentage increase since 1980.
[c]Includes a large proportion from aquaculture.
[d]Total does not include seaweeds, other algae, marine mammals, or recreational catches.

uses. The human food component (Table 10.5) comes from the 70% that is marketed fresh or processed into frozen, cured, or canned products. The part that is processed is mostly dressed, filleted, deboned, or otherwise separated into human food products and waste. The different products number many hundreds.

TABLE 10.3 U.S. Commercial Landings, 1992[a]

Species group	10³ MT	$ × 10⁶
Salmon	325	583
Shrimp	153	490
Crabs	283	471
Tuna	26	90
Menhaden	755	83
All others	4371	1961
Total	9367	3678

[a]From U.S. Dept Commerce (1993).

TABLE 10.4   Canadian Commercial
Landings, 1992[a]

Species group	10^3 MT	$\$ \times 10^6$
Cod	198	158
Herring	250	79
Salmon	64	193
Scallop	83	87
Lobster	42	317
Crab	41	73
All freshwater	39	70
All others	641	382
Total	1324	1471

[a]From Canadian Fisheries Statistical Highlights
(1992).

The waste from the larger fish-processing plants of the world is further processed into pet food or added to the supply of fish (30%) that is caught purposely for the so-called industrial uses. This waste and the industrial fish component are usually separated first into oil, water with dissolved solids, and solid materials. The oil is used for margarine, paints, inks, soaps, and many other industrial products. The solid materials are dried into fish meal, a high-quality protein supplement that is added to feedstuffs prepared for chickens,

TABLE 10.5   Disposition of World Catch, 1992[a]

	10^3 MT	Percentage
Total[b]	98,113	100.0
Human consumption	71,401	72.8
Marketing fresh	26,428	22.5
Freezing	23,689	24.7
Curing	9151	11.4
Canning	12,161	12.9
Other purposes	26,704	22.1
Reduction[c]	24,904	21.0
Miscellaneous	1800	1.5

[a]Source: FAO (1992). "Yearbook of Fishery Statistics," Vol.
73. FAO, Rome.
[b]Exclusive of marine mammals, aquatic plants, and aquatic
animals used for recreational purposes.
[c]Includes only whole fish, not waste from fish destined primarily for other purposes.

other farm animals, or fish. The water with dissolved solids, called "stickwater," is condensed and used for animal feeds.

The "miscellaneous" production, roughly estimated at 1,000,000 MT, includes sponges, shell products, corals, aquatic bird guano, and species of the live fish that are suitable for ornamental purposes. Some of these items, plus leather products from fish and reptiles, are manufactured into jewelry, apparel, buttons, fertilizer, and other industrial products that are eventually valued at several billion dollars annually.

The aquatic plant production (Table 10.6) is divided between a large fraction that is consumed directly by humans or animals and another large fraction that is processed to obtain colloids that are used extensively for thickening agents in foods, cosmetics, and paints.

The value of U.S. commercial fishery products was substantial in 1980. The major item was fresh and frozen fish valued at $2.11 billion, followed by canned fish at $1.81 billion and industrial products at $0.42 billion. The canned fish was predominantly tuna, $1.41 billion, followed by salmon at $0.40 billion.

A wholly different kind of product is recreation. No worldwide statistics are available, but fishery-based recreation in the United States is a major part of outdoor recreation. The U.S. Department of Interior estimated that 42 million fishermen fished about 500 million days in 1991 and spent about $24 billion for their fun (Table 10.7). About 84% of their fishing was in fresh water, the balance in salt. Other estimates suggest that recreational anglers contributed about 1.6 kg edible weight of fish per capita in addition to the officially reported commercial supply of about 6 kg.

Recreational fishing is valued not only for the fish caught but also for the quality of the experience. The judgment of quality varies greatly according to the preferences of individuals, ranging from the solitude, the companionship, the competition, or the excitement to the conviviality of evenings in the pub. It is popular enough to attract participation by about 20% of the population over 12 years old and induce the average participant to spend about 24 days and $650 annually. The total expenditures of $24 billion, which include travel, transportation, equipment, and fees, dwarf the total sales of less than $4 billion by the commercial fisheries, even though the number of days fishing has declined to about half the number in the late 1970s.

10.2.3 International Trade

About 40% of the seven major fishery commodity groups moves internationally (Table 10.8) and, in 1990, the exports were valued at nearly US$36 billion. The major commodities were frozen fish, of which nearly half was traded, and frozen crustaceans, of which about 80% was traded. About half of the animal foodstuffs and nearly two-thirds of the oils and fats were also traded. Much of this trade was between the lesser-developed producing countries and the developed countries and consisted of items that met international packaging

TABLE 10.6   Animal and Plant Production of Major Fishing Nations[a]

	Animals[b] (10³ MT)		Plants (10³ MT)	
	1980	1992	1980	1992
China	4240	15,007	1576	3553
Japan	10,410	8460	695	788
Peru	2731	6875		130
U.S.A.	3635	6843	163	84
Chile	2817	6502	75	127
USSR	9412	5611	20	30
India	2423	4175		
Indonesia	3080	3358	6	188
Korea, Rep.	2091	2696	317	604
Thailand	1650	2655	1	
Norway	2398	2549	106	189
Philippines	1557	2272	116	357
Korea, P.R.Dep.	1400	1750		
Spain	1240	1330	7	8
Canada	1305	1251	28	23
Mexico	1240	1248	35	62
All others	15,404	19,348	172	115
World total	71,040	98,113	3317	6179

[a]From FAO, "Yearbook of Fishery Statistics" (1980), Vol. 50; (1992), Vol. 72. FAO, Rome.
[b]Excludes marine mammal and recreational catches.

TABLE 10.7   Trends in U.S. Recreational Fishing[a]

Year	Total no. of fishermen (10⁶)	Total expenditures (US$ × 10⁹)	Total no. days fishing (10⁶)
1955	21	1.9	397
1960	25	2.7	466
1965	28	2.9	523
1970	33	5.0	706
1975	41	11.8	1058
1980	42	18.1	1002
1985	46	28.1	977
1991	36	24.0	511

[a]Source: "National Survey of Fishing, Hunting, and Wildlife-Associated Recreation." U.S. Dept. of the Interior, Washington, D.C.

and quality standards. Much smaller proportions of the traditional preparations were traded. These included the dried, salted, smoked, wet-salted, and pickled fish, which have long been prepared for local or national consumers.

The United States is a major trader. It imported products valued at $3.6 billion in 1980. The largest single item was $939,000,000 worth of nonedible fishery products, much of which was jewelry and apparel that included aquatic products such as pearls, corals, leathers, or shells. This item (in 1991) also included ornamental fish valued at about $30,000,000. The edible imports were valued at $5.7 billion, of which frozen shrimp were $2.2 billion, frozen fillets, blocks, and slabs were $704 million, frozen tuna (for canning) were $429 million, and lobsters were $270 million. The principal exporters to the United States were Peru, Canada, and Chile. Much of the rest was spread among the several dozen countries that produce frozen shrimp.

The United States also exported fish products valued about $6.9 billion in 1993. More than half of this was salmon, either canned, frozen, or as roe, and the major buying country was Japan. Canada has recently exported fishery products valued at about $1.3 billion annually, half of them to the United States. It imports products valued at about $400 million, about half of which are shellfish.

10.2.4 Major Fishing Nations

The major fishing nations are those that are fortunate in being located close to good fishing grounds plus those that, lacking good nearby grounds, have built distant-water fishing fleets that can fish off the coasts of other nations. The two

TABLE 10.8 World Production and Exports of Seven Fishery Commodity Groups, 1992[a,b]

Commodity group	Production (10³ MT)	Exports (10³ MT)
Fish—fresh, chilled, or frozen	14,523	5848
Crustaceans and mollusks—fresh, chilled, or frozen	2492	1989
Fish products and preparations—canned, wet-salted, or pickled[c]	5390	1334
Fish—dried, salted, or smoked	3604	510
Meals and solubles—animal feeding stuff of aquatic animal origin	6002	3297
Crustacean and molluscan products—canned, wet-salted, or pickled[c]	631	37
Oils and fats—crude or refined, of aquatic animal origin	1024	540

[a]Source: FAO (1992). "Yearbook of Fishery Statistics," Vol. 75. FAO, Rome.
[b]Value of exports equals US $38,657,000.
[c]Preparations may or may not be packaged in airtight containers.

leading nations, China and Russia, have both productive nearby grounds and large distant-water fleets (Table 10.6). China also depends on its intensive aquaculture (which is included in the FAO statistics). Other nations among the top 20 with large distant-water fleets include the United States, Norway, Republic of Korea, and Spain. These 20 nations produce about 80% of the world's aquatic food supply.

10.2.5 Fishing Grounds

Fish are caught in practically all of the inland waters of the world but in relatively limited parts of the oceans (Gulland, 1971). The bottom fish are available on the continental shelves or the upper part of the continental slope out to depths of about 500 m. The pelagic fish are available in the enriched areas, predominantly in upwelling zones in the temperate latitudes off the west coasts of the continents, close to the equatorial upwelling in the open oceans, in high-latitude mixing zones, and off the mouths of the great rivers of the world.

About two-thirds of the total comes from four limited areas of the oceans, the North Pacific rim between southern China and central United States, the North Atlantic rim between central United States and Spain, the west coast of South America off Peru and Chile, and the west coast of southern Africa from Angola to South Africa. The inland waters produce about 10% of the fish, a substantial and increasing part of which comes from aquaculture. Another eighth comes from archipelago areas of the southwestern Pacific and eastern Indian oceans. The rest of the oceans, which includes the majority of the tropical waters, produces only about one-eighth of the world total. Another statistic that illustrates the nearshore distribution of fish is that about 99% of the fisheries are within 200 miles of shore, a fact with major socioeconomic implications (see Chapter 13).

The United States is especially fortunate to control the fishing on rich grounds off New England and the Middle Atlantic region, and in the Gulf of Mexico, the Gulf of Alaska, and the eastern Bering Sea. Its adjacent areas of continental slope and upper continental shelf comprise about 20% of such areas in the Northern Hemisphere and equal about 30% of the land area of the United States. The annual per capita consumption in the United States increased from about 12 pounds in the early 1970s to about 13 pounds in the early 1990s.

Canada also controls large productive fishing grounds off Nova Scotia and Newfoundland in the northwest Atlantic and smaller grounds off British Columbia in the eastern Pacific.

10.2.6 Participation in Fishing

Any record of participation in fishing must include not only the fishermen, subsistence, recreational, and commercial, but all others who are involved in the processing and distribution of commercial products, and those who gain employment from the sport fishery. Worldwide statistics are not available, but in the United States commercial fisheries in the early 1990s, about 193,000 fishermen and 103,000 (average for season) shore workers were employed in processing and wholesaling. The fishermen operated about 18,900 fishing vessels of 5 net tons or more, and about 94,000 smaller boats.

The great majority of fishermen in the United States fish for fun. They numbered about 40 to 45 million in the late 1980s, or more than 200 for every commercial fisherman. It is a sport for both sexes, young and old (Fig. 10.6), although a little more than two-thirds of the anglers are men and a higher proportion of people less than 50 years old go fishing. Fishing is more popular among residents of small towns and rural areas, about 30% of whom fish, than among big-city residents, whose participation is about 14%. Fishing attracts people at all income levels.

10.3 TRENDS IN PRODUCTION AND USE

The first major step in the transition from subsistence fishing in the ocean to commercial fishing began with the use of sailing craft, navigation, and charts.

FIGURE 10.6 Angling in the United States is popular with men and women, young and old. (Photo courtesy of Washington Dept. of Game.)

These enabled people to fish much farther from home, to carry and use larger fishing gear, and to bring back their catches if they could preserve them. They eventually sailed thousands of kilometers to catch whales for the oil and cod for salting. As they did so they began to operate out of seaports instead of fishing villages and many of them became full-time fishermen. Such operations required substantial investments and expansion of markets beyond the fishermen's neighbors.

When motor-powered vessels became available during the latter part of the nineteenth century, another big step was possible. These vessels could tow large bottom trawls and return to port on a schedule with their catches temporarily preserved in ice. Here the fish could be sorted for shipment fresh, or preserved for longer periods by freezing, canning, salting, or other means. Since then, many technologies have been adapted to fishing. Now the most sophisticated vessels weigh perhaps 3000 tons and carry a crew of more than 100 who catch, process, package, and preserve their fish at sea. Since World War II the technological changes in vessels, equipment, and fishing methods have been almost revolutionary. The changes have led to major effects on the resources, the methods of conserving and allocating them, the entire international trade in fishery products, and the consumers.

10.3.1 World Catches

The catch data compiled by FAO, when considered along with the previous history of fishing and with the conservation and allocation measures of more recent decades (see Chapter 13), provide an extraordinary account of resource development and social change, while suggesting the likelihood of greater changes to come.

The first and most significant development was the trebling of world catches in two decades, the 1950s and 1960s, and the subsequent leveling off of the rate of increase. The circumstances surrounding this and the inferences to be drawn from it include the following:

1. About 90% of the catches came from the sea throughout these decades, with 10% from inland waters. Out of the proportion from the sea, about 99% was caught within 200 miles of shore—the zone where fishing now has been allocated to coastal nations.

2. The leveling off of the production from the sea was accompanied by collapse of production from several major stocks and a decline in the rate of discovery of new stocks of conventional species. The fishing technology enabled fishing on virtually all of the world's fishing grounds down to the depths that are likely to produce significant catches. It now appears unlikely that production of conventional species can be expanded significantly.

3. This situation, which developed within three decades after the millennia of fishing the sea, seems certain to foreshadow a worldwide change in our concepts of using the living resources of the sea. Major legal and social changes have already begun.

4. The annual rate of increase in catches during these two decades was about 6%, a rate about double the rate of increase in agricultural production from the land. This had led to hopes that the sea could feed more of the world's people—hopes that are now unlikely to be fulfilled.

5. Large nonconventional stocks are known, such as Antarctic krill, squids, and small mesopelagic species in the enriched zones of the oceans. Most of these are small animals and major problems of capture, processing, and marketing must be solved before these fisheries develop. Therefore their use seems likely to require capital-intensive, sophisticated organizations that will develop international markets. Small-scale fishing will probably be little involved.

6. The proportion of 10% from inland waters remained about the same during the 1950s and 1960s, which meant that the catches there were increasing at a rate of about 6% annually. But most of the inland waters had been for millennia within reach of subsistence fishermen who probably were using them to their limits. Further, an increasing proportion of the inland production was probably from aquaculture rather than capture fisheries.

The second development is the different trend in total catches of the developed countries of Europe and North America in comparison to the trends in the former Soviet Bloc countries and the developing countries of Asia, Africa, and South America.

1. The continents with mostly developing countries showed the most rapid growth in fish catches during the 1950s and 1960s and a definite leveling off or decline during the 1970s. The increase was due partly to major efforts by a few countries to build long-range fishing fleets, including Japan, Korea, Ghana, and the Soviet Bloc countries of the former Soviet Union, Poland, Romania, East Germany, and Cuba. By 1980, the Soviet Bloc countries had nearly 60% of the tonnage of the large long-range fishing vessels of the world. In addition, all coastal countries located close to good fishing, especially Indonesia, Thailand, Peru, and Chile, increased their efforts on nearby grounds.

2. The declines were due partly to collapse of some major pelagic fish stocks such as the anchovy off Peru and Chile and the northeast Atlantic herring, plus major declines in some regional bottom fisheries, such as the one off Northwest Africa.

3. In contrast, the catches of the developed countries in Europe and North America had a less rapid rate of increase during the 1950s and 1960s, and the North American catches during the 1970s were still increasing slowly.

Third, there has been a major shift in the marketing and consumption of fishery products that can be inferred from the trends in processing.

1. The quantities marketed fresh, which is the principal way that fish reach the poorer peoples in the developing countries, have increased little since the middle 1960s.

2. The quantities cured (dried, smoked, salted, pickled, etc.), which is the

other major form in which fish is consumed by poorer peoples, have increased at a rate of only about 2% annually since 1950.

3. In contrast, the quantities reduced to oil and meal increased at a rate of nearly 14% annually from 1950 to 1970, then declined slightly, due primarily to the collapse of the anchovy fishery off Peru and Chile. Nevertheless, during the 1970s and 1980s, nearly one-third of the world's fish production was reduced to oil.

4. The quantities of fish frozen have increased rapidly since 1950. The rate of increase was about 12% annually during the 1950s and 1960s, a little less during the 1970s to 1990s. This is the major form in which edible fish enter international trade.

Fourth, some significant socioeconomic consequences that have already begun probably will become more evident.

1. The leveling off of total fish production should, by itself, lead to higher overall average fish prices relative to other foods.

2. The increasing proportion of fish in international trade will probably also tend to increase average fish prices in the producing countries. The lesser developed of these countries will probably enter international trade by agreeing to joint ventures that supply the managerial and marketing expertise.

3. The increasing proportion of processed fish and higher prices will be beneficial in terms of product quality and less spoilage.

4. Higher prices will probably lead to upgrading from animal to human food of part of the fish that is now reduced.

5. The competition among fishermen and nations for the fish resources will probably intensify and make more urgent the development of fishery conservation programs.

10.3.2 Recreational Fishing

Recreational fishing has been increasing rapidly during recent decades in most countries of the world. At a conference that considered reviews of recreational fishing in 16 countries (12 from Europe), the following conclusions were reached (Grover, 1982):

> The size and basic social and economic dimensions of the recreational fisheries are no longer open to serious challenge: the recreational fishery is as big, if not bigger, than the commercial fishery in three of the largest and most industrialized countries represented at the Consultation. The country review papers revealed that the sport fishery is the dominant if not the only significant fishery in the inland waters of most of the countries represented at the Consultation. In conjunction with this it was recognized that the recreational fisheries are comparably important in many other of the more advanced countries which could not provide for one reason or another similar holistic data on their fisheries.

Surveys have been made of United States fishing, hunting, and wildlife-associated activity since 1955. The overall fishing experience is shown in Table

10.7. The total number of anglers is estimated to have doubled between 1955 and 1980 to 42 million, the days fished increased by a factor of 2.5 to about 1 billion, and total expenditures increased by a factor of 9 (although much of this was due to inflation) to about $18 billion. The proportion of fishermen in the population increased from about 14 to 20%, but the trend seems to have leveled out since the late 1970s and may be declining.

Recreational fishing is also important and increasing in Canada. A 1975 survey (reported in Grover, 1982) showed that the gross value of the expenditures by 6.1 million anglers exceeded $1 billion and their catches accounted for about 8% of Canada's total fish supply. And because Canada exported much of its commercial fish, recreational fishing accounted for about 44% of its domestically consumed fish.

The burgeoning recreational fisheries in most countries are generating reconsideration of government policies in order to strengthen the activity and collect much better information to guide decision making. Even in lesser developed countries, where fishing is predominantly subsistence, any potential recreational fishing is recognized as a possible tourist attraction.

10.4 FISHING SYSTEMS

A fishing system will be considered here to include the businesses and government agencies directly and indirectly involved in transporting the fishermen to the fish, in their actual fishing, and in their getting the catches to the consumer. After describing the large range of production levels, these activities will be discussed for each of the three main fishing systems, although the latter activity for commercial fishing may involve complex preservation, processing, and marketing, which will be considered in more detail in Chapter 12.

10.4.1 Production Levels

Fishing practices vary dramatically in catch rates, and many of the technical, social, and economic aspects of fishing can be better understood after comparing the general levels of catches by different kinds of fishermen. A comparison on the basis of tons/person/year of round weight is made in Table 10.9. This table is not based on averages (which are seldom available), but many fishing activities do in fact produce at the levels indicated.

This logarithmic progression in the catch/person/year illustrates two major differences among the fishermen. First is the very small catch per person by recreational fishermen relative to subsistence and commercial fishermen. Second is the great increase in catch rates of commercial fishermen as a result of the technological gains that have been achieved with vessels that can operate large trawls or purse seines and carry large volumes of fish. The catch rates can also be used roughly to estimate average annual earning levels of fishermen with the assumptions that low-value fish sells for less than US$200/ton, high-value fish

TABLE 10.9   General Production Levels of Various Fishing Practices

Catch (tons/person/year)	Fishing method
0.01	Occasional recreational angling
0.03	Average recreational catch
0.1	Regular recreational angling
0.3	Seasonal subsistence fishing with lines, traps, spears, or nets
1	Subsistence fisheries from manually operated boats
3	Subsistence or primitive commercial fisheries from small power-boats
10	Fishing from small coastal vessels for highly valuable fish
30	Fishing from medium to large vessels for moderately to highly valuable fish
100	Fishing from the best modern trawlers and purse seiners for moderately valuable fish
300	Fishing from purse seiners for low-value fish such as anchovies, pilchards, menhaden, or jack mackerel
1000	Fishing from the most efficient modern purse seiners for low-value fish when abundant

for $2000 to $4000/ton (at 1990 prices), and about half of the gross earnings from sales are used for the gear and vessel. The range is from a few hundred dollars for subsistence fishermen to $50,000 or more for a few fishermen with the best modern equipment.

10.4.2 Subsistence Fishing Systems

The ancient methods of subsistence fishing continue with little change among several million fishermen today. These are fishermen who live in small villages near rivers, lakes, or oceans and who fish for their own use and for barter or sale within nearby villages. Many of them farm as well as fish. They may fish individually with spears, lines, pots, or simple nets. They may also fish in teams that can handle beach seines or strings of gill nets. Most of their fishing craft lack power. Most of their catches are consumed fresh, but some may be preserved by salting or drying.

These fishing activities must be similar to those of the early Sumerians along the Persian Gulf and the rivers Euphrates and Tigris (Kreuzer, 1978). Here, before 2000 B.C. they fished with nets, traps, spears, and lines. They operated from shore or from small sailing vessels. They traded their fish for barley and wool and paid their taxes with them. Most of their catches were delivered fresh, but some were preserved by drying, salting, smoking, fermenting, or submerging in oil.

Similar practices by most of the world's fishermen probably prevailed into the nineteenth century. A majority of peoples almost everywhere were farmers, and farmers commonly fished nearby waters. Certainly they did in North America, where virtually all rivers and lakes were fished with nets until the fishing for all the more desirable species was allocated to recreational fishermen late in the nineteenth or early in the twentieth century.

There is, of course, no sharp distinction between subsistence fishing and simple commercial fishing. Both use similar methods, but the distinction becomes unmistakable as commercial fishermen work full time with costly gear and vessels and sell their catches to buyers who process and distribute them in distant markets. Nor is there a sharp distinction between subsistence and recreational fishing. Some anglers who fish primarily for fun also catch fish that are an important food for themselves and their neighbors.

10.4.3 Commercial Fishing Systems

Modern commercial fishing systems are groups of businesses that provide livelihoods to fishermen, vessel builders, gear builders, engine builders, supply and repair shops, fish buyers, processors, distributors, and retailers (Shapiro, 1971). The simplest are those that involve small motor vessels that deliver fish a few times each week to a buyer who may ice them and truck them perhaps 100 km to a retailer. The most complex are those that sustain fleet operations thousands of kilometers from home port, that process, package, and preserve the fish on board, and that deliver them to an international marketing system in lots of several thousand tons each. In between is a great variety of systems that have developed around the specific challenges of catching fish from each stock and supplying each group of consumers. Also included in the systems are the government services to the fishing industry and the government controls on the use of public resources, both to be discussed in later chapters.

No comprehensive description of the many unique systems is possible in this book, but some of the trends in their development and their modern role deserve special emphasis. Development trends since 1950 include:

1. Radical technological improvement in fishing vessels, especially those of larger size with more horsepower, better electronics for navigation and finding fish, more use of mechanical refrigeration and labor-saving equipment, more attention to comfort and safety, and design for multipurpose operations.

2. Enlargement of fleet operations to permit exchange of information on fish abundance and more efficient catcher operations involving support by processing and supply vessels.

3. Increasing bycatches and waste of fish. Recent estimates indicate that the global discards are probably in the range of about 20 to 40% of the landed catch. The amounts are small in the fisheries targeting with purse seines on off-bottom schooling species such as members of the herring family, but large in bottom trawl fisheries, especially those trawling for shrimp (Alverson *et al.*, 1994).

4. Continuing improvement and adoption of better preservation systems, especially freezing. The best-quality frozen fish begins the freezing process on board the vessel within a few hours of capture.

5. Sale of a larger proportion of the processed fish in convenient packages for household or restaurant use. Distribution by air freight is growing.

6. Increased capitalization and affiliation of fish companies with large food companies.

7. Increased international trade. World imports were about US$45 billion in 1992.

The role of commercial fishing varies in importance with abundance of fish stocks and the contribution of the catches to the economy, but the following circumstances are significant:

1. Many coastal countries such as Norway, Iceland, Peru, and Chile obtain a large fraction of national income from commercial fishing. A total of about 20 countries each harvests more than 1,000,000 tons annually. Their economies can be significantly affected by changes in their fisheries.

2. Fish is an important item in the diet of people in developing countries, where it is estimated that about 60% of the people derive more than 30% of their animal protein from fish.

3. Many of these countries are choosing to upgrade their fish products to increase foreign exchange. This action reduces spoilage of fish but increases prices and social stresses through competition between subsistence and commercial fishermen, thus placing increasing pressure on the resources.

10.4.4 Recreational Fishing Systems

Any recreational system delivers relaxation and enjoyment and the fisheries are no exception. The fish are essential, but fishing itself is not the single goal. People value the fishing experience for a whole complex of personal reasons, which include exercising a common right, being on or near the water, getting away from everyday routines, enjoying the surroundings, the feeling of freedom, companionship, closeness to nature, and travel. Their experience may be degraded by any of another complex of reasons, these usually including the presence of pollution, poaching or unsportsmanlike conduct, crowds, noise, and, of course, lack of fish.

A few wealthy anglers can obtain the experience by owning a piece of a stream or membership in a fishing club. But the great majority treasure the legal principle of freedom to fish in public waters held in common. They ask that their governments maintain the public right of fishing even though the fish are too few to go around, in which case they ask that the supply be allocated among user groups and enhanced by stocking. They also ask that the environment of their fishing be maintained in as natural a state as possible and that fishing be permitted only in sportsmanlike ways. They recognize that the fish resources may be insufficient to permit everyone to exercise his or her right and they accept restraints on time, place, and size of fish if so doing will reduce waste and distribute the catch equitably. Thus a major component in the recreational fishing system is government regulation (Chapter 13).

Any activity with as many participants who spend as much money as recreational fishermen generates a large business activity. People who fish for fun are vacationing and use equipment and expect service. Almost all of them own fishing gear. A large proportion own boats from which they fish. If not, they rent boats or buy passage on party boats. They travel, use automobiles, operate campers, stay in hotels, and eat meals. The result is the substantial economic development of organizations that provide equipment and services and of communities where such organizations are concentrated.

Additional fishing opportunities are provided by both governments and private firms. Governments may buy access for the public to choice fishing areas or construct fishing piers near urban areas. In many parts of the world, private fishing clubs may own fishing rights on streams and entrepreneurs may operate fee-fishing facilities.

Informational systems are vital to recreational fishing activities, especially for the fishermen who travel. They need to know where the fish are biting and how to catch them. Such information is provided by newspaper columns and fishing magazines. Governments may also publish information for anglers (e.g., the U.S. Department of Commerce annual "Fisheries of the United States").

Most recreational fishing is angling but certain other aquatic activities are pursued. Scuba diving to spearing fish is popular in estuaries. Spearing frogs is popular in a few places. Gathering pretty molluscan shells, harvesting oysters, or digging clams are popular sports in many saltwater areas. Even netting fish and trapping crustaceans, which are usually regarded as commercial activities, may be allowed as recreational activities in some places.

10.5 ROLE OF FISHERY SCIENTISTS

All of the businesses related to fishing depend on a supply of fish, and their existence depends on knowledge of where, when, how, and how much fish may be caught. Such information is, of course, similar to that required by governments for regulation of fishing (Chapter 13) or for government-stimulated fishery development (Chapter 15), but the presentation and use of such information may be completely different. An abundant stock that is difficult to catch may present no conservation problems to government but may provide a major opportunity for business. Or fishing on a heavily exploited stock under government regulation may require special long-range forecasts of catches as a part of investment strategy by business.

Fishery scientists are finding new roles in planning and policy development as well as contributing to many daily decisions by fishing businesses. Such activities are partly government services to business and partly private activities in close collaboration with upper- and middle-level management of the business organizations. The scientific role is to provide information in suitable format, at appropriate times, on how things can be done, at what cost, and with what benefits. Their role is usually advisory to business rather than the major administrative roles that scientists frequently play in government organizations, but an increasing number are managing fish businesses.

10.5.1 Commercial Fisheries

Successful fishermen know where and how to fish on the basis of their experience and that of their colleagues. But their experience is essentially a trial-and-error system that is based on inferences about the behavior of the fish and the efficiency of the gear used for catching. Such inferences can be reinforced or modified by interpreting the fishermen's success with reference to scientific studies. These are commonly made in collaboration with government institutions.

Almost all commercial fisheries pursue aggregations of fish—if they were randomly distributed, few if any would be caught. So a major requirement is knowledge of how and where fish aggregate as they grow, feed, migrate, and reproduce. This involves understanding of their sensory systems, their habitat preferences, and many aspects of their behavior (Bardach *et al.*, 1980). In addition, knowledge of their reactions to temperature, light, sound, or odor can lead to improvements in fishing. Such biological knowledge can be combined with knowledge of the interaction of fish and gear to improve fishing success. Useful studies include underwater observations of trawls and seines to discover how fish avoid them or statistical analyses of catches by gear modified in various ways.

But some stocks of fish are not well known because they have not been entirely located or they have largely evaded conventional gear. Study of these requires special exploratory vessels that can find the concentrations, assess their abundance, and try new methods of fishing.

Consideration of commercial requirements—for new vessels, for shore facilities, for new markets, or for mergers that involve significant changes in business practices—requires longer-term planning, frequently in five- to ten-year periods. Desirable information may include projections of supplies, methods, markets, competition, and government policies. Many public scientific reports will be available, but the final projections usually will be private.

Fishery scientists are finding important roles in such planning, as employees of either fishing companies or consulting firms. Their task is to work with financial analysts and company management to provide collective judgments about future opportunities. Included may be questions of fishing vessel design, plant location, potential yields from various stocks, new products, or new market possibilities (Chapter 12).

10.5.2 Recreational Fisheries

The role of fishery scientists in the recreational fisheries has been and remains predominantly governmental, in regulation and environmental management (Chapters 13 and 14). A limited role is in providing information on fishing and in development of fishing opportunities.

A rapidly growing role for both government and private fishery scientists is in policy formulation. The concern about the environment and the widening participation in recreational fishing are creating many complex issues. Use of water for fishing or other purposes heads the list. Then come issues of acid rain and the cost of pollution control, fair distribution of opportunities among city and rural residents, license fees and user charges, public versus private facilities, and many others. All government agencies that make decisions about the aquatic environment are involved, as are the hundreds of private organizations that represent fishermen or others concerned with the environment. Both groups depend heavily on scientific information.

A limited private role is in furnishing information to fishermen and assisting with planning of investments in facilities. A few become engaged in operation of recreational fishing facilities.

But almost all of them like to fish for fun themselves!

REFERENCES

Alverson, D. L., Freegurg, M. H., Murakowski, S. A., and Pope, J. G. (1994). "A Global Assessment of Fisheries Bycatch and Discards," Fish. Tech. Paper No. 339. FAO, Rome.

Bardach, J. E., Magnuson, J. J., May, R. C., and Reinhart, J. M., eds. (1980). "Fish Behavior and Its Use in the Capture and Culture of Fishes." International Center for Living Aquatic Resources Management, Manila, The Philippines.

Grover, J. H., ed. (1982). "Allocation of Fishery Resources: Proceedings of the Technical Consultation on Allocation of Fishery Resources Held in Vichy, France, 20–23 April, 1980." FAO, Rome.

Gulland, J. A., ed. (1971). "The Fish Resources of the Oceans." Fishing News Books, Ltd., Surrey, England.

Kreuzer, R. (1978). The cradle of sea fisheries. In "Ocean Yearbook 1" (E. M. Borgese and N. Ginsburg, eds.), pp. 102–113. Univ. of Chicago Press, Chicago.

Shapiro, S., ed. (1971). "Our Changing Fisheries." U.S. Govt. Printing Office, Washington, D.C.

11 | The Culture Fisheries

The challenge of aquaculture is to increase the potential supply of aquatic animals and plants that can be harvested at a rate of about 10 kg/ha/year from relatively unproductive areas to a maximum of about 200 kg/ha/year from the most productive waters. This challenge is met by discovering and alleviating the natural restraints on production in ways that cost less than the additional benefits expected to be realized. The methods range from such modest measures as reducing predation on the fish in public waters to a "farm" system of total control over all aspects of an organism's life.

11.1 PERSPECTIVES ON THE CULTURE FISHERIES

Aquaculture, like agriculture, is a farmer's approach to production of food or other things of value form the waters. Its concepts and practices are those of cultivators or herdsmen rather than fishermen. In fact, many aquacultural projects are closely associated with farming of livestock or field crops on privately owned lands and waters. Others are completely independent and may be located on inland or coastal waters that are leased from governments. However, all share the basic concept of farming. All are opportunistic, since all started with stocks and facilities that seem likely to be manageable, and provide a product for market at a profit.

11.1.1 The Scope of Aquaculture

Aquaculture (formerly spelled *aquiculture*) includes the culture of any kind of plants or animals. It may be called *husbandry* of aquatic organisms for the purpose of producing animals or plants for food, but it includes also the production of ornamental animals and plants in controlled waters. Aquaculture has even been defined to include hydroponics, the soilless cultivation of land plants, but that usage will not be considered here. Aquaculture may be best defined as that portion of fisheries production obtained through human intervention involving physical control of the organism at some point in its life other than at harvest.

The activities classified as aquaculture, however, remain uncertain with respect to the transition between subsistence aquaculture and modern habitat improvement. Both activities may include some control of desirable organisms and alleviation of environmental factors that limit populations or make difficult their capture. For example, some subsistence aquaculture consists of building a simple fish trap on the edge of a lake, feeding the fish with farm waste, and perhaps controlling predators. Another subsistence practice is to enhance the fishing by providing brush shelters in temporary oxbows cut off from a river during the dry season. These do not differ in principle from some habitat improvement activities (Section 14.5), but the latter have usually been undertaken to repair damage to the natural environment or to enhance new environments such as reservoirs.

Such simple subsistence aquaculture uses only a few of a spectrum of controls that may be exercised over aquacultural organisms. These simple practices, frequently called *extensive aquaculture,* may include transplantation of organisms to better growing conditions, construction of enclosures to protect them and make their capture easy, removal of their predators, and additions to their natural food supply. At the other end of the spectrum is *intensive aquaculture,* which may use an animal or plant that is especially bred for its domestic role in a completely controlled environment with a continuous production schedule. The animals and plants become products designed and packaged for the markets. The range of such practices might be compared to the range of agricultural production that falls between herding semiwild reindeer and raising poultry or vegetables on a modern farm.

Food, of course, is the principal product of aquaculture. The animals—fish, mollusks, and crustaceans—are grown mostly for human food, although in more developed countries some may be grown for ornamental purposes, and some for live bait for recreational fisheries. The plants, which are principally large marine seaweeds but include a few freshwater plants such as water spinach and watercress, are grown for both human and animal food.

A substantial part of the food fish production supports recreational fishing in developed countries. Much of this, notably in Canada and the United States, is public aquaculture that produces fish to stock public waters. But a large and increasing part of the production is private aquaculture that supports personal fishing in farm ponds and fee-fishing enterprises.

In addition to the food organisms, many kinds of valuable organisms are grown for ornamental use. A few thousand species of colorful fish are displayed in aquaria, and their production and sale are part of a large worldwide pet trade. Certain oysters produce pearls used for jewelry.

11.1.2 Recent Production

Statistics on aquacultural production have been largely ignored in national and international fishery statistical systems. Part of the production, as with oysters in the United States, may be combined with production from wild stocks. But even if the production of the organism is entirely from aquaculture, the statistics have frequently been included with the production from the capture fisheries. Furthermore, part of the subsistence production and most ornamental fish production has been omitted from the statistics or merely "guesstimated." Information on long-term trends in production has been, therefore, largely a matter of informed judgment, but in the 1980s the Food and Agricultural Organization of the United Nations (FAO) began to collect the statistics (Tables 11.1 and 11.2).

These statistics reveal the rapid growth in production, especially in Asia, where about 80% of the total is produced. Finfish, which are grown predominantly in freshwater ponds associated with irrigated agriculture, come mostly

TABLE 11.1 Aquaculture Production by Major Species Groups[a]

	1985	1992
	(MT × 1000)	
Carps, barbels, and other cyprinids	2995	6652
Salmons, trouts, and smelts	267	628
Tilapia and other ciclids	238	474
Jacks, mullets, and sauries	168	179
Shrimps and prawns	210	884
Oysters	910	954
Mussels	752	1086
Scallops	122	549
Clams, etc.	347	765
Brown seaweeds	2511	3640
Red seaweeds	644	1133
Miscellaneous aquatic plants	248	600
Total	9175	17,545

[a]Source: FAO Fisheries Circular No. 815, revised July 1994. FAO, Rome.

TABLE 11.2   Total Value of
Aquacultural Production[a]

	1985	1992
	(millions of US$)	
Africa	40	172
North America	527	968
U.S.A.	425	632
Canada	11	146
South America	311	1121
Asia	10,651.0	26,482
China	4953.5	11,642
Japan	2280.4	4665
India	507.5	1501
Indonesia	380.6	1983
Europe	1535.6	3296
World total	13,428.3	32,518

[a]Source: FAO Fisheries Circular No. 815, Revision 6, July 1994. FAO, Rome.

from Asia and Europe, where such aquaculture has been indigenous for centuries. Two countries, India and the People's Republic of China, produce about half of the world total, and most of the balance comes from the countries of eastern and central Europe (not detailed in the tables). A major part of all this production consists of carps (see Section 11.2.1).

Mollusks, which are grown mostly on small farms in coastal marine waters, come almost entirely from only about a dozen countries. The People's Republic of China produce more than half; Japan, the Republic of Korea, Thailand, the coastal countries of western Europe, and the United States contribute almost all of the balance. Part of the reason for the limited production of mollusks is the cultural aversion to them in Muslim countries. Crustaceans, especially shrimps, which are grown in coastal or estuarine waters, come mostly from Central America, India, Indonesia, and Thailand.

The regions that had the most rapid growth in overall aquacultural production between 1975 and 1992 were Asia and Oceania in molluscan and crustacean production, and Europe and the United States in finfish production. The regions with a low level of production and little, if any, recent growth were Africa, Latin America, and the Caribbean. These regions, with their many lesser developed countries, have had little or no cultural heritage of aquacultural production and, despite various assistance efforts, have seen no widespread development.

But overall trends in aquacultural production between 1975 and 1992 show a high rate of growth with encouraging prospects for the future. The annual growth rates were about 7% outside the African, Latin American, and Caribbean areas, a rate that is much higher than the rates of increase in population, food supply, or capture fishery production for the same period (Table 11.2).

Much of the rapid development in aquacultural production appears to be focused on relatively high-priced food. The crustacean culture, a large part of the molluscan culture, and the carnivorous fish culture (Section 11.2.2) of a few Asian countries, western Europe, and North America are all directed toward luxury products. Nevertheless, a substantial part of the actual increase in volume of finfish production comes from freshwater and brackish-water pond culture in countries that already had relatively high production of aquacultural products per capita. And there were no signs of leveling off in the group of countries with the highest production per capita.

Another widespread but almost completely unrecorded aquaculture business is the ornamental fish, or aquarium, industry. Axelrod (1971) estimated that retail sales of live fishes, aquaria, hardware, fish foods, and remedies were about $4 billion (U.S. dollars) annually, of which about half were in the United States. He estimated further that the business had grown at a rate of about 20% annually during the previous three years. Such a rate must have tapered off in the 1970s and 1980s, but the aquarium business remains at a substantial size.

11.2 MAJOR AQUACULTURAL SYSTEMS

Aquaculture utilizes animals or plants, water, land, food, labor, and money, all of which are organized by people to produce something for themselves or for a market. The usual discussion of aquaculture is arranged by species, but the hundreds of species that are grown around the world are produced in relatively few kinds of systems that make very different use of water, land, food, and labor. The following sections will discuss the principal systems that provide almost all of the production for markets. Emphasis will be given to major features of the systems and not the myriad of essential biological details, which are elaborated in references listed in the Bibliography at the end of this volume.

11.2.1 Omnivorous Animals in Ponds

Many animals are grown in stillwater ponds that are managed to produce all or a substantial portion of the food that the animals need. These animals include the numerous species of carps grown in eastern Asia and India, and the common carp that is grown there as well as in the countries of eastern Europe. Other animals are milkfish, grown in coastal brackish-water ponds of Southeast Asia; tilapia of several species, grown in either freshwater or brackish-water ponds; catfish, goldfish, and baitfish of several species, grown in fresh water in the United States; shrimp and prawns, grown in fresh or coastal salt water in various countries; and crayfish grown in fresh water in the United States.

Not all of these species are strictly omnivorous, but they are frequently grown in an omnivorous combination that makes use of all parts of the ponds. Such systems account for about three-fourths of the world production of finfish by aquaculture, and almost all of the crustacean production.

Many of the animals grown in this way have a high tolerance for water with low dissolved oxygen. Some can tolerate only 1 ppm, and this tolerance permits the stillwater ponds to be fertilized to produce food for the animals at a high rate. Most of such natural food is the mixture of algae and tiny animals that grows either on the bottom or in the water column. Some of the animals, such as the grass carp of China, will feed on rooted aquatic plants. Others, such as shrimp, may suffer when dissolved oxygen is less than 3 ppm and require supplemental feeding with a high-protein diet if high production levels are to be maintained.

Much of this aquaculture is conducted on saline lands or internal river delta areas, where agriculture is difficult or impossible. A fish crop may be more valuable than a field crop, and some land in the Mississippi delta area of the United States can either be used for field crops or flooded for fish. Milkfish are grown almost entirely in brackish-water ponds in coastal waters of Southeast Asia. A large fraction of the carps is grown in seepage ponds in or near river delta areas of China, India, or eastern Europe. Some animals are grown in ponds constructed in mangrove swamps. Annual Asian production in the early 1990s exceeded $20 billion in value.

Some of this pond culture is closely associated with agriculture. Ducks, chickens, and pigs are frequently grown on the edges of ponds and provide the fertilizer needed for the fish crops (Fig. 11.1). Field crop and food-processing wastes, such as cut weeds, rice bran, and oilseed cake, may be used as a fertilizer or even as feed for a few species of fish. Human waste is used in some places.

In some situations, fish can be grown as a part of irrigated agriculture using water and land that are too salty for field crops, because even freshwater fish will usually tolerate up to about 0.5% salinity. Rice production can be combined with fish or shrimp production if paddies are designed with deep-water trenches, but this system may be limited by the use of chemicals to control pests on the rice, which are toxic to the aquacultural animals.

Fertilization also produces rooted waterweeds, which are a constant nuisance and must be controlled by manual weeding, minimizing the area of pond water with depths less than 70 cm, using a fish such as grass carp to eat them, and fertilizing at a precise rate to produce plankton dense enough to reduce weed growth by shading, yet not reduce dissolved oxygen below the tolerance of animals.

Overall average production rates of intensively managed ponds in the various countries usually range from about 500 to 2000 kg/ha of pond, depending on the length of the growing season and the management methods. The growing season is much longer, of course, in the tropical countries, where the species grown are rarely inhibited by warm water. Production is increased by polycul-

FIGURE 11.1   A carp and duck farm in Egypt; designed by Egyptian engineers in collaboration with Chinese fish culturists and financed by private capital, it has several hundred hectares in ponds. (Photo by W. F. Royce.)

ture of several compatible species that will use different levels in the pond and eat various kinds of food with some feeding on the wastes of others. Other increases can be obtained by supplemental feeding or by precise fertilization that produces a maximum of food without undue depletion of oxygen, which can sometimes be mitigated if mechanical aerating devices are provided. A maximum annual production under experiment-station conditions appears to be about 7000 to 10,000 kg/ha.

A special advantage of all aquaculture is the opportunity to keep fish alive until the moment of processing or shipping to market. Most of the fish produced by small-scale pond culture are sold fresh in markets within easy trucking distance, so maintenance of quality is relatively simple. Large-scale operations can easily arrange an optimum combination of harvesting and processing. Shrimp may be supplied to the local markets if the price is good, but a large portion of them is usually processed for international trade according to international standards.

The actual production systems vary with the species, water, land, food, fertilizer, and associated agriculture, that is, within local conditions. Management of the ponds requires skilled day-to-day attention to control water quality, dissolved oxygen, fertility, weeds, diseases, and predators, and to harvest the crop at the right time. Pond culture is a relatively sophisticated kind of farming.

11.2.2 Carnivorous Animals

Carnivorous animals are those that are regularly supplied with a complete high-protein diet—a major part of the cost of their production. They include the salmonids that are grown in North America and coastal parts of western Europe, the yellowtail that are grown in Japan, and a variety of choice saltwater or freshwater species, such as basses, porgies, snappers, pikes, perches, eels, and lobsters. Shrimp and catfish, which are included in the omnivorous animals, might also be considered here since their culture may include a complete diet in the juvenile to adult stages.

Most of these animals require dissolved oxygen in their water at levels of about 4 ppm or greater. This means that they must be reared in cages through which water flows readily, in raceways supplied with running water, or in ponds with no fertilization. Some of them, notably the salmonids, usually require cool water that is less than 20°C.

Although pond culture of omnivorous animals started centuries ago, the intensive culture of carnivores is much more recent. It began during the early nineteenth century with the development of methods for the manual fertilization of trout eggs, the care of the eggs during incubation, the care and feeding of young trout, and the transportation of both eggs and fish. The techniques were widely adopted in Europe and North America in both public and private enterprises, during the 1860s and 1870s. During the 1980s the practice received much more scientific attention, and now it may resemble a "factory" type of operation (Fig. 11.2).

The complete diets that are required by intensive culture of aquatic animals may consist entirely of fish or other aquatic animals similar to their natural diets. If so, they require 6 to 8 kg of food per kilogram of weight gain. If the diet is artificial, it may be compounded from fish meal as well as a variety of grain meals, meat meals, and additives to supply minerals and vitamins as necessary. Usually such diets are composed of 20 to 35% protein, and between 1.5 and 4 kg of food (dry weight) are required per kilogram of weight gain.

The requirements of large quantities of running water and high-protein food mean that this intensive aquaculture must be located where both are readily available. If the diet is fish, the supply usually comes from scrap fish landed by the capture fisheries. If the diet is compounded from a variety of meals and brewery or slaughterhouse wastes, these must be available at reasonable prices, usually from within the country. Large quantities of running water must be obtained from streams of suitable temperature and water quality, preferably by gravity flow. Cage culture, on the other hand, can be located near any body of suitable water, salt or fresh, which may be a lake, slow river, irrigation canal, or estuary.

One of the variants of salmonid culture is called ocean ranching. Here, the fish are hatched from eggs, reared to juvenile stages in captivity in fresh water, and then released to their feeding grounds in the sea. When they are fully grown they return to their home streams, in this case their hatcheries. They can be

FIGURE 11.2   An aerial view of the Snake River Trout Co. at Buhl, Idaho. This "farm" can raise trout at a rate of about 400,000 kg/ha/year. (Courtesy of Snake River Trout Co.)

harvested along their migratory route if they are protected enough for some to reach the hatchery and contribute spawn for the next generation.

Public salmon ranching now supports an estimated 20% of the salmon fishing in the North Pacific and a substantial portion in the North Atlantic. Ranching by fishermen's associations in Japan supports most of the coastal fishing for salmon in that country. Private salmon ranching began in North America during the late 1970s and is expanding slowly in locations where the returning run is not subject to extensive public fishing.

All of the animals chosen for this intensive type of aquaculture are necessarily valuable because of the high cost of water, enclosures, and food. Their value can be further enhanced if they can be sold to the consumer alive, or if they are killed, bled, processed, and packaged to meet the highest market standards. Usually this is done, and large portions of such products enter international markets as high-priced or luxury foods.

11.2.3 Filter-Feeding Mollusks

Oysters, clams, mussels, and scallops feed on the tiny organisms in the water that they circulate through their shell cavities. Almost all of the cultivated species live in coastal marine waters in which tidal circulation brings their food. They are all immobile or only slightly mobile during juvenile or later stages, but eggs and larvae drift with currents for a time before becoming bottom dwellers. The best growing areas for filter-feeding mollusks are those that are naturally enriched by riverine or oceanic current processes.

But even though their food comes free with the water, their culture is complex. Almost all growing areas are public waters, which must be leased from the state and which are used for other purposes, some conflicting. Each species or race has its own tolerances for temperature, salinity, dissolved oxygen, wave action, and substrate. Each must also be able to survive predators (which are kept out of ponds and cages) and competitors, such as barnacles and other attached organisms. Of course, each must also remain fit for human consumption at the time of harvest. Suitable sites are limited, therefore, by proper ecological, sanitary, and publicly acceptable conditions.

Two methods of culture are in widespread use: bottom culture and hanging culture. Bottom culture, which requires much less labor and investment, is used when suitable environment is widespread. It involves natural seeding or spreading seed from hatchery or natural sources, protection of the animals from crawling predators, and eventual harvest. Oysters, the predominant mollusks grown on the bottom, are sometimes dredged after planting or natural seeding, separated if they are clustered, and replanted in superior growing areas.

Hanging culture involves attaching the mollusks to lines and suspending them from racks, rafts, or floats (Fig. 11.3). It has many advantages: the ani-

FIGURE 11.3   Racks from which strings of pearl oysters are grown in Ago Bay, Kakiojima, Japan. (Photo by A. Suomela.)

mals grow faster owing to better water circulation, the farming is extended to areas with unsuitable water depths and types of bottom, crawling predators are excluded, and entire floating rigs can be moved to take advantage of better environments at proper times for growing, seeding, hardening, or fattening.

An increasing problem in all coastal waters is pollution by sewage or toxic materials, and some good molluscan growing areas are slightly polluted by sewage. Most mollusks have a tendency to accumulate viruses or bacteria that may be harmful to people who eat them raw or only slightly cooked. They may also accumulate toxins that are dangerous to people. Consequently, most governments inspect molluscan supplies and the water in which they are grown for sanitary quality (Section 12.3.5).

Mollusks will purify themselves of human disease organisms when they are held in clean water for as little as 24 to 48 hr. This may be done in a circulating system in which the water is purified by filtering, ozone, ultraviolet light, or other means. Another purification method is to transplant the mollusks to a clean growing area for a suitable time. The process must be carefully controlled to maintain good water for the mollusks while removing viruses and bacteria.

Most mollusks are marketed whole in the shell or as raw meats after shucking. Scallop meats may be frozen, but the others frequently develop undesirable flavors after freezing or canning. Consequently, most of them are marketed fresh within economical transport distance, and great care is taken to avoid spoilage and contamination.

Mollusks are usually high-priced, in spite of the fact that their food is at the bottom of the food chain and they simply filter it from the water passing by. The edible portions are only 10 to 20% of body weight, so shipping is costly, and the special handling necessary to assure a sanitary product is an additional cost.

11.2.4 Aquatic Plants

Almost all of the aquatic plants used by humans are the marine algae, which are of two very different kinds. The tiny planktonic forms are essential food for many larval animals and are reared in a secondary aquaculture to feed seed animals (Section 11.3). The attached leafy algae, or seaweeds (marine macrophytes), are eaten by humans, fed to stock, added to soils as a fertilizer or conditioner, or used as a source of vegetable gums that thicken foods, cosmetics, and paints. Several dozen genera of red, brown, or green seaweeds have been harvested for centuries, but as demand has increased a few species have been cultured in large quantities.

Culture of seaweeds grew out of attempts to improve the seaweed habitat by providing suitable hard-bottom materials for their attachment. This added to the supply but was only a temporary solution for the increasing demand. Then it was discovered that it was practical to grow some species much like mollusks, from cuttings suspended from rafts or floats. Next it was discovered how to control the sexual reproduction of several major kinds of seaweed and how to attach the young thalli to lines for suspension. Now nearly 4,000,000 metric tons of red, brown, and green seaweeds are produced annually in eastern Asia (FAO, 1993).

An important genus of red seaweeds is *Porphyra*, or "nori," which began to be cultured in large quantities in Japan when methods of controlling the reproductive cycle were discovered. The complex cycle starts with the production of macroscopic carpospores in early spring. These are squeezed out of the sporangia on the leaves of nori into water with clean shells, which the spores penetrate after germination. This starts the *conchocelis* phase (a generic name given when the phase was thought to be a new plant), and the shells are suspended in water with controlled temperature and light. After about 5 months, when the tiny plants become visible to the unaided eye, they produce monospores, which adhere to nets placed in the tanks. The nets are either placed in the sea immediately or stored temporarily to spread out production through an annual cycle. The plants mature at a length of 15–20 cm after 2 or 3 months of placement. Mature nori is harvested from about December to March, washed thoroughly, chopped into tiny pieces, suspended in water, and dried to a paper-like product on fine screens.

Other kinds of algae are cultured for dried human food or for their gums. A few rooted freshwater plants, such as watercress, are grown for human consumption. Most of the current production of aquatic plants is based on species that have not been domesticated. But recent studies have led to improved methods of controlling reproduction of the plants, and new varieties have been bred that possess exciting possibilities.

11.2.5 Ornamental Plants and Animals

Maintenance of aquaria with displays of plants and animals has become an increasingly popular personal hobby as well as a public service in many parts of the world. Axelrod (1971) estimated that about 20,000,000 home aquaria were kept in the United States at an annual cost for each of about $100 for apparatus, livestock, food, and remedies. He estimated further that there were about 450 aquarium fish farms and importers in the United States, who handled more than 200,000,000 fish. The fish and supplies were distributed by about 6000 pet shops and 10,000 variety stores. The business in the United States appeared to be about half of the worldwide business. The outstanding displays are, of course, in the great public aquaria that can maintain tanks of either fresh or salt water up to sizes large enough for small whales.

The aquarium supply business has probably doubled in recent years. It deals in animals that are worth far more than animals grown for food. They are usually sold on a unit basis, but if compared to food animals on a weight basis, the prices are usually ten to hundreds of times greater. Unusual display species may be valued at $1000/kg or more.

Such valuable fish justify much more investment in habitat and care per individual than do food fish. The habitat may be a closed circulating system in a large public or business installation in which the water is treated biologically, chemically, and mechanically to maintain optimal quality. It may be a saltwater system far from the sea, which uses synthetic seawater. Individual tanks can be temperature controlled and aerated as necessary.

The widespread development of the hobby is relatively recent. Fish were kept as pets many centuries ago, and public aquaria were established in Europe and North America during the latter half of the nineteenth century. At first they were limited to displaying local species, but colorful tropical species were soon discovered and techniques of shipping were developed. Now, with aerial shipment common, several hundred species are available through dealers in many parts of the world. Many species of wild freshwater fish are caught in the headwaters of the Amazon and Orinoco rivers and shipped out of Peru and Colombia to international markets. Many supplies of exotic cultivated fish are produced in Singapore, Hong Kong, Thailand, the Philippines, Taiwan, and Japan. United States production has been centered in Florida, where several hundred growers operate.

11.2.6 Culture-Supported Capture Fisheries

Several of the freshwater and anadromous species are produced by public aquaculture, especially in North America. These are the trouts, salmons, basses, pikes, and perches that are reared to stock natural waters in support of recreational fishing or, in the case of salmon, as mitigation for loss of spawning grounds needed to support both recreational and commercial fisheries.

Concern for the decline of public fishing in streams and lakes led to efforts to replenish the stocks during the middle of the nineteenth century. Private aquaculture had long been practiced for food production, and discovery of improved methods of culturing trout let to public programs for restocking natural waters. The first public hatchery was constructed at Huningue, France, in 1852, and claims were made that all of the streams in France could be replenished with fish at negligible cost.

This attractive concept spread to North America, where the first trout hatchery was built at Mumford, New York, in 1864. Many others were soon built by government and by private operators who sold eggs or fish to the governments. Culture of other species soon started, as with Pacific salmon and rainbow trout at McCloud River, California, in 1873 and various marine species such as cod, flounders, and lobster at Boothbay Harbor, Maine, and Woods Hole, Massachusetts.

The practice of fish culture became politically popular, and it was the major part of the fish conservation movement into the 1940s. The number of public hatcheries grew to 68 in 1900 and to 629 in 1948. By then, their value was being questioned, and the numbers declined to 502 in 1965, 92 of which were operated by the U.S. federal government, and 410 of which were run by the state governments [U.S. Bureau of Sport Fisheries and Wildlife (BSFW), 1968]. They produced 10,100 MT of fish for stocking, of which 95% were salmonids.

Most of the salmonids were trout, which were predominantly grown to catchable sizes before release by the hatcheries at a national average rate of 113 g/person-day of fishing.

Warm-water species were also grown but were released predominantly as fry because of the cost and difficulty of rearing them to larger sizes. These were mostly catfish, largemouth bass, bluegill, red-ear sunfish, northern pike, muskallunge, walleye, and striped bass.

The practice of fish culture continues in the United States, although comprehensive statistics on production are not available. The federal government operated 88 hatcheries in 1980, which produced 2800 MT: 95% by weight were salmonids and 54% by number were "warm-water" species (U.S. Fish and Wildlife Service, 1980). Fragmentary statistics on state production from about 400 hatcheries (personal communication from R. Martin, Sport Fishing Institute) indicate some expansion, so overall U.S. production was probably about 10,000–12,000 MT in 1980.

But the growth of private aquaculture eclipsed public aquaculture activity in the United States during the 1980s. Overall production of fish was about 56,000 MT in 1980, but during the decade the total production of fish, mollusks, and crustaceans reached about 300,000 MT. It has continued to expand in the 1990s.

Some private recreational fisheries are also supported by aquaculture in the United States. Following the discovery of improved farm management methods, especially stocking with a combination of largemouth bass and bluegill, the number of ponds increased from an estimated 20,000 in 1934 to well over 2,000,000 in 1965 (Swingle, 1970). These ponds were multipurpose and were used for conservation or stock watering as well as to grow fish. Their construction was assisted by extension programs and by state supplies of fish for stocking in some places.

The interest in such pond fishing led to another recreational fishing service, fee-fishing or catch-out ponds. Many of these are maintained by private operators who charge for the fun of fishing or for the fish caught, or both.

11.3 SEED PRODUCTION

Throughout much of the aquacultural world the term *seed* has been borrowed from agriculture to apply to young aquatic animals or plants at or before the stage when they are planted in growing enclosures or attached to racks or floats. At this stage, they are usually large enough to be identified by the unaided eye and have passed through the stage requiring extra special care. Many other terms are in common use, such as *fry* and *fingerling* for young fish in North America, *spat* for young mollusks, and *postlarvae* for various crustaceans.

Seed production is the major turning point in the domestication of plants and animals. First, it assures the opportunity for cultural operations to produce plants or animals at a higher rate than they can be produced in the wild. Second, when parents of the seed can be selected, the genetic development of superior cultured organisms can begin.

Production of aquacultural seed differs from production of agricultural seed in the sophistication required. Whereas domestic mammals are large at birth and are cared for by their mothers, young fish are usually only 3 to 5 mm long at hatching, and they are extremely delicate. Salmonid fry, which range from about 15 to 30 mm long at hatching, are the exception and are relatively easy to care for. Molluscan or crustacean eggs and larvae are even smaller, usually microscopic, as are the spores of macroalgae (see the description of handling the spores of *nori*, Section 11.2.4).

These tiny organisms require environmental conditions of water quality, temperature, and light that are tailored for each species. They also need special foods, which, for each species of animals, must usually be particular living microorganisms. And because they grow rapidly, perhaps doubling in weight in a few days, each organism may need a different kind of food from week to week. Production of suitable microorganisms requires a secondary aquaculture that can provide the proper food at the right time.

The ideal food production system will require maintaining a brood stock selected for desirable characteristics and controlling reproduction to produce seed in the quantities and at the time needed. The fragile organisms must be reared to seed size with laboratory-type care. Such seed production is usually undertaken by specially equipped organizations as a service to farmers who grow the animals or plants for market.

The commercial practice of seed production is seldom easy. Brood stock for some animals either cannot be kept in captivity or will not spawn. These have included milkfish, which are widely grown in eastern Asia; yellowtail, which are grown in Japan; and even mullet, which are grown in many countries. The young of these animals have been collected in the sea and transferred to the growing ponds. Penaeid shrimps migrate offshore to spawn and the females carry the fertilized eggs, so the seed has been obtained by collecting gravid females and holding them in tanks until they release their young. Most bivalve mollusks broadcast eggs and sperm in the water. Their fertilized eggs drift for a time, hatch, grow, and then descend to the bottom, where they become attached. Their seed, or spat, is obtained by placing suitable collectors at the right time in areas where setting occurs.

Diligent work by fishery scientists in many parts of the world is bringing the reproductive processes of more organisms under control (Fig. 11.4). The methods usually begin by conditioning the adults through control of environmental factors such as light, temperature, water currents, diet, segregation, or other factors that are related to the maturation of the gonads in the particular species. Then, if eggs and sperm cannot be manually removed, or the animals will not

FIGURE 11.4 Mature carp being transferred to breeding ponds in Indonesia. (Photo by E. Schwab, courtesy of the FAO.)

spawn by themselves, gonadal or pituitary hormones may be used to induce spawning.

Injection of pituitary hormones to induce spawning is revolutionizing some major aquacultural activities. For example, the production of most of the Chinese and Indian carps formerly depended on collection of naturally spawned seed. But almost all of the species will spawn after pituitary injection, and special seed-producing facilities are being built to supply the farmers.

Desirable genetic changes were accomplished long ago for a few animals. Common carp has been bred into numerous varieties for characteristics such as greater meat production, resistance to disease, and tolerance of different water conditions. Rainbow trout have been bred for rapid growth rate, better food conversion on artificial diets, spawning at most times of year to permit continuous production cycles, and disease resistance. More recently, trout intended for stocking have been hybridized to produce interesting variants, and some domesticated strains have been crossed with wild ones to improve their survival after stocking. But perhaps the most spectacular modifications are in the colorful goldfish and Oriental carp called *koi,* which are now available in hundreds of varieties.

11.4 HEALTH AND QUALITY MAINTENANCE

Any suitable organism, whether it is an interesting catchable, attractive display, or vigorous bait organism, must be healthy. Yet all aquaculture crowds and stresses organisms in ways that enhance the possibility of disease. Closer spacing, greater bodily contact, and increasing contact with body wastes make the physical transfer of disease organisms easier. Confinement is never a state of life toward which wild organisms have evolved, and stress always tends to increase with abnormal diet, abnormal activity, accumulation of wastes, interference with reproduction, or increased fighting.

11.4.1 Scope of the Problem

Control of diseases of aquacultural organisms is an immense challenge. Hundreds of bacterial, viral, mycotic, protozoan, and metazoan diseases are known, and many more probably exist. Little scientific attention has been given to the problems of fish disease, and few people with the special skills necessary are available. It was estimated some years ago [U.S. National Research Council (NRC), 1973] that the United States had fewer than 100 full-time disease specialists, including those occupied with research, diagnosis, and teaching. Yet it was estimated at the same time that 10 to 30% of production costs in both private and public operations in the United States were due to disease.

Every aquaculturist must seek the best balance between the cost of production and the risk of disease. The costs go down as the organisms are crowded and fed with cheaper food, but the risks go up; it has been asserted by fish disease specialists that every aquaculturist must expect a disastrous epidemic sooner or later.

11.4.2 Prevention

The best disease control is avoidance. First, avoid stress and sources of infection. Most diseases are waterborne, and the ideal water supply should have no disease organisms in it. Such supplies may come from underground sources or from treatment such as chlorination plus dechlorination. When surface water must be used, it is desirable that wild organisms of the species being grown not occur in the supply, as transfer of disease is usually easiest among individuals of the same species.

Another tactic is containment. Aquacultural ponds should be arranged so that each drains independently of others. Then an epidemic in one pond will not automatically spread to the ponds below. This tactic must include a vigorous avoidance of the transfer of disease organisms on nets or workers' clothing if an epidemic occurs. It may also include sterilization of the pond by drying or by chemical treatments such as lime. A last resort may be the slaughter of diseased and exposed animals.

A third tactic is the use of disease-resistant strains. All domesticated plants and animals have evolved or have been selectively bred to be better adapted to the growing systems. A few strains of fish have been developed and some oyster populations have been rehabilitated by breeding the survivors of epidemics. But this tactic is not possible, of course, without methods of controlling the reproductive cycle.

The primary tactic used by aquaculturists is meticulous attention to water quality. They must ensure that dissolved oxygen is adequate at all times and that body and food wastes are recycled in the pond or flushed out. They must be sure that suitable temperatures prevail or, if temperatures are temporarily adverse, that stress is reduced by oxygenation or change in diet. Ponds must be clean and free of plants or animals, such as snails, that may carry the vectors of disease.

Not to be overlooked are predators (including humans) and other possibilities of physical injury. Many nondomesticated animals fight among themselves, and most choice predatory fish will eat their own young. Fish-eating birds, mammals, and snakes may find small captive fish especially attractive. Handling fish during transfer among ponds, or when sorting for size, also requires great care.

A further insidious threat is contamination. This may come from pesticides that are harmful to the organisms or that concentrate in ways that make them harmful to humans. Another possibility is the development of bad flavors due to blooms of certain algae in the ponds or contamination by petroleum products.

11.4.3 Therapy

The primary control of diseases of aquacultural animals is, like preventative medicine, based on good nutrition, good water, sanitation, and avoidance of overcrowding. The food must supply essential nutrients, but should not transmit disease organisms, which is likely if raw fish is included in the diet without pasteurization. The water must have temperature, dissolved gases, and pH suitable for the species at all times, and it must not have toxic materials. Infection must not be spread from one tank or pond to another through common use of nets or transfer of animals. Overcrowding, which happens quickly when young fish are growing rapidly, leads to cannibalism and excess competition, with large animals getting most of the food.

When disease strikes on the fish farm, the operator must usually act quickly on the basis of the general appearance and behavior of the animals, without the assistance of laboratory procedures. The operator will seek first to correct any aquatic environmental problems and then may apply a treatment.

As the pathology of plants and animals has become better known, more techniques have become available to the aquaculturist, but even the best of them are not cure-alls. Traditional treatments for external parasites have included chemical baths that use common salt, formalin, malachite green, potassium permanganate, or copper sulfate (although some may be illegal for food fish). These may be administered at low concentrations for long periods in flowing water, or at high concentrations as a quick dip in a special tank.

Diseases caused by internal parasites or systemic bacteria may be treated by drugs in the food. Wide-spectrum drugs of the sulfamerazine, nitrofuran, and other groups are used for bacterial infections. Calomel (a mercurial), carbazone (an arsenical), and others are used for parasites. Other methods for control of bacterial and viral diseases are being vigorously investigated. They include injection of antibiotics as well as immunization through injection, inclusion in the food, or immersion.

11.5 SOME PUBLIC POLICY ISSUES

Aquaculture is broadly accepted by the public in most countries, because private aquaculture supplies food and public aquaculture helps to maintain fishing in public waters. The issues related to government support of such activities are not whether to support them, but in what way and how much (Chapter 15 and Section 11.2.6). Nevertheless, various aspects of these and other aquacultural activities cause public concern.

The age-old perception of fish as a publicly owned resource is at the root of much conflict. Fishermen are comparable to hunters, whereas aquaculturists are farmers. The fishermen may see rising aquacultural production as increasing their competition and may resist such growth in political ways. They may also have fished for wild animals, such as oysters, on grounds proposed for lease and regard such a lease as a violation of their rights. And private oyster culture, when carefully controlled, may produce 7 to 20 times as much per unit area as untended public grounds, thus exacerbating the competition.

Many other people oppose the lease or sale of public waters for aquaculture for a wide variety of reasons. Coastal marine waters are especially controversial because of their many uses. Molluscan or algal culture may involve use of racks or rafts that interfere with boating or merely affect the ambiance of the landscape. Fish pens and cages may also be considered unsightly or interfere with other uses of the waters.

Intensive aquaculture of carnivorous species produces organic wastes that may be undesirable. The amount of such wastes from any facility is a little less than the total food fed; they consist of the animal body wastes and lost food. Treatment of wastewaters may be required if dilution in the receiving water is not adequate. On the other hand, many coastal waters are so badly polluted that they are unsuitable for food production. Some people will argue that if the waters are used for oysters, there will be greater incentive to keep them clean.

Most aquaculture involves moving live animals among farms, among watersheds, and even among countries. This means a constant danger of transfer of animal or plant diseases, of alteration of native gene pools, and of introduction of nuisance species. There are many unfortunate examples. Water hyacinth has been introduced in many waters, where it has become so abundant that it is considered an ecological and economic disaster. Live bait fish have been released in ponds where they did not live before and have changed the entire population balance. Fish and molluscan diseases have been transported with consequent damage to both cultured and wild stocks. Exotic ornamental fish have escaped and become established in many warmer climates and had impacts on local fish. Yet there is always a possibility that the introduction of an exotic species may have great benefit, as in the case of Pacific salmon now thriving in the Great Lakes of North America.

Although their production is not really considered aquaculture, the fisheries for wild ornamental species in many tropical countries catch fish for the aquarium business. Some of these fish are unique and rare, and through overfishing are becoming threatened or endangered species.

Aquaculturists in the United States complain of overregulation that drives up their costs. Some have noted that they must satisfy the requirements of about 40 different agencies, local, state, and national, and that many agency representatives have no acquaintance with or sympathy for aquacultural problems. Clearly, there is a widespread need for better development of policy and law about aquaculture to safeguard the public interests and to avoid unnecessarily inhibiting the aquaculture industry.

11.6 ROLE OF FISHERY SCIENTISTS

If the optimistic projections of demand for aquacultural production materialize, the supply of aquacultural scientists may be a major constraint on meeting the demand (Section 2.5). Much more basic research will be needed, especially on physiology, nutrition, and pathology of several dozen animals and plants that are on the path to domestication. A few of these will probably be genetically modified to become better suited to each of the major kinds of producing systems. Simultaneously, applied research should expand on many complex aspects of water quality management, disease control, seed production, diet formulation, and production strategy.

Such research will require a major public commitment, perhaps similar to the experiment-station concept of agricultural research in the United States. Other public activities will be needed for planning aquacultural development and resolving the public problems associated with allocation of water, control of the spread of fish diseases, and conflicts with the capture fisheries.

But the major opportunities for employment of fishery scientists may be related to production. Private producers will need supervisors, marketing experts, and managers, as well as the advisory services of nutritionists, fish veterinarians, and water quality experts. Such producers may rely on highly sophisticated firms for their seed supplies. Credit agencies will need expert advice on all aspects of aquacultural financing. Engineering firms will need broadly experienced experts on production systems.

At this time, such projections are speculative. Aquacultural production is growing rapidly, but its further acceleration will probably depend on the national and international commitments to the scientific base and on the availability of suitable land and water for the farms.

REFERENCES

Axelrod, H. R. (1971). The aquarium fish industry—1971. *Tropical Fish Hobbyist*, November, pp. 19–23.

Food and Agriculture Organization (FAO) (1993). "Aquaculture Production. 1981–1991," Fish. Circ. No. 815, Rev. 6. FAO, Rome.

Swingle, H. S. (1970). History of warmwater pond culture in the United States. *In* "A Century of Fisheries in North America" (N. G. Benson, ed.), Spec. Publ. No. 7, pp. 95–105. Amer. Fish. Soc., Bethesda, Md.

U.S. Bureau of Sport Fisheries and Wildlife (BSFW) (1968). National Survey of needs for hatchery fish. *Res. Rep.—U.S. Bur. Sport Fish Wildl.* **63**, 1–71.

U.S. Fish and Wildlife Service (1980). "Propagation and Distribution of Fishes from National Fish Hatcheries for Fiscal Year 1980." U.S. Dept. of Interior, Washington, D.C.

U.S. National Research Council (NRC) (1973). "Aquaculture in the United States; Constraints and Opportunities." National Academy Press, Washington, D.C.

12 | Food Fishery Products

When a finfish dies after being caught it becomes raw material for a product. It consists of many tissues, including trunk musculature (the part that is normally human food), skin, skeleton, fins, blood, head, and viscera. The viscera may include large ovaries or testes that may be prized for human food and a liver that may be sought for its oil or vitamins. Its whole body immediately begins to undergo microbiological and chemical changes that must be reckoned with. It probably will not be consumed whole, so processing will begin with division into different parts according to expected use. Each part may be preserved appropriately for a product that will be distributed and ultimately consumed.

Other fishery food animals such as crustaceans and mollusks have different body structure but undergo similar conversion from living animal to multiple products. Most of the products are destined for human or animal food but some important by-products are used for industrial purposes and a few others may be used as drugs.

This chapter introduces the student to the application of fishery science to the dynamic fish business. The complex management and engineering aspects will be covered little if at all, although fishery scientists will usually work closely with managers and engineers. A statistical summary of major products and international trade was given in Sections 10.2.2, 10.2.3, and 10.3.1, and these sections should be reread during use of this chapter.

12.1 CHARACTERISTICS OF FISH FLESH

Fish flesh differs markedly from the usual mammalian and avian flesh in the characteristics of its composition, nutritive content, and contaminants. Even in one country, one or two hundred species may be available, with a few major species being marketed in many different convenience products. It usually provides a unique and valuable addition to human and animal nutrition, and unfortunately it is a highly perishable product.

12.1.1 Composition

Fish flesh (defined as the muscular and fatty tissues of the trunk), which is the part usually destined for human consumption, varies from about one-fourth to three-fourths of the total weight. The differences are largely associated with body shape, nutritional condition, and skeletal characteristics. Species with long trunks and small heads, such as tunas, salmons, carps, cods, and herrings, are usually 50 to 70% flesh (Kizevetter, 1973). Species with short trunk and large head, such as rockfishes and sculpins, may have only 30 to 40% flesh. Most of the popular carnivorous fish such as basses, perches, groupers, snappers, and porgies have an intermediate proportion of flesh.

Not all of the flesh can be recovered in the usual processing operations. Some of it lies among bones, so that perhaps 10% less than the preceding figures can actually be obtained in fillets or portions prepared for canning. Of course, since some products such as steaks and canned herrings may contain the bones and some modern machines remove minced flesh from bones almost completely, the actual portion available for market is quite variable.

Other aquatic animals usually have even less flesh. Oysters commonly have only 10 to 15% meat, the rest being shell. Shrimp, crabs, and lobsters usually have between 15 and 25% meat.

Thus few aquatic animals contain more human food than half of their live weight, and in many it is much less. The balance is usually wasted if the fish are marketed whole, or merely gutted and headed, as they are in many communities. If they are further processed into fillets, steaks, or any of the many convenience packages, the balance frequently can be economically used as animal food or industrial products in either fresh or further processed form. Tuna cannery wastes, for example, may be incorporated into canned pet food and fillet plant wastes may be dried into fish meal. Oyster shells, which are largely calcium carbonate, may be returned to the oyster grounds as culch, or ground up to be a calcium supplement in avian diets.

The major chemical components of fish are water, lipids, protein, and ash. The water and lipids (principally oils) together usually make up about 80% of total weight, the protein about 17%, and the ash about 3%. The proportion of ash is, of course, lower in the flesh (about 1%) than in the whole fish with all of its bones.

The most variable major component in fish flesh is the oil, which varies inversely with the water from about 1 to over 20% of body weight depending on species, seasonal fatness, and status of the gonads. It also varies within the fish, dark flesh having more oil and liver as well as gonads sometimes being very rich in oil. The variability of oil in the flesh is a major factor in flavor and quality, with the more oily flesh usually being preferred. The quantity of oil frequently influences processing methods and price.

12.1.2 Nutritive Value

The kinds of oils, proteins, and ash in fish make it as good a high-protein food as the flesh of mammals and birds for the general human diet and a superior food for various special diets. The many kinds of fishery-produced foods provide highly attractive variability.

Fish is a desirable food primarily because of its protein, which is a major proportion of its dry weight and is of high quality. Fish proteins, like those from mammals and birds, contain as components relatively high proportions of lysine, threonine, and tryptophan, amino acids that are scarce in plant proteins and that are needed by animals to avoid malnutrition. Thus fish in the human diet and fish meal in other domesticated animal diets may be essential if those diets lack other animal proteins. And despite the diversity of fish available for the diet, they are all similar in the quality of their protein.

The oil content is also of nutritional importance from the standpoint of both quantity and quality. The oil proportion of 5 to 15% in the flesh of moderately oily fish such as herring, mackerel, and salmon is about the same as the fat in lean meat. The fish with less than 5% oil such as cod, carp, flounders, most of the perchlike fishes, and all shellfish have a much lower calorie content than meat or poultry. Thus the fish in the diet can be chosen according to whether moderate or low caloric content is desirable in combination with the high-quality protein.

Fish oil (except crustacean oils) is quite different in quality from that of other animal fats, principally because of a higher proportion of polyunsaturated fatty acids. These have been shown to be desirable in human diets because they tend to lower blood cholesterol, and they also oxidize more easily than other animal fats and cause more rapid chemical change in fish flesh.

The other major characteristic of fish flesh is the usually low proportion of carbohydrates. These are rarely above 1% except in a few shellfish.

12.1.3 Bioactive Compounds

Ancient dietary taboos such as not eating mollusks or fish without scales probably arose because of a perceived relationship between their use and human health. Some unfortunate incidents must have arisen from contamination, but others probably came from intrinsic hazards in the organisms. Ancient medical practitioners perceived benefits in various extracts of aquatic plants or animals that they prescribed.

Now, when almost all aquatic organisms have been identified and can be captured in quantity, some are known to be naturally toxic or venomous and to be avoided for those reasons. But the toxins or venoms may have useful pharmacological properties as may other compounds found in aquatic animals.

The *toxic* animals appear to be those that have biologically magnified toxins in their diet or their environment to quantities that are hazardous to humans. The most serious natural toxins in edible animals are of three kinds: ciguatera, which is found in many species of inshore tropical fish between latitudes 30°N and 30°S; paralytic shellfish poison, which is found in several species of mollusks at many scattered locations in northerly latitudes; and puffer fish toxin. Both ciguatera and paralytic shellfish poison are widespread and erratic in their occurrence, and both appear to be related to the presence of toxins in the organisms eaten by the fish or shellfish. Neither is destroyed by cooking. Both are neurological poisons that cause tingling and numbness and both are occasionally lethal. Puffer fish toxin is more potent (and more frequently lethal), but the toxin occurs primarily in the viscera. Puffer flesh is usually only slightly toxic and is eaten in some places for the intoxicating effects.

Other, less hazardous natural toxins occur occasionally. At least two species of fish, one called castor oil fish, have a purgative effect. A species of trunkfish has toxins in its skin. Some species accumulate heavy metals, notably mercury, from natural sources to the extent that regular consumption should be avoided.

One form of toxin develops occasionally after capture of fish in tropical climates. Scombroid poison may occur in tuna, mackerel, mahi-mahi, and a few other fish if they are not chilled promptly after capture. Proper chilling is difficult with large catches of large fish, which may lie on deck too long and whose temperature may have been 25°C or higher at capture. The exact nature of the toxin is not known but it is unaffected by canning or freezing.

Venomous animals occur almost everywhere in the shallow seas, less commonly in fresh waters, but most of them are relatively innocuous. They include jellyfish, sponges, anenomes, corals, and others with stinging cells that are used for capturing food and many kinds of animals with spines for protection, which may be equipped with poisons. Most of the fish with poisonous spines belong to the stingray, weever fish, stargazer, and scorpion fish groups. More dangerous though rarer animals are venomous sea snakes and a few species of snails called cone shells.

The study of aquatic organisms for potential drugs extends the search for natural drug products from terrestrial sources, where many useful drugs have been discovered during the past half century. Various extracts or isolates of aquatic origin already have been discovered to have antimicrobial, cytotoxic, cardiovascular, or psychotropic effects and the search has been accelerating during the 1980s and early 1990s.

12.1.4 Contaminants

Fish flesh may become contaminated with microorganisms pathogenic to humans or may accumulate chemicals that, like the naturally occurring toxins, may make them inedible. The effect may be immediate with raw sewage and petroleum spills or long delayed because of storage and cumulative processes. For example, pollutants may accumulate for many years in bottom muds from which they can be released by dredging or floods. And some organisms may be unable to excrete chemicals that do not harm them but that are dangerous to humans. Such organisms may become dangerous to eat after months or years of exposure. (Contamination through spillage, a fundamentally different process, will be considered in another section.)

Domestic sewage, if not treated, frequently contains two types of bacteria and a virus that can cause serious human health problems. The most troublesome bacteria are of the *Salmonella* (not in any way related to the salmonid fish) and *Shigella* groups, which cause food poisoning, typhoid, and dysentery. The virus is the one causing infectious hepatitis.

These organisms are most commonly transmitted to humans when consumed by filter-feeding mollusks that are then eaten raw or only slightly cooked. The dangers of such contaminated mollusks are recognized in most countries, which test and set standards for the quality of water in which the mollusks are grown or cleansed as well as for the mollusks themselves. Finfish products are much less likely to be implicated in these health problems because they are usually well cooked. But there is danger that handling the contaminated fish can lead to transfer of the organisms to other foods, and the dangers are increasing with the use of human sewage in farm fish ponds.

Heavy metals, notably mercury, cadmium, lead, and zinc, are of special concern, although serious poisoning has occurred only from mercury. Mercury occurs naturally in all oceans and, as noted earlier, a few large, high-seas fish accumulate enough to be of concern. Much more serious has been mercury from industrial pollution, which has caused lethal poisoning episodes and indefinite bans on fishing in certain waters. A common regulatory limit on mercury content of fish flesh is 1.0 mg/kg wet weight for consumption not more than once weekly and 0.5 mg/kg for regular consumption.

Organic chemicals, especially the chlorinated hydrocarbons such as DDT, aldrin, dieldrin, lindane, polychlorinated biphenyls (PCBs), and others with industrial or agricultural uses, have increased in some aquatic environments to the point at which they have damaged populations of plants and animals and made some animals dangerous for human consumption. The prevalence of these has led several countries to monitor routinely the water and animals in certain places to ensure their safety.

Petroleum, even in small amounts, taints the flesh of fish, but such fish are so easily identified and unlikely to be eaten or even marketed that they are not a human health hazard.

Unusual food may cause color and flavor changes in fish flesh. Pink flesh may be related to a diet of crustaceans. Green flesh may also occur. Flavors that can be described as weedy, muddy, iodine, oily, or sulfide may occur in certain species, seasons, or places. The undesirable quality can be controlled only by avoiding capture of the animals with such conditions.

12.1.5 Disease Organisms

All fish carry disease organisms, but those that are seriously ill are usually consumed by predators before they are caught or they are so obvious that they can be discarded. Of special concern, therefore, are fish that look well but carry diseases transmittable to humans. Disease organisms that occur solely in the viscera, which is discarded or dried to animal-grade meal, are of little concern.

Human diseases that fish can transmit (other than as a carrier of a contaminant) are restricted to a few worm parasites that inhabit a fish host at one stage in their lives and can develop at the same or another stage in humans. They include the broad tapeworm, *Diphyllobothrium latum,* the adults of which grow in human intestines from larvae that occur in the flesh of many species of northern freshwater fish. Another is the lung fluke of humans, *Paragonimus,* which has intermediate hosts in crayfish and freshwater crabs of Southeast Asia. A third is a marine roundworm, *Anasakis,* which infests several species of fish and can grow in the human stomach or intestines. These are hazards only occasionally when fish are consumed raw.

Offensive fish diseases are much more diverse. Microorganisms including bacteria, fungi, and protozoans cause pustules and soft spots in the flesh. Larval stages of flatworm and roundworm parasites live in the flesh or skin of many different fishes. An external crustacean parasite attaches itself with an anchor in the flesh of a few fish. Such offensive diseases are usually avoided by rejecting infected fish, although a few parasites or soft spots in fillets of valuable fish can be located and individually cut out by "candling" over a bright light. Offensive diseases in farm fish can be positively controlled. If infections occur, treatment must start immediately.

12.1.6 Deteriorative Processes

Fish probably make up the most diverse and the most perishable group of human foods. Each species has unique flesh characteristics that usually vary with size, sex, season, and locality. Changes start with the first struggle after contact with fishing gear and, of course, accelerate after death. Then changes are associated with each of the many possible ways of handling between capture and consumption. Not all changes should be considered spoilage, because many are readily acceptable to consumers, but their ideal is usually strictly fresh fish and most changes are regarded as deteriorative.

The *struggle* during capture, especially with large fish on a line, causes an accumulation of heat and lactic acid in the body of the fish, both of which change flesh quality and may contribute to more rapid changes after death.

Death is soon followed by rigor mortis, which is associated with enzymatic changes in the flesh. This lasts for various lengths of time depending on species and temperature but it is a period when mechanical injury to the stiff body is especially easy. After rigor, the deteriorative changes accelerate as physical, microbial, and chemical factors act together.

Physical changes after death are usually caused by handling and may contribute to subsequent microbiological spoilage. These changes include crushing and bruising in nets, on decks, or in storage; puncture wounds due to forks or careless guttings; or failure to bleed promptly. Other changes may be caused by undue exposure to sun or drying. Later on, changes occur in the firmness and texture of the flesh, which are associated with both the prior condition of the fish while alive and the progressive microbial–chemical changes during storage.

Microbial changes are inevitable after death. Prior to death the skin keeps microbes out, but soon afterward the skin and flesh can be invaded, especially through bruises and cuts and in association with chemical changes. The sequence of change is usually marked by odors that start with the natural species odor, then become stale, sour, or acidic, and progress to bitter, ammoniacal, and fecal. If the fish happen to be packed out of contact with air, a different group of microbes may develop sulfurous or rotten egg odors. The sequence is also marked by changes in appearance as colors fade, as skin and gills become cloudy, and as eyes change from round and bright to sunken and opaque.

Chemical changes follow rapidly. Usually the first are changes due to the fishes' own digestive enzymes acting on the dead flesh. Then come changes in flavor due to action of flesh enzymes. Later come major changes in protein and fat composition from the combined effects of body and microbial enzymes. Other changes include loss of water and gradual oxidation of fats to produce rancid flavors.

12.2 METHODS OF PRESERVATION

If fish are not to be consumed soon after death they will spoil unless preserved. The period of acceptability varies with climate, species, consumer, and condition of the animals. It may be less than an hour for some crustaceans, and it is seldom longer than two days for finfish in warm climates.

Thus, if a fisherman catches more than about a day's ration for himself, family, and friends, he needs to preserve the catch for future use or for a market. If he is a commercial fisherman, he will probably sell his catch to a processor who will prepare a product to be sold ultimately to consumers at a profit. But every commercial fishery is unique. It is a system of resource capture, processing, transport, and marketing that has brought together a supply of food and a set of consumers. The methods of preservation have been developed to fit many circumstances and produce food varying from the full fresh flavor to diverse kinds of relishes. Sources giving detailed descriptions of some methods are listed in the Bibliography at the end of this volume.

12.2.1 Live Fish

The best possible method of preservation is to postpone death whenever practical. It is commonplace in many clam, oyster, and lobster fisheries in which the animals are kept alive until plunged into the consumer's cooking pot. Many crabs, which spoil very quickly after death, are kept alive on board vessels until they can be delivered to processing plants, which cook them immediately.

A few finfish are marketed alive in some localities. These are usually species that tolerate crowding in tanks during transport or display. Included are carp, eels, and a few tropical species. In some places anglers are allowed to catch their fish to be cooked for them immediately or to be taken home.

Fish farmers have the advantage of keeping their fish alive until they are in prime condition for market and then killing and processing them in ways that minimize deterioration (see Section 12.3.1).

12.2.2 Curing

Curing includes the traditional methods of preservation by salting, drying, smoking, pickling, and fermenting. The practices have been widespread for millenia and they vary greatly according to the raw material, the traditions of the communities, and the markets. The products include the lowest-priced dried fish and the highest-priced caviar.

The preservation principle is to reduce the rate of deterioration and create flavors that are attractive to consumers. The biological principle of the various methods is to decrease the water content or to increase the acidity of the fish and thereby change the microbial flora. It is usually impossible to eliminate the microbes, because even heavily salted fish will support halophilic bacteria or molds. Rather the processes have developed through trial and error to control the microbes and produce salable products, some of which are highly prized.

Many curing practices are wasteful of the raw materials. Poorly controlled smoke drying may overcook the fish and cause large losses of protein quality. Similar protein losses may occur during the wet salting and pickling processes. Air-drying, which is common in many countries, is difficult to control because of weather changes and the likelihood of insect infestation. Even after careful processing, the products may spoil due to increase in moisture content or infestation by insects during storage.

It is not surprising that some of these varied and widespread practices create health hazards. Many products carry risks because of dysentery-type organisms and are unsuitable for people at risk of dysentery, especially young children. A few carry risk of botulism.

The overall proportion of world catch that is preserved by curing has decreased since 1950 when it was about 26%. It is currently only about half as much (Fig. 10.8). In contrast, the growing markets for dried fish meal and frozen fish now receive much greater proportions of the world catch. Although they are not usually included in the curing processes, most marine plants are dried to a form suitable for human food or as a preliminary to processing.

12.2.3 Chilling and Freezing

Fresh fish, if it lives up to its name, is almost always fish that has been chilled and kept that way until it is consumed during the few days that it retains its freshness. Chilling with ice is the predominant way of reducing the rate of deterioration, but this method has its limits. Cod, for example, can be expected to remain acceptable for 15 days at 0°C, 6 days at 4.4°C, and less than 3 days at 10°C.

The period of acceptability also varies greatly with species. Large firm-fleshed fish such as halibut will remain acceptable up to about 21 days. Small soft-fleshed fish such as hake, even at 0°C, may not be acceptable for more than about 5 days. Oily fish and the oily dark-meat parts of fish spoil more rapidly than the less oily portions.

Chilling with ice is, of course, the common way of temporarily preserving fish on board vessels until the catch can be landed for final processing. But the final processing can never reverse any deterioration, so the chilling must be done promptly and with care to chill all parts of the catch uniformly without crushing or bruising the fish.

An alternative method of chilling now becoming widespread is the use of refrigerated seawater to which additional salt may be added. This has advantages such as reducing physical damage and disadvantages such as salt penetration into the flesh of small fish.

If chilling reduces the rate of deterioration, freezing should be better, but there are added problems. Each freezing, storage, and thawing cycle further reduces quality, even if only slightly. Thawed fish differ in flavor and texture from fresh fish in various ways according to species and the frozen experience. A few, such as oysters, become unacceptable to most consumers.

The changes in flavor and texture are related to the initial quality, the freezing process, the storage conditions, and the storage period. The initial quality must be better than just acceptable and the better it is the longer the product can be kept in storage. The freezing process should be quick, which is defined in some places as cooling from 0°C to −5°C in less than 2 hours and continuing the process until −20°C or lower is reached. The product being frozen at this speed must usually be less than 15 cm thick. Microbes usually cease to multiply at lower than −10°C, but they live at lower temperatures and resume activity as soon as the temperature rises to tolerable levels.

The storage conditions that most affect quality are temperature and packaging. Temperatures of −30 to −40°C are recommended for long-term storage of many species. But at any temperature, the species with less oil keep their quality longer than the oily species. Airtight packaging is essential to prevent drying, which is called "freezer burn," a condition of spongy fibrous flesh with off-flavors. The packaging should not only be airtight but not even slowly permeable by oxygen, which will react with fats and create rancid flavors. Additional protection for fish to be packaged can be supplied by a dip in a gel solution, and large fish can be protected for short-term storage by glazing, that is, dipping the newly frozen fish in water. The best practical storage conditions still result in steady loss of quality. The period with acceptable quality is less than a year for the most amenable species and less than 6 months for the least.

12.2.4 Heating

Many cooking processes are used to prepare fishery products. The basic principle is to slow or stop the action of microbes and all enzymes. Temperatures throughout the products can be raised to pasteurization levels of about 80 to 85°C or to sterilization levels of about 115°C.

Products cooked to the pasteurization levels are boiled, fried, or steamed in a wide variety of ways. Whole gutted fish can be boiled in brine, cooled, and distributed immediately. Shrimp can be boiled on board the trawler to arrest deterioration immediately and deliver them in better condition to the packaging plant ashore. Fish portions are increasingly prepared as convenience foods by breading, partial cooking, packaging, and freezing for distribution to consumers.

Perhaps the most widespread processing of fish products by heat is canning, packing the fish in hermetically sealed containers that are sterilized to virtually inactivate all microbes and enzymes. Such products have the added advantage of being immune to contamination until opened by the consumer. Major canned fishery products include shrimp, tuna, salmon, herrings, and mackerels. Most are for human consumption, but a substantial fraction is for pets.

Another common process that uses heat is manufacture of fish meal and oil for animal food and industrial use. The raw material is waste from fish-processing plants, species caught incidentally during trawling but unsalable for food, and small abundant species such as various herrings, anchovies, and mackerels. The raw material is cooked, pressed to remove oil and water, and dried to a meal in hot air or high-temperature steam dryers. The oil is separated from the water, which contains many desirable solubles. This "stickwater" is evaporated to about half solubles and sold in liquid form or dried completely and added to the meal.

12.2.5 Irradiation

Gamma irradiation can be used by itself or in combination with other methods to delay deterioration in fishery products. Many studies during the past two decades have determined the effects of various dosages on different products. Irradiation will inactivate or destroy bacteria but will not stop the continuing chemical changes in flesh. Higher doses that are necessary to destroy bacteria will also produce flavor changes. Ultraviolet irradiation can be used to sterilize water and studies have been made of its use on fishery products. No extensive commercial application to fishery products exists as of the early 1990s.

12.3 QUALITY

The quality of a fish product is ultimately a concept of consumers that embodies an immense range of conditions. Consumers may consider satisfaction, nutritive value, flavor, appearance, safety, presentation, confi price. The fisherman and the intermediate owners who catch, proces transport, and store the product en route to the consumer will each h limited concept based on the quality required by and price expected from the next buyer. Each will be inclined to deal with only a part of the consumer's expectations. Most concepts of quality include judgments difficult to quantify but a few are broadly agreed upon and enforced by governments or effected by industry in its general interest. These include factors of health, safety, truthful labeling, correct weight, lack of contaminants, and lack of offensive substances.

12.3.1 Quality Control

Quality control includes the measures taken to provide the consumer with an acceptable product. It is much more difficult with wild fish than with meat from domestic animals. The latter are usually killed instantly at an optimal age, in a known condition, and in quantities that can be immediately sold or processed. The sanitary conditions of slaughter and handling can be maintained at high levels. Consequently, the postmortem deterioration can be minimized during any periods of processing, transport, and storage.

Wild fish, on the other hand, are usually caught throughout the year, the condition varying by season, and the quantities by seasonal abundance rather than according to market demand. The locality of capture and death is usually on a vessel or on a beach far from a sanitary processing facility and the catches are necessarily handled extensively between death and preservation. Consequently, the postmortem deterioration starts and may be extensive before processing can take place. Further, it may vary greatly within the catch of a vessel during a single trip according to species, day of capture, and conditions of stowage.

After arrival at the processing plant, the fish are under control of a single organization that can more easily ensure proper care, but subsequently the product will usually be transferred to the consumer through processor, wholesaler, and retailer under widely varying handling and storage conditions.

Thus control of fish quality becomes a very complex problem. The diversity of fish and the products, the inherent susceptibility of all of them to spoilage, the variability of quality within catches, and the many people involved in handling all add to the complexity. Essentially the quality received by the consumer is a function of the care and skill of all the people involved from the fishermen through the retailers.

One approach to quality control is through combined industry effort, which may be broadly supported when publicity about trouble with one company's product can damage the sales of all companies making the product. An example is the National Food Processor's Association (formerly the National Canner's Association) of the United States, which maintains research laboratories that deal with many fishery product problems. It publishes recommended process times for established canned products, researches processes for new products, helps train cannery workers, and requires adherence to a code of ethics for members. It helps establish and may participate in joint government–industry quality control programs.

12.3.2 Quality Assessment

The ultimate judge of quality is the consumer and a major method of testing is the organoleptic test by a taste panel. A variety of methods are used, some with untrained test panels and others with trained panels. The untrained panels are commonly presented with pairs of samples to be compared or with samples to be ranked according to particular characteristics. More complex descriptive tests with trained panels are desirable for product development or research. These might consider ranking a sample according to each of the seven major quality factors of fish: (1) general appearance, (2) appearance of the flesh, (3) texture of the raw fish, (4) odor of the raw fish, (5) odor of the cooked fish, (6) flavor of the cooked fish, and (7) texture of the cooked fish.

A variety of chemical tests for fish spoilage have been tried, but few have widespread application. Those that are most used are measures of volatile bases and trimethylamine (TMA) because of the production of ammonia and amines during spoilage. The TMA test is most useful for codlike fishes because the precursor of TMA, trimethylamine oxide, occurs in variable amounts in most other species of marine finfish and is usually absent in freshwater species. Other tests are being tried such as measurement of electrical resistance as an index of change.

A variety of other chemical tests are used where toxins are suspected. These are, of course, designed according to the suspected material. Two of the most widespread are the mouse test for the naturally occurring toxins of paralytic shellfish poison and a test for histamine, which, although not toxic, is an index of scombroid toxin when present in large quantities.

Microbiological tests are used extensively in research on fish quality and for routine quality control on most processed fishery products. They are essential to identify the role of the various microbes in the deteriorative process and in the development of human health hazards. They are, however, usually too time-consuming for inspection of fresh fish products. Rather the relationships between the microbial changes and the organoleptic, chemical, and physical changes are used as a basis for quality judgments.

Microbiological tests are also used for routine assessment of the quality of bivalve mollusks and the water in which they live. This is provided by determination of the number of coliform organisms, which is an index of pollution by human waste.

12.3.3 Quality Standards

Standards are usually product specifications that are voluntary or mandatory, reasonably complete, and widely applied. They may be the minimally acceptable attributes in the product or the attributes of two or more grades. The standards have two general purposes. They assist the industry through description of uniform products and development of confidence in them. When accompanied by inspection systems, they protect the consumer by preventing the sale of harmful or substandard items.

Many countries have voluntary or mandatory standards for a wide variety of fishery products. An example of government–industry collaboration in use of standards occurred in the United States following the introduction of frozen fried fish sticks during the 1950s. Sales grew rapidly at first, then decreased due to some low-quality products that could not be identified at time of purchase by the consumer. Government and industry organizations together promulgated minimum quality standards for many frozen fish products. Those meeting the standards could carry a government certification of grade or a statement that they were packed under continuous inspection.

Standards for bivalve mollusks include both the amount of paralytic shellfish poison (PSP) and the numbers of coliform organisms. In the United States the standard for PSP is set by the Federal Public Health Service at a maximum of 80 μg/100 g. The bacteriological criteria for fresh and frozen shucked oysters at a satisfactory level is set by the same agency at a fecal coliform density of not more than 230 MPN (most probable number) per 100 g and a 35°C plate count of not more than 500,000/g. Even though these standards are met, the oysters must be identified as having been produced under the general sanitary controls of the National Shellfish Sanitation Program (see Section 12.3.4).

International standards have been under discussion by more than 30 countries since the 1960s under the Joint FAO/WHO Food Standards Programme called the Codex Alimentarius Commission. In 1978, 116 countries had become members of the Commission. Minimum standards for a few fishery products have been formulated, the provisions usually covering such matters as: (1) name of the product, (2) forms of product included, (3) description of all forms of product, (4) essential composition and quality factors, (5) food additives permitted, (6) potential contaminants, (7) hygiene, (8) weight and measures, (9) labeling, and (10) methods of analysis and sampling.

Unfortunately, few if any of these standards have been accepted by governments, but they have provided an international view of many matters of hygiene, contaminants, and defects. Agreement on matters of degrees of staleness, deterioration, and contamination has been very difficult because of immense differences in market requirements among countries. More widely used are criteria suggested by the International Commission on Microbiological Specifications for Foods (ICMSF), which have been incorporated into the regulations of a number of countries.

Standards also are being established for fish seed used by fish farmers. These include attributes such as correct identification to species and strain, freedom from animal diseases, and (in the case of oyster spat) freedom from unwanted invertebrate animals.

12.3.4 Codes of Practice

Codes of practice are process specifications or guidelines. They are usually advisory to industries that produce similar products or use similar facilities. They may include recommendations for type of construction, mode of facility operation, methods of cleaning plant and equipment, personal hygiene, packaging requirements, and handling of the raw materials at all stages. They are commonly prepared in sufficient detail for use by all employees who handle the raw materials.

International codes for several fish products have been prepared by the Codex Alimentarius Commission. The one for frozen fish contains several hundred recommendations that start with raw material requirements and fishing vessel design and end with retail practices.

A few mandatory codes of practice are enforced by governments. An example in the United States is the Code of Federal Regulations relating to human food manufacturing, which is enforced by the Food and Drug Administration. Some of the provisions apply to specific fish products.

Another example is the National Shellfish Sanitation Program of the United States government. The national program was started in 1925 and has been revised periodically. It details the codes of practice for ensuring sanitation of shellfish growing areas and for the harvesting and processing of shellfish. It assigns responsibility to the states and requires compliance by the states with respect to all growing areas, harvesters, persons, and facilities.

Government regulatory standards or codes are always an interference with free trade and require a pragmatic approach to the questions of what can be done and what it will cost. There is no possibility of complete consumer protection unless questionable products are removed from the market, such prohibitions serving neither the consumer nor the industry, and it is equally difficult to try to mandate uniformity for products as highly variable in character as fish. Standards and codes must be based on sound, fair, and widely understood management practices and devised through collaboration between industry and government.

12.3.5 Inspection Systems

Quality control of any product must include monitoring by inspection. The purpose is to evaluate how well the production system has met the quality objectives, to identify any problems, to assist in their correction, and to ensure delivery to the consumer of a safer, more valuable product. Inspection is always a responsibility of facility management and supervision. It may also be conducted by an industrywide organization or by a government agency.

Japan and most European countries have mandatory inspection of raw fish as landed at dockside where the fish are displayed for auction. Any fish found to be unsuitable for human consumption are condemned and used for fish meal. Other countries and localities may also inspect the raw fish at landing.

The inspection procedures include process inspection to ensure adherence to a code of practice and end-product inspection to ensure attainment of product standards. The process inspection should start with the fisherman, who sorts and preserves the catch, and continue with each subsequent owner. The end-product inspection can occur only when deterioration has been effectively stopped and the product protected by packaging and processing against further deterioration.

An example of voluntary government inspection in the United States is the provisions of paragraph 197 of the Federal Code of 1981 with respect to canned and frozen shrimp. The provisions include: (1) exclusion of uninspected shrimp from plants, (2) general requirements for plant and equipment, (3) general operating conditions for the plant, (4) code marking procedures for each lot of shrimp, (5) detailed processing procedures, (6) sampling procedures for examination after processing, (7) labeling requirements and permission to state on labels "Production supervised by U.S. Food and Drug Administration," (8) certification and record-keeping procedures, and (9) inspection fees. This system is a combination of process and end-product inspection. Similarly detailed provisions are given in other sections of the code for many other fishery products.

Another inspection system that is an example of a pragmatic approach to a very complex problem is inspection of bivalve mollusks in the state of Washington, United States, for PSP. The clams and oysters are caught commercially as well as for personal (sport) use. The occurrence of PSP varies with location, species, and time of year. The only valid test procedure is a bioassay that has been performed by the State Public Health Laboratory in accordance with federally prescribed procedures (Houser, 1965; Jensen, 1965). Samples are taken from the waters by the State Department of Fisheries and, as necessary, commercial fishing area closures are announced and enforced by the Department of Fisheries. Personal use fishing areas are under control of the county health departments, which set regulations as they deem necessary.

This complex system has worked. Despite the occurrence of lethal concentrations of PSP every year in certain species, locations, and times, there have been no cases of poisoning from commercial harvests since 1957 and no serious cases from sport harvests.

12.4 MARKETING STRATEGY

Fishery scientists have been predominantly concerned with fish resources, and the marketing function is regarded as part of business. Yet food fisheries exist because markets bring fish to the consumer, and scientists have begun to take an increasing role in marketing strategy.

Marketing is a function of every business and the consumer may know little of the business except the marketing. But marketing is especially important in fish businesses, which have such diversity of products, such volatile supply, and such difficulty with maintaining quality. The successful fish businesses are those in which the managers have developed effective marketing despite such difficulties. In fact a major activity of most managers of fish businesses is marketing.

Marketing strategy in the dynamic fish business involves setting fundamental objectives. These may include targeting a share of products in existing markets; trying new products, services, and distribution systems; entering into new markets; or abandoning old products and markets. Each will be considered on the basis of the ability of the business to perform and the relative attractiveness of each opportunity. Evaluation of the attractiveness requires knowledge of the consumers and a forecast of their buying habits, information that is gathered through research.

12.4.1 Long-Term Trends

A worldwide pattern has been evolving in the fisheries for several decades. The surge in world fish production following World War II (Fig. 10.1) and the rapidly increasing proportion of the catch marketed as frozen or fish meal products (Fig. 10.8) have involved major changes in distribution and marketing. The most remarkable change in distribution during the surge in production was the increase in international trade in fishery products (see Table 10.6), from about 20% of catches in 1950 to about 40% today. In terms of volume, the catches increased about three and a half times, and the international trade about seven times.

Much of the increase has been due to improvement of the freezing process. During the 1930s, fish were frozen primarily to extend their shelf life and thawed before selling (when they were commonly an inferior product). During the 1940s, a few fish products were sold frozen to consumers. Then, during the 1950s, convenience products and packages were introduced for family and food service institutions and these have continued to grow in popularity and volume.

One of the key factors in this growth of frozen products was an increase in consumer confidence. Frozen fish were first regarded as inferior (and still are in some countries) until major efforts to control quality were started. These worldwide efforts have also contributed to the increasing proportion of fish in international trade. Further improvement of quality is continuing as more fish are processed and frozen at sea within hours of capture and as tougher inspection procedures are instituted by companies and governments.

Another change has occurred in distribution patterns. The specialized wholesaling necessary for small lots of fresh fish was not necessary for large lots of frozen fish in which quality was more certain. Food service and retail institutions began to buy directly from processors when they could be assured that every shipment met quality standards. Still another change has occurred in company size. Frozen fish sold in larger lots, to different markets, and in international trade has involved greater risks and required more capital. This has led to more mergers and larger organizations.

Canned fish production has increased at a rate scarcely greater than the increase in population. Unlike the quality of frozen fish, there has been no major increase in the consumer's perception of quality, which is associated primarily with the species canned. And the supply of the major canned species—herrings, tunas, and salmons—has been limited by the sustainable yield of the resources. The present markets seem to be largely composed of traditional consumers who are familiar with the products.

The supply of fish meals and solubles increased dramatically during the 1950s and 1960s, from about 1 million metric tons (MMT) to about 5.5 MMT (product weight). This was used primarily in domestic animal feeds, at first as a routine additive, but later more flexibly according to supply and price relationships with plant proteins. The supply during the 1970s declined somewhat with the collapse of the Peruvian anchovy fishery and seems unlikely to increase to the previous peaks. Markets are now dependent on the changing animal feeding practices and the supply of competitive proteins as well as the supply of fish.

Fish marketing in the future seems likely to be at least as dynamic as in the past. With the supply of conventional species nearing a limit and with improvements in quality control of frozen fish, the price of fish relative to competing food products should rise and cause changes in consumer buying habits. Part of the challenge will be to market the new product forms, especially the rapidly increasing production of minced fish and simulated products based on it (e.g., "crab legs").

12.4.2 Market Intelligence

Information for business forecasts, either for routine day-by-day decisions about sales and prices or for special occasional decisions related to new activities, derives primarily from market data and analysis. The market data come from company records, news sources, trade association statistics, and government statistical services. The latter are a primary source in many countries and their collection and presentation are a substantial activity of fishery agencies.

An example is the Federal fishery statistical service of the United States. Commercial fishery statistics are collected through about 45 Resource Statistics Offices of the National Marine Fisheries Service and with assistance of other state and federal agencies (e.g., U.S. Department of Commerce). Most basic production data are derived from copies of the first sale. Other data on processing, inventories, imports, exports, vessels, employment, or businesses are derived from a wide variety of sources.

Market news useful principally for day-to-day decisions is published three times weekly from five major offices. It includes data on landings, market receipts, cannery receipts, and prices at ex-vessel, wholesale, processor, importer, and broker levels. Data on imports, exports, cold storage holdings, and cannery packs are published at weekly or monthly intervals. In addition, many pertinent news items are included in the bulletins.

Market statistics more useful for special decisions are assembled and published annually. These include most of the preceding kinds of data summarized for each year with appropriate comparisons over the previous five or ten years. These are accompanied by other comparative data on foreign fishing, marine recreational fishing, or employment in fisheries. A datum of special importance is the long-term trend in consumption per capita of each major fishery product.

12.4.3 Consumer Preferences

Specialized market research may include structural analysis of consumer groups, for example, a stratified sample of consumers may be asked about purchasing habits and preferences. The data are then analyzed to determine underlying motives according to regional and sociological characteristics. An analysis of trends may be obtained through recurring surveys of a consumer panel composed of a representative sample of consumers.

Information on acceptance of certain brands may be obtained from consumer panels or from retail trade panels (samples of retailers from whom information on trends in purchases of certain commodities is obtained). The potential acceptability of new products may be estimated by special sampling surveys among consumers or by furnishing them examples of proposed products.

Such special surveys are usually performed by market research institutes and are far more expensive than use of routine statistical reports. They are commonly conducted, therefore, only for large businesses or trade associations.

12.5 ROLE OF FISHERY SCIENTISTS

The continuation of changes in supply, products, and markets seems certain to involve many more fishery scientists at all steps between capture and consumer. Especially challenging will be the problems associated with increased production of nonconventional species. Many of these are small and their use will require far more research on their flesh characteristics, processing methods, potential products, and market development. Use of such small fish will probably accelerate development of minced fish products, such as fish cakes and sausages, and of fish protein concentrate suitable for human use. Reduction of waste through utilization of bycatch from shrimp and other fisheries and through reduction of spoilage in small-scale fisheries is another major opportunity to increase supply.

Improvement of quality will remain a primary objective of most fish handling. Despite the gains of the past three decades that have been associated with the increase in frozen fish products and international trade, more can be done and the improvements extended to fresh and cured fish, which still comprise the majority of products in many countries. Basic researchers in universities and government laboratories will extend knowledge of the chemical and microbiological factors. Applied researchers will work in quality control laboratories. Broadly trained inspectors will be used by more businesses and government agencies to ensure adherence to standards.

Market intelligence systems will probably be extended to cover many more countries and to provide current data on the international markets for more products. Design and supervision of such systems require very broad knowledge of worldwide fishery practices.

The trend toward larger and more vertically integrated businesses will probably include a larger proportion of private employment for fishery scientists. Some will be able to advance the quality control through adherence to the international standards, and others may assist with marketing and investment strategy. Some of their work will be associated with improved processing at sea and its integration with the fishing.

Codex Alimentarius Commission Recommended International Standard (CAC/RS) for:

Quick Frozen—Gutted Pacific Salmon. CAC/RS 36-1970.
Fillets of Cod and Haddock. CAC/RS 50-1971.
Fillets of Ocean Perch. CAC/RS 51-1971.
Fillets of Flat Fish. CAC/RS 91-1976.
Shrimps or Prawns. CAC/RS 92-1976.
Fillets of Hake. CAC/RS 93-1978.
Lobsters. CAC/RS 95-1978.

Recommended International Code of Practice for:

Fresh Fish. CAC/RCP 9-1976.
Canned Fish. CAC/RCP 10-1976.
Frozen Fish. CAC/RCP 16-1978.
Shrimps or Prawns. CAC/RCP 17-1978.
Molluscan Shellfish. CAC/RCP 18-1978.

REFERENCES

Houser, L. S., ed. (1965). "National Shellfish Sanitation Program, Manual of Operations. Part I. Sanitation of Shellfish Growing Areas." U.S. Dept. of Health, Education, and Welfare, Washington, D.C.

Jensen, E. T., ed. (1965). "National Shellfish Sanitation Program, Manual of Operations. Part II. Sanitation of the Harvesting and Processing of Shellfish." U.S. Dept. of Health, Education and Welfare, Washington, D.C.

Kizevetter, I. V. (1973). "Chemistry and Technology of Pacific Fish." U.S. Dept. of Commerce, National Technical Information Service, Washington, D.C. (Translated from Russian.)

13 | Regulation of Fishing

Fishing has been a human activity since the first hunters and gatherers appeared on earth, and it is still important to a large fraction of our populations. The regulation of fishing, which has evolved with public support almost entirely during the past century, is one of the three major activities of fishery management. The others are environmental management and artificial stock enhancement. Usually all three are discussed together, but they differ in many respects and thus they are treated separately in this volume. A discussion of regulation follows the earlier coverage of stock enhancement, and environmental management will be discussed in Chapter 14.

The first and major difference is that each is regarded by the public in a very different way. A fishing regulation restricts people from the centuries-old traditional use of a public resource that many regard as a right. It is a legal action and, if it is to be effective, it must be acceptable to a majority of the people who are being restrained. Proposals for regulation are usually controversial and must be supported by sound information on their need. Proposals must be discussed in detail with people who will be affected or with their representatives. After a regulation is promulgated it must be enforced and monitored.

The other two activities are much less controversial. The purpose of environmental management is partly to improve the aquatic environment, which most people support. It also serves partly to prevent abuse of water by pollution or use for purposes that interfere with fisheries, which may be contested by only a few people. Artificial stock enhancement through public aquaculture is widely supported politically and economically by people who benefit from it.

The second difference among fishery management activities is in the scope and kind of scientific studies that support the decisions. Fishing is regulated on the basis of recurring assessments of the condition of the stocks and condition of the fisheries. If it is a major oceanic stock, the assessment may require a large, continuing scientific investigation; if it is a minor domestic stock, it may require only an opinion survey among the people using it. The aquatic environment is managed on the basis of limnological or oceanographic studies and detailed studies of the life of the organisms involved, plus studies of the feasibility of alternatives for change. Artificial stock enhancement through public aquaculture requires support from aquacultural sciences related to rearing the fish and from ecological studies of receiving waters as a basis for stocking policy.

This chapter begins with a discussion of the common principles of public action that apply especially to regulatory activities. It includes a summary of how agencies regulate for conservation and other purposes and an account of the new system of oceanic fishery regulation.

13.1 ORIGINS OF PUBLIC POLICY

The present functioning of fishery resource agencies developed basically from public attitudes that have evolved over a very long time. It must be presumed that the earliest people who subsisted largely by hunting and fishing originated the concepts that the game and fish belong to no one until after they are killed and physically possessed. Such wild animals were important to everyone who depended on them, yet, with a few exceptions, they could not be restrained and possessed while alive. If someone with unusual ingenuity managed to possess or claim the wild animals, he would have been considered to be reducing the resources for other people.

Apparently such natural law with respect to fish has been followed through the ages. In early Roman law, fish were placed in the category *res nullius*, things that belonged to no one, on the basis of the argument that this classification conformed with natural or moral law. Similar principles obtained by law in medieval times in Europe continue to the present. In most countries fishing laws are still based on the assumption that the fish belong to no one until caught, except for certain government waters and for private aquaculture.

Since fish have commonly been considered to belong to no one, fishing has been regarded by everyone as a right. In ancient China, all waters were free and open to fishing, with the exception of a few imperial reserves. Under Roman law, the sea and public waters were open for fishing by anyone. However, private waters were recognized and granted by governments in situations such as coves, backwaters, small lakes, and aquacultural ponds. Exceptions were also made in some places in favor of the fisherman who first occupied a site with fixed gear, such as a trap. He was allowed exclusive use of the shore or water for a reasonable distance around his gear even though he did not own the fish until he caught them. Today, most domestic waters remain open to fishing by all citizens of the country having jurisdiction. In countries with large numbers of recreational anglers, this right may be exercised by a large proportion of the population, who cause major reductions in the resources, and who support regulatory and stocking activities.

Those who fished in international waters, namely, the sea and a few large boundary lakes or rivers, went beyond the limits of domestic authority into an area that, in early times, was shared universally. It was an area of no law, an area of freedom from all domestic authority. But since A.D. 1300 that area has been gradually reduced and now we have a new Law of the Sea in which control of the fisheries is an important part.

13.2 PRINCIPLES OF PUBLIC ACTION

Almost all of the action that involves the common property resources embodies three fundamental principles:

1. Every social change produced by an individual, company, or government decision gives an advantage to some person or group and a disadvantage to others. Even when the purpose is to correct a disadvantage, the benefits cannot be assumed to be uniformly distributed.

2. Every ecological change caused by people in their use of water or the living resources gives an advantage to some organisms and a disadvantage to others, as long as the changes are within the physical and chemical limits tolerated by some living organisms. Such opposite effects are frequently overlooked when action is taken to improve conditions for something we value.

3. Advantages and disadvantages to segments of either society or the organisms in the environment seldom become evident simultaneously; this fact complicates the decision process when either benefit or damage is long postponed. Further, our ability to predict the consequences of delayed reactions rapidly declines with the length of the delay, and some decisions may be irreversible.

These principles are cited to emphasize the complexity of the decision process. Seldom are the issues black or white in either a social or environmental context. Seldom will action be taken solely on the basis of technical information, which never will be as complete as some people will desire or demand. And always the technical and social issues, both short and long range, will need to be evaluated together in a search for an optimal solution. Thus the decision maker's problems become the research problems: biological, technical, economic, legal, organizational, and political.

In a different sense, another set of practical democratic principles governs the action in regulation of fishing:

1. Most of the people being regulated must agree about the need for regulation and they must understand how the regulation is supposed to work. This requires knowledge of the resources and a respected forecast of the likely consequences of the obvious alternatives.

2. The regulation must be enforced or else the action of a few violators will destroy any confidence in its effectiveness.

3. Legal authority for the action must be secure and, if more than one political authority is involved, firm cooperation must be the rule. If the stock migrates between political jurisdictions, only joint action will be effective.

13.2.1 Government Decisions

Government control of common property resources is divided between executive and legislative agencies with varying degrees of satisfaction to the people. Currently the control methods are in turmoil as public aspirations change and as agencies vary in their ability to cope with the complex problems.

The complexities arise from the natural variability in the condition of the resources, from the unevenness of social and ecological impacts, and from the ease with which fish and water cross political boundaries. The natural variability in the condition of the resources and the unevenness of the ecological impacts require a technical basis for decision. The unevenness of social impacts, especially when some people are excluded from an activity that they regard as their right, requires a political basis for decision. And movement of fish stocks across political boundaries means that neighbors must agree if either is to make effective decisions.

The resulting arrangements are different in every political unit. Legislatures specify certain requirements by laws and delegate authority for specific kinds of actions to agency directors. The agency director may be responsible to his or her chief executive or to a permanent commission. New controversies are frequently dealt with by appointment of special study groups or temporary commissions. When large segments of the public become involved, the decision may be made by a referendum. And when existing laws conflict (as they usually do), the decisions may be made in the courts.

When the resource is divided among political entities, the first step toward collective decision making is usually an agreement among them that identifies the problems, specifies goals, and creates an organization with defined responsibilities. Usually the organization's responsibilities are primarily technical, that is, to determine the facts to be considered and to coordinate the negotiations. Authority for making regulatory decisions is usually reserved to the governmental partners in the agreement.

The extension of national authority to 200 miles under the new Law of the Sea is certain to require many new international fishery agreements. These will involve extension of national boundary lines, allocation of fishing on stocks shared by countries, permits for foreign fishing, and investigation of the status of the resources.

13.3 CURRENT REGULATORY OBJECTIVES

Fishing is regulated for many reasons, of which the oldest are probably allocation of the right to fish and protection of public health. The more recent goals have arisen from the realization that the resources are limited and that the public will benefit if waste is prevented, if the stocks are conserved, and if fishing is orderly.

13.3.1 Conservation

Since the occurrence of widespread public acceptance of the facts that fishery resources are exhaustible and that controlled fishing is essential for conservation, fishery scientists have been asked how much and what kind of fishing should be allowed. They have been studying the dynamics of wild animal populations since early in this century (Chapter 8). The most useful concept they have developed is Maximum Sustainable Yield (MSY), a value that can be calculated in various ways if the fish stock is assumed to be in a steady state.

After successful application of MSY to some relatively stable fisheries, the concept was widely accepted as an objective for fishery regulation. But regulation according to the concept does not always protect the stock or satisfy the users. Some very large stocks have collapsed while being regulated under the concept; the assumption of a steady state was invalid. Some commercial fisheries have become uneconomical even though the stocks have been sustained at maximum levels. And maximum catches do not reflect the concept of a quality experience for most groups of recreational fishermen.

The objective of regulating fisheries for MSY has been abandoned, although MSY continues to be a useful computation during the assessment of the condition of fish stocks. Instead, MSY is modified by biological, economic, social, and political values to produce the maximum benefit to society (Roedel, 1975). The resulting objective, called optimum yield (OY), may be (1) equal to MSY in some stable commercial fisheries used entirely for food, (2) zero catch for an endangered species or a species in a fragile environment, (3) near zero if the species is an essential food for a more desirable species, (4) a moderate fraction of MSY in order to produce trophy fish for a recreational stock, and (5) a catch rate greater than MSY if the species is an unavoidable component of a multi-species fishery.

International acceptance of OY as the central element of conservation came in the Convention on Fishing and Conservation of the Living Resources of the High Seas that was negotiated in Geneva, Switzerland, in 1958, but this was only a step toward a comprehensive Law of the Sea that required many more years to negotiate. Throughout the decade of the 1970s a new and comprehensive Law of the Sea was negotiated and the fishery provisions of that Law have been widely accepted (Section 13.6 and Appendix B). The major thrust of the Law gives coastal states authority over fishing and fishery resources out to 200 miles and the obligation to conserve the resources on the basis of scientific studies and in collaboration with neighboring states.

National fishery laws preceded the new Law of the Sea by hundreds of years with many objectives other than conservation. The conservation objectives that have been developed around the concept of OY in the Law of the Sea have been or will be adopted by most countries. But the other objectives of fishery regulation are constantly before the fishery agencies and may be their major activity. These will be discussed in approximate order of their importance.

13.3.2 Allocation of Fishing Rights

Allocation means reducing the right of some nations or people to fish where and how they please. It is a dominant feature of fishery law in most countries simply because there are not enough fish for all fishermen to fish in the way they please. It is always controversial when begun. The decisions about who gets the rights are political decisions. The laws are difficult to devise in a democratic manner and difficult to enforce (Stroud, 1980).

Allocation started centuries ago with kings' decrees to reserve fish for themselves or to protect their subjects from foreign fishermen. Now the new Law of the Sea gives coastal nations the right to control access to their 200-mile fishing zones and to allocate the catches between their nationals and foreigners.

Allocation within countries is widespread. In North America, most freshwater fish stocks have long been allocated to recreational anglers. Treaties with the North American Indian tribes have provided special fishing rights for them. In addition, many laws that set closed seasons, closed areas, catch quotas, and restrictions on gear are really for the purpose of both allocation and conservation.

Some allocations among nations and among fishermen using different kinds of gear have been established and accepted for a long time, but as commercial stocks become fully exploited, further allocation is needed. So many fishermen and so much gear enter the fisheries that many fishermen barely make a living and conservation becomes difficult. Consider, for example, a fishery for roe herring in Alaska in which an annual quota may be taken in less than one hour's fishing.

This condition is so common that a senior British fishery scientist (Graham, 1943) stated a Great Law of Fishing: "Fisheries that are unlimited become unprofitable." He used the word unlimited in the context of the amount of effort that must be reduced if the fishery is to be economically rational. This is attempted through a form of allocation called limited entry.

The causes of the condition are not entirely clear. Part of the difficulty arises from the misleading trends in the landings during the early period of the exploitation of a stock or stocks. This is a time of exploration and innovation as new grounds are discovered and gear is adapted. A large proportion of the catch from a newly fished stock consists of large, older fish that may persist in the fishery for several years. The nearby grounds that become depleted are easily bypassed in favor of more distant and productive grounds so that total landings and perhaps even catch per vessel tend to increase. It is a time of good profits that quickly attract fishermen and capital.

The reckoning comes with little warning or understanding. New grounds that produce exciting catches can no longer be found but hope remains that they may still exist. The accumulated stock of large fish has been caught, and catchable fish become available only as they grow above the minimum size. The catch per trip declines with the abundance of the stock, but hope remains that this decline is only a natural fluctuation. The large new vessels and plants that were built because of the prosperity a short time earlier are in financial trouble that can be corrected only by still greater catches.

These trends that deceive the fishing people and the money lenders are partly explicable by the uncertainty of whether the fluctuation in stock size is caused primarily by natural events or by fishing. If it has been caused primarily by natural events, such as those that cause unsuccessful year classes, then a reversal can be expected. On the other hand, if it has been caused by the application of so much fishing effort to the stock that its abundance has been depressed to the level of maximum sustainable yield or below, then a reversal cannot be expected without a reduction in fishing effort.

The reduction is accomplished by licensing only some of the fishing people and, in effect, allowing them to own an exclusive right to fish. Such limited entry has been tried in numerous fisheries around the world, usually with a mixture of social, economic, and conservation goals. When limits have been imposed on a fishery during its period of growth they have been reasonably successful, but attempts to impose them after fisheries have become unprofitable have created difficult social, legal, economic, and political problems (Rettig and Ginter, 1980) that have usually prevented achievement of the objectives of limited entry.

13.3.3 Orderly Fishing

Many provisions of fishery law help to identify fishermen or to avoid conflict. Licenses or regulations may specify locality, period, or kind of gear in ways that keep rival fishermen apart or that avoid physical interference such as destruction of crab pots by trawl gear. Licenses identify fishermen or dealers and the privileges that they have received. Laws and regulations may specify enforcement procedures or record-keeping requirements.

13.3.4 Prevention of Waste

Farmers' experience with raising animals probably led to early fishery regulations that were designed to protect spawning females and young fish. Now, in

North American recreational fisheries, fishing is usually prohibited during spawning seasons and young fish are protected by minimum size limits. Common sense led to regulations to prevent capture of animals in poor condition, such as some crustaceans immediately after molting. Other regulations attempt to prevent gluts of fish in excess of the ability of facilities to process them or catches by recreational fishermen in excess of their ability to consume them. An increasingly common regulation in some sports fisheries is a requirement for barbless hooks, which allow release of small fish with less risk of injury.

13.3.5 Protection of Public Health

Centuries ago, after people became ill from eating fish, some leaders established taboos to forbid consumption of animals such as fish without scales or certain mollusks. Now it is known that a few fish and shellfish may carry toxins or human disease organisms, and catches may be inspected and controlled by fishery agencies (see Section 12.1.3). For example, mollusks from polluted waters may transmit gastrointestinal diseases. Mollusks and fish may contain deadly toxins due to natural causes or to pollutants. A few aquatic animals may be venomous and should be avoided. Still others may transmit parasites to humans if not well cooked.

13.4 REGULATORY DECISIONS

The process of making regulatory decisions depends on the existence of a legal basis, the objectives and technical reasons for the proposed decisions, the experience of people being regulated with past decisions, the perceptions of people about how the proposed regulations will affect them, the nature of the stocks, and the administrative structure of the regulating agency. Let us discuss each of these factors in general with respect to the regulation of fisheries and then describe two quite different examples.

Some legislatures decided centuries ago that passage of detailed fishery regulations was impractical and they delegated restricted authority to fishery agencies. This is common practice today, but many new laws in the form of intergovernmental agreements are needed among political entities, for example, coastal countries, that share resources. So the first decision may concern the adequacy of existing authority and whether it must be modified to permit effective regulation. If the law permits, a proposed regulation with clear objectives and a technical rationale is prepared, reviewed, and distributed to the people concerned.

If the people have had experience with the past decisions of the agency that gives them confidence in its effectiveness, they will usually accept the basis for the decision. If not, they may question the objectives and supporting material in great detail. If the regulation is merely a minor modification of a previously acceptable regulation, it may not be questioned at all.

If some fishermen or companies think they will be adversely impacted they will contest the proposed action and try to modify it. This will usually begin in public hearings and, if a modification is not obtained or if they are not convinced that their concerns are unjustified, they may go to court. Vigorous dissent, including court action, may result if any people feel that their rights are being abridged.

Several aspects of the stocks must be considered during the regulatory process. If consistent regulations on many small stocks are needed, as in most domestic freshwater habitats, the regulations should be tailored to average conditions rather than to individual stocks. If a single large oceanic stock is being regulated, the regulations must be tailored to it. If the abundance of the stock is unpredictable, regulation of the catch may be modified on the basis of within-season experience. If the stock migrates into another political jurisdiction, regulation should be based on agreement with the other government.

Regulation is more important for the recreational fishing system than for either the subsistence or commercial systems. The subsistence systems have usually developed around traditional local activities with little or no government involvement. The commercial systems, if they do not catch fish sought also by recreational fishermen, involve competitive businesses that use fish stocks and that hurt few besides themselves if they abuse them. The recreational systems, on the other hand, involve far greater numbers of people who not only engage in fishing but treasure the environment and conditions under which they fish.

The government must not only act in the interest of the fishermen but must be perceived to do so. It must know the condition of the fish stocks and their environment, protect them from abuse, and allocate them fairly. The allocations involve use of the water for all its competing purposes as well as use of the fish stocks by all kinds of fishermen.

All of these activities require decisions that are primarily political. The decisions rely on the best possible technical information on the fish and their environment, but most issues involve values that are difficult to quantify especially because environmental values are perceived so differently by people. Also difficult is a definition of optimal yield from fisheries shared by subsistence, commercial, and recreational fishermen. Much effort has been expended by resource economists to develop economic rationale for the allocation or OY decisions, but although the results help make decisions they do not substitute for political insights or judgments.

The type of administrative structure and overall competence of the agency also will influence the decision process. If the director is responsible for the decisions, public hearings may be needed to deal directly with the concerned people. If the agency has a commission with broad public representation, the commission may be able to make decisions with a minimum of public hearings. The decisions will also be shaped by the ability of the agency to enforce, monitor, and assess the effects of regulations because the process is continuous, usually with an annual cycle.

13.4.1 Multiple Domestic Fisheries

Recreational fisheries in the individual states of the United States provide examples of regulatory actions on stocks within a single political entity and with a long accepted allocation of the right to fish. Here commercial fishing for the recreational species has been prohibited for many decades.

Many stocks of different species in rivers and lakes with widely varying degrees of exploitation are involved. Usually the actions taken are only an annual "fine-tuning" of previous regulations. In general, the procedures (Smith *et al.*, 1979) consist of an annual schedule, preparation and review of proposals within the agency, publication of proposals for public comment, public hearings and/or commission meetings, further consideration of proposals, adoption of regulations, and promulgation. The whole process usually requires several months, although provision is commonly made for immediate action in emergencies.

The resource information base for such regulation may vary from collection of fishermen's opinions about trends in catches to continuing statistical systems such as routine "creel" sampling in the field or an annual report required as a condition of a license. The stocks are usually far too numerous to permit individual attention and the role of the fishery scientist is usually judgmental rather than analytical. Detailed field studies of sample stocks may be undertaken and used as a guide to regulation of a group of similar stocks.

The characteristics of the fishing people are at least as important as the condition of the resources. The number of fishing people is very large (an estimated 25% of U.S. citizens go recreational fishing), although most of them fish only occasionally. They fish for pleasure and do not want to risk breaking the law. The regulations must be simple, easy to follow, and widely applicable to different waters if they are to be acceptable.

13.4.2 A Large Oceanic Fishery

At the other extreme in terms of complexity, time required, and cost is regulation of a large oceanic fishery. Regulation for conservation that is required by the new Law of the Sea must be based on continuing scientific assessment of the condition of each stock and the effects of fishing on it. Regulation must be consistent on the part of all political entities that share in the fishing (most large oceanic stocks migrate across political boundaries) and, in addition, regulation must meet all of the normal needs of flexibility according to the variations in social and environmental conditions, enforceability in all jurisdictions, and practicality within organizational and budgetary restraints. Such requirements can be met only through development of suitable laws in each political entity, a reliable statistical data base, an experienced scientific and administrative staff, and confidence on the part of the people being regulated that the system is fair and effective.

But if the fishery has been long established it will have attracted large investments in catching, processing, and marketing facilities. Any attempt at regulation will require estimates of the economic consequences of alternatives as well as estimates of the sustainable yields. Such attempts in large established fisheries have frequently failed because of vigorous opposition by the participants who expect large economic losses. Therefore, if effective regulatory action is not taken early in the development of a commercial fishery, the most probable action will be only a gradual modification of the practices.

13.5 ESTABLISHMENT OF A NEW FISHERY REGULATORY REGIME

The experience of the United States in establishment of an oceanic fishery regulatory system consistent with the new Law of the Sea provides an example of the procedures required to change complex legal, administrative, and scientific activities.

Few of the oceanic fisheries off the United States had been regulated prior to the early 1970s, when a general recognition developed that foreign fleets were catching more fish in the waters off the United States than were U.S. fishing people, and that the stocks in those waters (some of which were depleted) were larger than the stocks off most other countries. Law of the Sea negotiations had resumed at about the same time and general agreement on fishery law had been reached. But, with no end in sight for completion of the Law of the Sea negotiations, the immediate issue was whether the United States could control the fishing off its coasts in order to conserve the resources and allocate more of the catch to U.S. fishing interests—an issue similar to that faced by most coastal countries.

13.5.1 Policy Development

After vigorous expressions of public concern about the impact of foreign fishing on stocks of fish off the U.S. coasts, the next step, taken in 1974, was a thorough study of issues that concerned people in the diverse fisheries of the United States. Discussions continued for about two years with representatives from various fisheries and state and federal agencies. The results were, first, a much better understanding throughout the fishing communities and government of the diversity of issues involved and, second, a set of recommended policies (U.S. Dept. Commerce, 1975) that were stated as follows.

> A. The Federal Government will have the primary responsibility for management of fisheries in the 3- to 200-mile zone. This will be undertaken with the fullest possible participation and cooperation of the states, recreational and commercial interests, other special interest groups, and the general public.

B. Management programs will be developed and implemented by the most localized political entity with sufficient competence and jurisdiction, so that local social and economic considerations are fully recognized in the management actions, consistent with sound biological considerations and national priorities and policies.

C. User or consumer groups with interest or concern will advise government on fisheries management policy.

D. The United States will control access to fisheries resources in the 3- to 200-mile economic fisheries zone and to anadromous fish resources under U.S. jurisdiction.

E. The Federal Government will allocate fish stocks among domestic and foreign users for the benefit of the United States, under the concept of optimum utilization of the total biomass in each ocean region (i.e. taking into account biological, economic, and social factors).

F. The United States will encourage expansion of its domestic recreational and commercial fishing industries to the extent that it is in the national interest and economically efficient to do so.

G. The United States will facilitate the access of U.S. distant-water fleets into the economic zone of other countries.

H. The U.S. Government will provide assistance to displaced U.S. distant-water fishing fleets.

I. Present international fisheries institutional arrangements will be altered or abolished; new ones will have different operating principles. The Federal Government will be responsible for enforcement of fisheries regulations developed under exclusive jurisdiction and will utilize the services of the states as appropriate.

J. The Federal Government will be responsible for enforcement of fisheries regulations developed under exclusive jurisdiction and will utilize the services of the states as appropriate.

K. The Federal Government will create a national fisheries management regime featuring strong and active state cooperation and partnership.

In addition to these recommended policies, other issues were identified but were unresolved with respect to policy recommendations. These included the criteria for applying economic and social factors in determination of optimum yield, allocation of fishing rights, and numerous economic matters associated with foreign fishing, license fees, foreign investment, and indemnification of U.S. fishing people in case of loss. Other problems were noted with respect to the new organizational structure that would be required and how to budget for it.

13.5.2 Enactment of Legislation

Following the careful development of policy and widespread discussion of its implications, bills were introduced in Congress that resulted eventually in the Fishery Conservation and Management Act of 1976 (FCMA). Its purposes were:

(1) to take immediate action to conserve and manage the fishery resources off the coasts of the United States, and the anadromous species and Continental Shelf

fishery resources of the United States, by establishing (A) a fishery conservation zone within which the United States will assume exclusive fishery management authority over all fish, except highly migratory species, and (B) exclusive fishery management authority beyond such zone over such anadromous species and Continental Shelf fishery resources;

(2) to support and encourage the implementation and enforcement of international fishery agreements for the conservation and management of highly migratory species, and to encourage the negotiation and implementation of additional such agreements as necessary;

(3) to promote domestic commercial and recreational fishing under sound conservation and management principles;

(4) to provide for the preparation and implementation, in accordance with national standards, of fishery management plans which will achieve and maintain on a continuing basis, the optimum yield from each fishery;

(5) to establish Regional Fishery Management Councils to prepare, monitor, and revise such plans under circumstances (A) which will enable the States, the fishing industry, consumer and environmental organizations, and other interested persons to participate in, and advise on, the establishment and administration of such plans, and (B) which will take into account the social and economic needs of the States; and

(6) to encourage the development of fisheries which are currently underutilized or not utilized by United States fishermen, including bottom fish off Alaska.

The FCMA became effective March 1, 1977. (It should be noted that the Act pertains almost entirely to fishery regulation in spite of use of the word *management* in its title and text.)

13.5.3 Implementation of the Regulatory System

The primary organizational requirement of the new Act was establishment of eight Fishery Management Councils: New England, Mid-Atlantic, South Atlantic, Caribbean, Gulf of Mexico, Pacific, North Pacific, and Western Pacific. These were designed to bring together national, state, and local interests in each region for preparation of fishery management plans that were required to be prepared with specified factual contents and to meet a set of scientific standards. The primary administrative requirement of the Act was assignment of authority to the Secretary of Commerce to approve the fishery management plans and implement them. The primary enforcement authority for any regulation was assigned to the Coast Guard. The primary responsibility for international fishery agreements was assigned to the Secretary of State.

Immediately after passage of the FCMA in 1976, preparations began for the fishery regulation that was to become effective on March 1, 1977 (U.S. Dept. Commerce, 1979). Well before then the councils were chartered, their members appointed, their scientific and statistical committees organized, and their advisory panels from the industry named. Council staffs were appointed and standards established for Council operations. Council headquarters were established and work began.

A major concern of many people was control of foreign fishing and this received top priority. All international fishery agreements were reviewed and new agreements were negotiated in compliance with the Act. Preliminary management plans were prepared by teams of federal and state fishery scientists. Some were reviewed and approved by Councils and the Secretary of Commerce before the effective data of the Act and several more soon afterward. Governing International Fishery Agreements were concluded with fourteen nations and permits issued for nearly 1300 foreign vessels to fish. Enforcement by vessel and aircraft patrols began soon after the effective date. United States observers were placed on foreign vessels for about one-fourth of the days that they fished.

By 1980, the scope of the fishery regulatory problems was becoming apparent (Office of Technology Assessment, 1977). Foreign fishing had been brought under control but domestic fishery regulation had scarcely changed. Out of nearly 300 fishery stocks that had been identified, management plans were being developed for 76, but scarcely any more had actually been implemented than those that had been approved in 1977. The latter included, of course, some of the largest stocks in U.S. waters.

Acceptance of regulation by domestic fishing people was slow for several reasons. They had expected a bonanza with the reduction in foreign fishing without realizing, or accepting, that their own fishing had been a factor in the overfishing of many stocks. Naturally, they resisted restraint on their activities. Previous regulation, if it had existed, had not needed as rigorous a justification as that now required by FCMA. Council procedures brought out the full complexity of the fishery interests. Council procedures also were extremely slow, requiring up to two years or more between the decision to proceed with a management plan and its implementation. And most fishery management plans have been subject to annual review and revision.

But the FCMA, with its requirement of scientific standards for fishery management, has also brought a radical change in the supporting research system. Many researchers from both federal and state agencies have become involved full-time with support of Council operations. In many cases this has meant a reduction of long-term research that is essential for obtaining a better understanding of complex interspecies relationships and ocean climate impacts on the stocks because the researchers have been assigned to short-term monitoring of stocks and delivery of information to the Councils. Many new needs of the Councils for information have been identified, not only on the condition of the stocks but also on the social and economic structure of the fishing industry.

The passage of FCMA and its implementation to date have been just the beginning of a new and comprehensive system of oceanic fishery regulation for the United States. Many people from government and industry have become involved with a complex new law, and are trying to make it work for themselves and the country. Much remains to be done in the 1990s to bring more overfished stocks under regulation and to speed up the decision process. But the major obstacle to optimal use of the resources is no longer the lack of a mechanism for control of the fishing. It is the effectiveness of the regulation and the information on which it is based.

13.6 IMPLEMENTATION OF THE NEW LAW OF THE SEA

All coastal countries have a recognized right to use and regulate the fishing in their coastal zones. By a 1980 count, 97 countries had claimed jurisdiction over 200-mile-wide zones.

The new Law of the Sea (see Appendix B) created a massive reallocation among nations of the right to fish. The 200-mile zone along coasts and around islands contains about 99% of the world's traditional ocean fish stocks and it includes an area nearly equal to the area of the world's land. Much of the catches from these stocks had been taken by the distant-water fishing fleets of about two dozen countries that traditionally had fished in many areas without restraint and without even maintaining the records of catch and effort that are essential for tracking the condition of the stocks. Particularly damaging had been the practice of "pulse" fishing, in which some fleets fished an area intensively until the fishing became unprofitable and then moved to another area.

Now coastal countries have control of their resources. They have the opportunities to obtain optimal yield, to allow their citizens to fish them, to lease any part of the fishing rights to foreign fishing people, and to use access to the resources as part of a bargain, such as a joint venture in which the country receives a fish-processing facility and part of the catch in return for the fishing rights.

Along with the opportunities come obligations. The Law requires control of fishing to ensure conservation, which requires scientific knowledge of the resources and statistical information on the fishing. The Law requires granting of access to foreign fishing people if the coastal state does not harvest the optimum catch.

Taking advantage of the opportunities and complying with the obligations will require most countries to form new or enlarged fishery regulatory and development organizations. Regulatory organizations (see Chapter 3) will need competence to perform research, collect statistics on the fishing, negotiate with foreign countries about fishing and fishing boundary zones, make decisions about regulations, and enforce regulations. Development organizations will need competence to perform economic, social, and organizational planning, as well as knowledge of the business of fishing, processing, and marketing (see Chapter 15). All countries will have boundary problems with their neighbors and will probably join regional fishery organizations. Implementation will be a long and continuing process.

13.7 ROLE OF FISHERY SCIENTISTS

Fishery regulation has been, for several decades, a major field for application of fishery science, especially with respect to large fish stocks that can be regulated on a unit basis. The fishery biologists describe the populations that comprise each stock and estimate the sustainable physical yields with varied amounts of fishing. The fishery economists add economic concepts to the biological models and estimate the sustainable economic yields with varied amounts of fishing. Both kinds of fishery scientists have contributed greatly to effective regulation of fishing on the stocks that come close to satisfying the inherent assumptions of the theories. These have been the large single-species stocks in which abundance is strongly dependent on their own density.

Other stocks, such as those whose abundance is strongly dependent on environmental conditions, have not been satisfactorily regulated. The environmental conditions may be physical, such as temperature, or biological, such as abundance of essential food or of a competing species. These stocks typically have occasional, very large year classes and their abundance is not self-controlling, at least much of the time. Several anchovy and herring fisheries of this type have collapsed without adequate warning from the scientists. Still other stocks that comprise a mixture of species as in a bottom trawl fishery have been difficult to regulate for optimal yield. It is not possible to maximize the yield from more than one species at a time and the interactions of the several species have proven to be very complex.

A similar difficulty occurs in tropical reef fisheries, where dozens of species may be harvested by a large number of fishermen using simple gear. Each species presents a different reaction to the fishing, and statistics on catch and fishing effort may be nearly impossible to obtain. Multi-unit stocks, which typically support recreational fisheries in fresh waters, are also difficult to regulate for optimum yield. The regulations must be relatively simple and consistent among the many units in different bodies of water—hence optimization is complex. The attempts to regulate these complex fisheries have sometimes resulted in regulations that could not be justified to the fishing people and, as a consequence, the credibility of the regulatory program has been reduced.

Clearly, the major professional challenges in the 1990s are to forecast reliably not only the abundance of stocks under the diverse kinds of fishing and environmental changes, but also the accompanying sociopolitical changes. Good forecasts will establish the credibility of the basic information about the stock and permit better analysis of the alternatives. Such an effort should start very early in the exploitation of a stock and include explicit recognition of the uncertainties.

Forecasts plus all of the desirable descriptive information about the stocks, catches, and fishing provide an essential basis for allocating the fishing among the users, but never a sufficient basis. Users are also concerned about the equity of the regulations and will raise many issues that will require political solutions.

Fishery scientists will need to be increasingly aware of the many biological, environmental, economic, social, legal, and other factors involved in regulatory decisions. All of these factors will be changing more and more rapidly as our human population grows and increases its use of its environment (Murdock *et al.*, 1992).

REFERENCES

Graham, M. (1943). "The Fish Gate." Faber & Faber, London.

Murdock, S. H., et al. (1992). Demographic change in the United States in the 1990's and the 21st century: Implications for fisheries management. *Fisheries* 17(2), 6–13.

Office of Technology Assessment (OTA) (1977). "Establishing a 200-Mile Fisheries Zone," Assessment Report. U.S. Congress, Washington, D.C. (Six working papers in a single volume.)

Rettig, R. B., and Ginter, J. C., eds. (1980). "Proceedings of a National Conference to Consider Limited Entry as a Fishery Management Tool." Univ. of Washington Press, Seattle.

Roedel, P. M., ed. (1975). "Optimum Sustainable Yield as a Concept in Fisheries Management," Spec. Publ. No. 9. Amer. Fish. Soc., Bethesda, Md.

Smith, W. B., Lewis, S., and Graff, D. R. (1979). "Procedures for Adopting Fishing Regulation Changes. Proceedings of the Eighth Annual Meeting of the Fisheries Administrator's Section, 28–30 May 1979." Amer. Fish. Soc., Bethesda, Md.

Stroud, R. H. (1980). Evolving fishery resource allocation. *Sport Fish. Inst. Bull.* 316, 1–4.

U.S. Department of Commerce (1975). "Fisheries Management under Extended Jurisdiction: A Study of Principles and Policies," Staff Report to the Associate Administrator for Marine Resources, U.S. National Oceanic and Atmospheric Administration. U.S. Dept. of Commerce, Washington, D.C.

U.S. Department of Commerce (1979). "Operational Guidelines for the Fishery Management Plan Process." U.S. National Marine Fisheries Service, Washington, D.C.

14 | Aquatic Environmental Management

Fishery scientists have become increasingly involved with management of the fish's habitat as humans intrude and alter the natural environment. People's concerns about both fish and water have been transformed into law and organizational responsibility. The earliest fishery agencies were faced with pollution problems, but now society perceives their environmental work in two ways. First, fishery agencies have responsibility to preserve and improve the environment for the benefit of the fish. Second, because the condition of the fish is an index of the condition of the aquatic environment, the agencies are asked to be judges and guardians of environmental quality. Many fishery agencies, especially those concerned primarily with inland waters, find that their major activity is aquatic environmental management.

Water, of course, is the predominant resource, being vital for all life. It is another common-property resource and its use has generated controversy throughout history. Water has long been allocated to its various uses and managed under complex national and local laws by many water agencies.

Waters have had two major categories of use: instream, or flow, and offstream, or withdrawal. The instream uses have been for power generation, navigation, waste disposal, fish habitat, and recreation. The offstream uses have been for irrigation, thermal electric power plant cooling, municipalities, and manufacturing.

This chapter will summarize the aquatic environmental issues and the current activities of fishery scientists as they try to ensure a productive environment for fish and a quality aquatic environment for people.

371

14.1 TRENDS IN WATER USE POLICY

Throughout recent centuries water development has been a major factor in economic development. Water transportation along rivers and estuaries has been augmented by dredging, damming, and construction of canals. Irrigation water has been a key to development of many arid regions, where expanded agriculture has produced centers of trade and industry. Flood control and other water storage dams have made possible the use of rich river bottomlands and production of hydroelectric power. All available water is now claimed in many regions.

Use of waters for recreation and production of fish is also a long-established use, but one that was regarded as incidental until recent decades. Now it is recognized as another major use that may or may not be compatible with other uses. Clearly, optimizing use of water is a major challenge. Its diverse uses generate intense competition, its conservation is vital, and older concepts are no longer acceptable to many people. One approach of recent decades is through formation of river basin commissions that can plan more efficient use of the water, reallocate some of the water according to modern public needs, and coordinate diverse government actions.

The two major aspects of water use policy that are important to fisheries are the allocation systems, which may determine whether water will be available, and control of quality, which may determine whether fish can live.

14.1.1 Allocation of Water

Water is allocated among users by water law that has evolved over centuries, but that recently in numerous countries has been tending toward a system called prior appropriation. The system is governed by three principles: (1) the state claims ownership and control over water, allowing private persons to acquire rights only by virtue of a state permit; (2) permitted uses of water must be reasonable and beneficial; and (3) the water users have property rights protected against infringement from later users, rights existing in perpetuity if used and in most states transferable by the owner to another person (Trelease, 1970).

State ownership is a key to planning the initial allocation of water or its reallocation through purchase if different uses are in the public interest. Such interest is now focusing more on fisheries and recreational uses of water and, in the United States at least, efforts are under way to allocate minimum stream flows for these purposes (Orsborn and Allman, 1976).

14.1.2 Environmental Quality

For thousands of years, laws about water quality were directed primarily at prevention of human diseases through control of pollution and purification of water supplies. Only within the past century have the laws been expanded to cover other uses, and only within about the past two decades have the laws been extended to more adequately protect the recreational, fishery, aesthetic, and other quality values of the waters.

The major step in overall environmental quality control came with passage in the United States of the National Environmental Policy Act (NEPA) of 1969. This followed the greatly increased concern about quality of the environment and the Act embodied important new policies and administrative practices. It is intended "to create and maintain conditions under which man and nature can exist in productive harmony, and fulfill the social, economic, and other requirements of present and future generations of Americans." Its administration requires "that presently unquantified environmental amenities and values may be given appropriate consideration in decision making along with economic and technical considerations". This is done, first, by requiring collection of all pertinent environmental, economic, and technical information in an Environmental Impact Assessment (EIA) and, second, by exposing the decision process to scrutiny by all concerned agencies and people. The NEPA is applicable in the United States to several executive orders and laws pertaining in some way to fisheries, including Protection of Wetlands, Floodplain Management, Coastal Zone Management Act, Wild and Scenic Rivers Act, Fish and Wildlife Coordination Act, Fisheries Conservation and Management Act, and the Endangered Species Act. Other acts, in particular the Clean Water Act, are exempt from NEPA because they have similar requirements for EIA and public involvement. NEPA applies to Federal actions but this is deemed to include state and local actions where federal funds are involved.

Concern about environmental quality is worldwide and many countries have adopted some of the procedures of NEPA—in particular the practice of preparing Environmental Impact Statements (EISs) (United Nations Environment Programme, 1980a, 1980b). Additional steps seem certain in the near future as the concern is transformed into administrative action suitable for each country.

14.2 ENVIRONMENTAL IMPACT ASSESSMENT

Environmental Impact Assessment has no universally accepted definition, but it can be described as a comprehensive evaluation of a proposed project's impact that covers all aspects with which people may be concerned. It should contain pertinent quantitative information about social, economic, and technical matters, as well as qualitative information about aspects such as historical, aesthetic, health, recreational, and other values that cannot be readily quantified. The project may be a proposed law, policy, program, procedure, or structure.

EIA grew out of the deficiencies of cost–benefit analysis (CBA), which cannot adequately quantify many of the values that people recognize. It benefits the decision process by providing all interested parties with a full compilation of facts and judgments about the project that will lessen the likelihood of controversy and mistakes. An EIA is not a decision, but it should provide the decision maker with a detailed view of the implications of each alternative course of action.

The particular requirements for EIAs vary with country and with type of project. Further, they have been changing frequently with efforts to make the process more workable. Numerous guidelines are available, but they are too diverse to summarize here. The results are even given different names, including Environmental Impact Statements or Reports, but the process in some form has become common all over the world.

Fishery scientists become involved with preparation or review of EIAs according to each government's procedures. EIAs are required in the United States for all federal fishery management actions and for some state actions. They are also required for many construction projects, both public and private. A large proportion of the latter have an impact on water quality and fishery agencies are asked routinely to comment on many of them.

14.3 THE SCOPE OF AQUATIC ENVIRONMENTAL CHANGE

Every species modifies its environment, and humans more than any other. Even centuries ago, through their use of fire and intensive herding of grazing animals, humans changed the landscape, the rivers, and perhaps even local climate. Now, with their increasing population, increasing use of all resources, especially energy, and increasing production of nonbiodegradable organic materials, they are changing virtually all ecological aspects of the earth including the oceans and entire climates (Gucinski *et al.*, 1990).

Much of this change is reflected in the waters. The water itself is a vital resource and a large fraction of continental and coastal waters is used and abused in some way. In addition, what humans do on the land and to the atmosphere may also modify the waters. In consequence, the impacts on fish and fisheries are exceedingly varied. Every change will have a subtle if not a dramatic effect on the ecosystem and the organisms in it. Even identifying and measuring impacts is a challenge, particularly because of their diversity and the presence of normal seasonal changes. The following summary will include only major changes of general concern.

14.3.1 Physical Changes

Bottoms and Basins   Harbors are dredged; swamps and estuaries are filled. Rivers are dammed, streams are channelized and confined in culverts under highways. Sewage sludge and other solid wastes are dumped in the water. Normal sediment transport of streams is increased by runoff from agricultural, urban, and deforested lands.

All of these activities change the living area of all organisms that they touch. The waters may be deeper or shallower, faster or slower. The bottoms of channels or culverts may be concrete or rubble instead of gravel. The sludge and wastes and sediments accumulate where the current slows to change the bottom to silt or mud.

Flow   Offstream use of water always reduces the flow immediately below the diversion and usually all the way downstream, because only part of the water is returned after domestic or agricultural use. And some diversions may transport water for hundreds of miles. Reservoirs change the flow regime and, in some climates, lose significant amounts of water through evaporation. Agricultural and forest practices may change the soil's water retention and, in turn, the runoff pattern. Harbor developments may change the mixture of salt and fresh water.

Reduction of flow simply reduces the living space for fish and limits populations during low-flow periods in summer or winter. Even changes in flood flows may force a change in population of a species that has adapted its life to the floods (e.g., young salmon migrating to the sea).

Temperature   Almost all changes in flow are accompanied by changes in water temperature. Lower summer flows will allow water to warm more rapidly. Lower winter flows in cold climates may allow streams to freeze. Reservoirs change the unstratified temperature of a stream to a layered lake, warm on top, cold on the bottom. And the water released from them may be cooler in summer and warmer in winter.

Offstream use of water may add large amounts of heat. Thermal electrical plants and industrial processes that use water for cooling may return the water at temperatures too high for any living thing. Irrigation water may also be returned at a much higher temperature than when diverted. Opening the forest canopy in sunny climates may raise summer temperatures.

Temperature changes associated with flow changes or removal of the forest canopy usually mean a change in the fish populations that favors warm-water species, which may not be as desirable as the cold-water species they replace. Return of warm water used for cooling may cause fish to avoid the immediate area of entry as well as cause a change in species composition over a large area of warmed water. On the other hand, such changes may attract desirable fish that may be easy to catch and be regarded as beneficial (Gore and Petts, 1989).

Suspended and Settleable Solids   The predominant source of the fine materials that are carried by the waters is the natural erosion of soils caused by rivers and ocean waves. But also included are the particles originating from agriculture, mining, logging, road construction, and a few industrial processes, notably paper making. They are carried by currents, deposited as velocity is reduced, and carried again with increased velocities. These materials change light penetration and consequently reduce the photosynthesis in the water. They may also increase temperature, causing the surface to be warmer. In addition, they may adsorb toxic materials and concentrate them on the bottom, from which the toxins may be released if the bottom if disturbed.

Increases in these solids cause a variety of changes that are detrimental to fisheries. Especially important is the increased blanketing of the bottom, which, when extreme, may smother fish food organisms or molluscan populations, or prevent successful salmonid reproduction. A few solids such as certain wood fibers from pulping operations may harm fish because they irritate the gills, although most inorganic soil particles are not directly harmful except in very high concentrations.

Entrainment   As water is withdrawn for offstream use it will contain fish that are almost certain to be harmed. Obviously they will be killed if the water is used for irrigation or treated with chlorine for a city supply. But they may be injured even when the water is passed through a riverbank turbine used for power generation. The impact is worst when very large volumes are withdrawn from tidal waters for cooling and when the fish are tiny. The tidal exchange moves the water back and forth by the intake and fish eggs and larvae cannot withstand suction against a screen.

One solution is a barrier that will pass water but not fish. Sonic and electrical barriers have been tried but screens are best. Much engineering design has been expended on them and they work with larger fish and small volumes of water. But many installations cannot be screened satisfactorily.

14.3.2 Addition of Wastes and Chemicals

Nutrient Materials   The category of nutrient materials includes body wastes from humans and animals, organic wastes from food-processing plants, and materials with high nitrate or phosphate content, notably detergents. Most of these are commonly included in sewage, but urban sewage frequently contains toxic materials that may have different effects. All of these materials increase the ability of the waters to support plants and animals—up to a point. The changes are somewhat analogous to adding fertilizer to a garden. If the waters are relatively barren, additions will increase fish production, but if the waters are already enriched by nutrients, the additions may be damaging, a process called eutrophication. The enlarged populations will increase turbidity and reduce oxygen to levels that some organisms cannot tolerate.

Toxicants   Toxicants include a large and rapidly increasing assortment of chemicals that may damage fish, humans, or both.

Heavy metals pose special problems because they usually accumulate in organisms at successive trophic levels and may reach concentrations thousands of times as great as in the water—a process called biological magnification. For example, mercury from pollution or even from entirely natural sources may accumulate in fish to levels dangerous to human health. Zinc, copper, lead, cadmium, selenium, and many others may harm fish.

Pesticides, including herbicides, insecticides, and fungicides in thousands of different formulations, are being used in ever larger quantities in agriculture, public health, and for household or garden purposes. Each compound must be considered individually because they have different rates of degradation in the environment and different toxicities in organisms. Many tend to be adsorbed on suspended sediments and accumulated in bottom deposits from which they may be released by dredging or floods. Some degrade very slowly and some, notably DDT, are biologically magnified in fish flesh to concentrations dangerous to humans.

Phenols range from extreme toxicity (pentachlorophenol is a highly toxic wood preservative) to low toxicity, but they usually taint the flesh of fish at concentrations lower than lethal levels.

Acids and Alkalis   Acids and alkalis change the water's pH. Fish usually live at pH levels between 6.0 and 9.0, although they may not tolerate a sudden change within this range. Acids usually have been of greater concern because they originate in large quantities from mines and various industrial processes. But these sources have largely been controlled and worse damage has been occurring since the early 1980s from acid rain, which originates largely from burning coal and oil. The process produces mixtures of sulfuric and nitric acids in the atmosphere and these may accumulate and drift hundreds of kilometers before "raining" at a pH frequently less than 4.0.

Oil   Petroleum products have received much attention as pollutants because of high visibility and several large, spectacular oil spills. These have killed most plants and animals in the intertidal zone and a few organisms in deeper water as the oil has emulsified and sunk. The spills of refined oils such as diesel oil and gasoline have been more damaging than spills of crude oils because the former contain more water-soluble materials, some of which are acutely toxic to fish. All of the oils contain substances that may taint fish flesh temporarily. None of the large, spectacular spills has had a measurable impact on commercial fishery populations other than mollusks and crustaceans in the immediate area of fouling or poisoning. Small recurring spills that are common in commercial harbors will, however, completely change the populations of organisms within a harbor.

14.3.3 Induced and Synergistic Effects

The impact of physical change or a pollutant on the environment is seldom isolated or immediate or constant over a large volume of water. The impact on fish of some changes may be additive to others or exacerbated by them. Still other changes may be cumulative or indirect in their impacts. The following are some major aspects that have received attention.

Mixing Zones   When something transportable is added to water it is progressively diluted in a mixing zone. The boundary of the zone is where the organisms of concern can withstand long-term exposure. It has been officially defined (U.S. EPA, 1976) as "an area contiguous to a discharge where receiving water quality may meet neither all quality criteria nor requirements otherwise applicable to the receiving water." Such zones are permitted when a decision has been made to sacrifice an area of water because it is not practical to treat the effluent further before discharging it. Many such instances involve discharge of heated water or domestic sewage.

Definition of such zones must be done on a case-by-case basis. It will usually involve consideration of physical, chemical, and biological characteristics of the effluent and receiving waters as well as the probable pattern of exposure of important species.

Oxygen Depletion   When nutrient materials are added to water they increase the activity of the entire trophic system. Sewage and food plant wastes are biologically degraded, which requires oxygen immediately, and later the degraded products fertilize the growth of plants and animals in the water, which also requires oxygen. Runoff from agricultural lands and urban areas will also fertilize the waters. Even sewage treatment that removes solids and sterilizes the effluent leaves the fertilizers in the effluent. The consequence is eutrophication of the receiving waters.

A little eutrophication produces more fish, but an almost inevitable result is production of plants and animals that exceed the capacity of the waters to sustain them. An early and common symptom is occasional oxygen depletion that kills both plants and animals. A later symptom is transformation of the aerobic system to a persistent anaerobic system that produces foul odors.

Fish vary in their oxygen requirements somewhat according to their accustomed level. Those that live near the surface of the ocean or in cool tumbling streams, where the waters are nearly saturated with oxygen, suffer much sooner than those species that live near the bottom of eutrophic waters. Salmonids usually require more than 4 mg/liter of oxygen, whereas several of the species of carps used in stillwater aquaculture live comfortably at 1 mg/liter.

The saturation level varies according to temperature and salinity. It is about 14 mg/liter in fresh water near freezing but only about 6.8 mg/liter in seawater at 25°C. The low saturation level in warm seawater means a limited capacity of such waters to receive nutrient materials without damage to surface species of fish.

Excess Dissolved Gas   Waters supersaturated with air (to about 110%) occur occasionally and harm fish through formation of internal gas bubbles. Such waters occur naturally in a few underground sources that have been used for aquaculture and occasionally in the midst of algal blooms in quiet ponds during sunny afternoons. But they are an increasing hazard associated with dams and heated effluents, as well as eutrophication.

Water dissolves more gases at higher pressures and lower temperatures. Water dropped over spillways carries air into the plunge basin below, where it dissolves under the water pressure at some depth. Then the gases in the water are transpired by fish, which develop internal bubbles (similar to the bends of divers) as they return to near atmospheric pressure close to the surface. A similar supersaturation occurs when saturated water is warmed in a power plant or near the surface of a quiet pond because the solubility of gases declines.

Multiple Changes   Near any urban area, multiple changes are the rule rather than the exception. Usually the streambed or the harbor is a changed environment, the water flow is modified, the water is warmed, and diverse things are added to the water. All of these are in addition to natural changes, whose normal seasonal variation is usually substantial.

Impacts become much more difficult to predict under these conditions. Organisms that are stressed by higher temperatures or low oxygen may become more susceptible to a toxin. One pollutant may affect spawning, another migration. Some toxicants have effects that are additive, others have minimum concentrations below which the effects are not additive, and still others have synergistic effects in which a combination of materials is far worse than the sum of separate activities. Some effects are cumulative or long delayed.

14.3.4 Community Changes

All of the changes discussed here affect the fish, as has been mentioned, but they also cause changes in the entire biological community of the waters (Hart and Fuller, 1974; Perkins, 1974). Such changes are important to the fish because populations of fish food organisms, either insects or small fish, may be changed and thus the populations of their food organisms may be changed.

Changes occur not only in the relative abundance of the different species but less tolerant species frequently disappear. This leaves more living space for the species remaining and the abundance of some of them may increase greatly. Such changes are usually gradual over long periods of time as newly abundant populations develop and become genetically adapted to their changed environment. Then such populations have their own impact on the environment and induce further changes.

A well-known example of changing abundance of fish occurs in newly flooded reservoirs. Typically these cover land containing substantial amounts of organic material that fertilizes the water and the fish, which are relatively scarce at first, but then grow rapidly. Only after five to ten years do the fish populations become relatively stable and then only if control of the water in the reservoir is consistent from year to year.

Some ecological changes may be highly undesirable, such as the accidental introduction of the zebra mussel, a small mollusk from eastern Europe, into the Great Lakes in 1985. It probably was carried in the ballast water tanks of vessels coming from Europe, and discharged as the vessels changed their ballast water. Apparently none of its traditional competitors or diseases live in this area, and it has spread very rapidly throughout the Great Lakes, the Mississippi River drainage, and the Hudson River. An example of its prolific reproductive capacity, in the absence of its natural competitors and predators, is the discovery of a 3-inch-thick growth of them on an auto body pulled from Lake Erie in 1989 (Ross, 1994).

Unfortunately so little is known about the complex interactions of the many species involved in an aquatic community that it is difficult to predict the eventual species balance. The best guide is actual experience with each kind of aquatic environment.

14.4 STRATEGY FOR PROTECTION

The protection of waters in favor of their quality and suitability for fish is a continuing process of social choice. So is regulation of fishing, but that policy can be made largely on the basis of broadly accepted objectives and authority for decisions is regularly delegated to executive organizations. Water quality policy, on the other hand, depends much more on judgment and provision for meeting the diverse objectives of vested interests of water users. A particular difficulty is comparison of uses to which a dollar value can reasonably be estimated (e.g., irrigation) and other uses that cannot be evaluated in dollars (e.g., many aesthetic and recreational uses). Representative organizations are essential for policy formulation and delegation of authority for decisions to executive organizations is less satisfactory.

14.4.1 Policy and Organizational Development

Development of any water policy involves changing existing policies or practices in favor of new goals and objectives. The proposed changes should be based on full consideration of all technical and organizational aspects of existing policies and identification of all the issues.

The process can begin with appointment of a study group that will be charged with accumulating basic information on the legal, political, social, economic, and environmental aspects of the problem. Included in the aspects should be the history of the studies, the decisions, the traditions, and the mistakes, as well as the need for additional information. The group may be a task force, commission, or legislative committee. It should include or hear representatives of all user groups and obtain scientific information unbiased by political judgments as well as the value judgments of the people who may be affected.

After the facts and value judgments have been accumulated the group can define the reasonable alternatives for action and predict the probable impact of each on the users (see Section 13.2). Then it can recommend policies and organizational development designed to meet both immediate and long-term needs. But the predictability of much environmental change is so poor that recommendations must usually include expanded research and monitoring. The action can then be implemented with appropriate legal or administrative steps.

A frequent problem in such action with respect to water quality is the multiplicity of authorities having jurisdiction. These may include, in the United States, several agencies from each of the local, state, and national levels of government. Each will have its own turf to protect, its own constituency, and its own procedures. Each may resist change of authority and, consequently, there may be pressure to take no action. Getting action will be expedited by broad and accurate public understanding of the alternatives. Scientists can help by making sure that their findings are distributed widely and fairly to the people who are concerned. But the ultimate success of changed policies depends on changing administrative action in accord with public understanding.

14.4.2 Allocation of Water Use

An obvious strategy for protecting the waters is to allocate the conflicting uses as much as possible. It is as absurd to expect to maintain pristine water quality in the midst of a busy harbor as it is to permit dumping waste on a unique coral reef.

The allocation of water use requires consideration of each body of water and each use on an individual basis. The objective will be identification of multiple nonconflicting uses, of uses that can be concentrated in a restricted location, of preserves that may be needed for aquaculture or endangered species, of opportunities for mutual benefits such as using waste heat, and of uses that can be located elsewhere.

The possible mechanisms for allocation start with the marketplace in which the uses have long been traded and with the government organizations that are already in place. Uses can be changed by referenda in which voters indicate their preferences. They might be changed by court action under existing laws, which would force agencies to change their practices. A new agency or law may not be needed.

Given workable laws and organizations, there are many possibilities for action other than enforcement of regulations. These include strategies such as low-interest loans or tax forgiveness, restrictions on permits, grants of funds for specific objectives, and agreements among agencies to help each other. Examples of fishery bargaining include siting a marina a given distance from an oyster farm, requiring a marina to maintain pump-out facilities for boat holding tanks, or persuading a federal and a local agency to support each other's water use regulations.

14.4.3 Criteria and Standards

Most water uses are important, and in most situations multiple uses prevail. Each will require a different amount and schedule of water of different quality. But every natural aquatic ecosystem has evolved a balance among the organisms and between those organisms and their environment. Any man-made change will cause some change or series of changes in the system that will favor some organisms over others.

Some people argue that it is necessary to preserve entire communities but this is impossible except for a unique single use, such as an endangered species or critical habitat, so a compromise is usually reached to protect and enhance the organisms that are harvested while accepting moderate changes in the other species and the environment. This involves establishment of criteria for satisfactory quality and quantity needed for each specific use. If several uses are expected, the criterion will be established for the most demanding use. Criteria initially reflect scientific judgments about what must be identified and controlled for each specific use. They must be tested by monitoring during use and corrected if necessary.

Every water management program is a practical compromise between cost and effectiveness and thus more stringent criteria should be established as the demand multiplies and as reserve capacity is needed for future use. Once criteria have been accepted, specific standards for each aspect of change are established by law or regulation. Standards for contaminants need to include guidance on environmental chemistry—on toxicity of various forms, transformation among forms, bioaccumulation, sampling methods, and analytical methods. Standards for instream flows should include minimum or maximum flows that will maintain suitable conditions for feeding, rearing, spawning, and migrating according to the stream and the fish desired.

14.4.4 Biological Tests

Criteria and standards are determined in part on the basis of tests of the effects of changes or additives on the environment. These include before-and-after surveys of water chemistry and biological changes in either populations or individuals of valued or index species, as well as similar surveys of the accumulation of toxicants in the environment and the bodies of animals—the latter being especially important with respect to public health.

One of the oldest tests that is still in widespread use for estimating the amount of oxidizable organic material in wastewater (e.g., human body and food wastes) is determination of *biochemical oxygen demand* (BOD). This is a laboratory test done by inoculating a diluted sample of wastewater with microbes, placing it in an airtight container, and incubating it for 5 days at a temperature of 20°C. The dissolved oxygen is measured before and after and the amount consumed is an index of the amount of biodegradable organic matter. This test has been used so long that it has become a standard, but determination of *chemical oxygen demand* (COD) or *total organic carbon* (TOC) may be used in its place because they can be done more quickly.

Another common kind of laboratory test is the *bioassay*. These are used to determine (1) toxic levels of substances to given species, (2) environmental stimulation by nutrients, (3) estimated concentrations of materials in water or animals when chemical tests are unsatisfactory, and (4) suitability of natural waters for different organisms, as well as various less important purposes.

A widespread bioassay is determination of acute toxicity to fish. The procedures are carefully standardized and usually involve placing a group of similar fish in each container of a series having a gradient of toxicant concentrations. The series of containers is held under constant conditions of temperature, dissolved gases, daily light cycle, and toxicants insofar as possible, usually for 96 hours. The result, if the gradient of toxicant concentration is chosen correctly, is survival of all fish in at least one container and the death of all fish in another. The determination that is usually reported is the estimated concentration at which half of the fish die—the median lethal concentration, or LC50.

The 96-hour LC50 value is, of course, intolerable as a level to be permitted, but it is modified with an *application factor* (AF) that is based on judgment and experience. Tentative factors for materials that are nonpersistent or noncumulative are concentrations less than 0.1 of the 96-hour LC50 at any time or place after mixing with the receiving waters and a 24-hour average less than 0.05. For persistent or cumulative toxicants, the tentative corresponding maxima are 0.05 and 0.01, but long-term tests of many of these are being undertaken to refine the AF. If two or more toxic materials are present in the receiving water, their effects are considered to be additive and the permissible concentrations are modified accordingly.

Such an indirect approach to determination of a permissible level is necessary because (1) determination of a safe level for each combination of organism, environment, and toxicant may require very long-term experiments, even throughout the lifetime of an organism, (2) decisions cannot wait, and (3) the multitude of toxicants have varied effects in different receiving waters.

Better predictability of the impact of the multitude of new toxicants is urgently needed. Many of these are slowly biodegradable or nonbiodegradable and have subtle but significant impacts in low concentrations. The study approach has been to devise sequential tests in which the first step is to compare the possible environmental concentrations with concentrations that produce known biological effects. If there is a wide gap and the environmental chemistry of the toxicant is well known, a simple application factor may be adequate. If not, further studies will be undertaken to determine effects throughout the life of an organism and on its environment and to determine the environmental chemistry. Special attention will be paid to biodegradability and bioaccumulation. If the toxicant is suspected of being a carcinogen, even more rigorous testing will be done, especially if it might become a contaminant in human food.

Concerns about aquatic pollution are being developed into protective laws in many countries, but there is wide divergence in kind and scope of concern. Where people have suffered from disease or poisoning (e.g., mercury accumulation in fish), the laws tend to focus on human health. Where there is widespread fishing, the laws focus on the species being used, especially if they have been stocked. Where there is widespread recreational use, the laws protect against contamination, bad odors, weed growth, and other aesthetic degradation. Increasingly, people are concerned about the health of ecological communities.

So the testing serves two purposes: first, pragmatic compliance with the law, and second, development of better understanding of the total impact of changes in the waters.

14.4.5 Waste Management

Waste management is a burgeoning activity that involves almost everyone who uses water and has an effluent pipe. It is also a rapidly growing occupation of people employed by public and private polluters, who are commonly trying to find the optimal choice among costs, compliance with the law, uses for waste materials, less use of water, and better treatment for unavoidable discharges. The occupation involves many specialists other than fishery scientists, but the latter have substantial involvement with some aspects of the process other than the overall condition of the receiving waters.

Fishery waste management began with a break in the tradition of discarding the waste at sea, or dumping it from the processing plant into the harbor. Efforts are being made, first, to produce more human food products from the fish such as fish sausage or dried fish products; second, to produce more animal feeds such as canned pet food, dried fish meal, or liquified fish (by acid processing); and third, to develop industrial or agricultural products such as oils, fertilizers, or adhesives. Such changes have been slow to come because a large proportion of fish processing is done by small-scale operators who find that production and marketing of new products is too costly and technically complex.

Aquacultural wastes arise from intensive fish rearing in which the fish are fed a complete diet. Their body wastes and the wasted fish food affect streams just like similar amounts of sewage and require similar control.

Thermal pollution by waste heat has potential for enhancing aquaculture and improving fishing in the mixing zone of the effluent. In many locations the water supply for growing fish or mollusks is too cold at times for optimum growth or reproduction and fish farms are being built to use the warm water. Similarly, warming natural waters may attract fish in numbers and species that anglers favor.

Animal and plant wastes may also be used profitably in tropical aquaculture. This is done in stillwater ponds that are fertilized to produce food for the fish (see Section 11.2.1). The fertilizer may be human, poultry, mammalian, or field crop waste that is added to the ponds in carefully controlled amounts that will optimize fish food production while still maintaining a suitable environment in the pond for fish. Sewage can be used if it does not contain toxins, but these are common in urban effluents.

14.4.6 Organic Waste Treatment

Sewage and food plant wastes that do not contain too many toxins are commonly treated by a process that simulates, enhances, and isolates a decomposition phase of the natural ecosystem. In the activated sludge system the water to be treated is aerated in tanks where the flow and temperature are controlled in ways that maximize the growth of microorganisms. The end product is a sludge consisting largely of the bodies of microorganisms in a solution of nutrient materials. Another aerobic system, trickling filters, allows the waste to trickle over a porous medium such as a rock filter in ways that are favorable to microorganisms.

An anaerobic system is similar to the natural process that results in natural gas. It involves microorganisms that produce organic acids and methane gas. The organic solids in suspension are held in tanks from which air is excluded and under temperature, flow, and acidity conditions that enhance the process. The end products are sludge and gas.

A third system, the use of oxidation ponds, is similar to the management of tropical fish ponds. The ponds are shallow and are kept aerated to the bottom in order to avoid the bad odors of anaerobic processes. The inflow, recirculation, and aeration are carefully controlled. The end products are nutrient-rich water and sludge, which must be removed occasionally from the empty pond.

14.4.7 Acid Rain

Atmospheric pollution has begun to cause major fishery problems in some fresh waters. It is occasioned when the combustion of fossil fuels (especially coal) releases sulfur dioxide (SO_2) and nitrogen oxides (NO_x) into the atmosphere, where they are transformed into sulfuric and nitric acids that are transported long distances with the prevailing winds.

Such releases were not identified as the cause of serious environmental degradation until the late 1960s, and there were several reasons for this. One reason was that the sources not only were far from the areas of impact but were, in fact, in substantial compliance with laws that required release of gases from tall smokestacks in order to avoid local contamination of the ground. Another reason was that natural sources of acid reduce the pH of much precipitation to levels near 5.0, a value that happens also to be about the minimum tolerated by the more resistant species of fish and other aquatic organisms. A third reason was the presence, in many areas exposed to such rain, of soils containing limestone and waters with substantial hardness that buffer the acids. And there had been no noticeable impact on the ocean, which usually has a pH of about 8.2.

But during the 1970s abundant evidence accumulated of the depression of rainfall pH to levels of 4.2 to 4.7 in large areas downwind of major industrial centers (Haines, 1981), which had acidified lakes and streams in soft-water regions. Some of these were already acidic waters with pH close to 5.0, and the natural populations of organisms were adjusted to the acidity. With the increases in acidity the aquatic ecosystems have undergone major changes. Fish have been eliminated from hundreds of lakes in Scandinavia, southeastern

Canada, and northeastern United States. The problem has been complicated by troublesome amounts of toxic metals that have been included in the rain or remobilized from sediments.

The remedy is reduction in the acidic precipitation. Mitigation of impact on the lakes through addition of limestone helps, but such treatment is expensive —it ranged up to $500 per hectare in the late 1970s. Introduction of acid-tolerant fish (trout) has been promising, and other techniques are under study.

14.4.8 Reduction of the Risk of Accidents

Large impacts on the environment occur occasionally by accident due either to human errors or to infrequent natural events. Reducing the likelihood or impact of such events is an essential part of any protective strategy and a few examples especially important to the fisheries should be noted.

Oil Spills   Large and widely publicized spills have occurred in recent years from wrecked tankers and from uncontrolled oil wells. Minor spills are commonplace during routine oil transport and transfer operations. In addition to spills, large quantities are discharged with ballast and bilge water at sea. The impact on the environment varies with the kind of oil and the kind of cleanup operations.

Crude oils are usually relatively nontoxic but kill by fouling animals and plants with which they come in contact. The light refined fractions such as gasoline and diesel oil are more toxic and kill many animals, including fish eggs and larvae, at very low concentrations. In addition, the oil may taint fish and shellfish that survive and make them at least temporarily inedible.

Oil spills on beaches may be attacked by dispersants or detergents. These are toxic in themselves and they may render the oil more toxic. If the fisheries are of primary concern, the use of detergents in the cleanup should be avoided. Smaller oil spills while still on the water may be mopped up. Special floating equipment is available to separate oil and water, retain the oil, and tank it for later disposal.

Repeated unattended minor oil spills that prevail in many harbors have a cumulative effect as does the substantial quantity of oil that is discarded in urban sewage systems. Part of the oil sinks as it ages and fouls the bottom as well as the shore. The recurrent toxins and repeated fouling cause major changes in the local aquatic ecology that are eventually more serious than the impact of a major large spill.

Protective strategy must center largely on prevention or minimization of impacts. Ship traffic can be better controlled to reduce the risk of accidents. Ships can be better designed and ballasting procedures modified to minimize loss of oil when discharging ballast. Oil transfer operations can be surrounded by temporary floating barriers—a practice now required in some places. Clean-up equipment and strategy can be prepared in advance. A powerful incentive for improvement is legal responsibility for the cost of any cleanup and damage to the environment.

Storms and Floods   Although irregular in occurrence, storms and floods commonly cause pollution problems that may aggravate environmental changes caused by the event itself. They blow away or wash away topsoil that has been exposed in agriculture, construction, or logging operations. The top-soil usually gets into streams, where it adds to the sediments being transported during the high waters as a result of the natural stream erosion. The floods may cause highway or logging road failures—the latter a major source of sediments resulting from logging operations. The accumulation of water in unstable soils on steep slopes, and perhaps as snow on top of the soil, may cause a slope failure or avalanche that puts debris in the stream below.

Stormwaters may be a major problem. In urban areas with extensive paved surfaces the runoff into streams may be very rapid. Or if domestic sewers are combined with storm sewers the surge of water may overload the treatment facilities and require dumping of untreated sewage. The solution is to separate domestic and storm drains as much as possible. Stormwaters also trigger release of toxins from waste storage areas on land. Toxins may be washed off the surface of industrial waste storage areas or leached from general-purpose land-fill dumps.

Minimization of the impact of such events requires better planning of construction and operations to withstand unusual events. A common approach is to use the frequency distribution of rain or height of flood waters in past years as a basis for predicting the probability of occurrence during any period of time. Frequently 100 years is chosen and the expected maximum during that period is used as a basis for the design.

Carelessness   Many accidents occur as a result of ignorance of waste management practices, equipment failure, or simple carelessness. Something may be spilled or released in excessive quantities at a time when the assimilative capacity of the stream is low. The evidence may be a fish kill. A common provision of fishery law is a financial penalty for such kills. Frequently the size of the penalty is based on a determination by fishery scientists of the number and value of fish killed.

14.4.9 Incremental Effects

Incremental effects are the little events that produce additive effects—events that are minor in themselves, but that may in combination eventually transform the environment. They include the steady expansion of all structures and activities associated with growing populations. They also include nonpoint source pollution, which creates what are probably the most pervasive and ubiquitous water quality problems (Ischinger, 1979).

Some of the changes associated with population growth or economic development must be accepted as inevitable. Few people regard protection of the environment as worth any cost; indeed, many people around the world regard the environment as something to be used or conquered.

But the damage to the fisheries caused by many practices has risen alarmingly. Most of the changes identified in Section 14.3 have been associated with incremental effects as well as overt impacts. The raw sewage from the first small community is added to by other communities and, even though it may eventually be treated, the organic load of the receiving waters has been compounded. The first toxicants are scarcely noticed, but eventually they produce major fish kills.

The water courses themselves are changed. A stream is first channelized and diked through the community, then through the farmlands. A dam that first changes the flow regime is followed by other dams that may eventually completely regulate the river. The landing place on the beach is followed by a pier, then by dredging and more piers and eventually a harbor. The shallow bay where fish spawned is dredged and filled. The several communities grow into a megalopolis.

Control of the human activities that have these impacts depends on public awareness of the problems and on better planning for development. Fishery scientists can do much to describe the impact of past practices and to predict the result of current trends. The planning must develop and incorporate the concept of a *best management practice* (BMP) for each of the many development activities. Each BMP will be a compromise (perhaps not explicitly) between the direct cost of the action and the indirect cost of not taking it. The use of BMPs must have the force of law and their application may be assisted by incentive or bargaining arrangements. Further, their use must be subject to updating as they are found to be inadequate or unduly restrictive.

14.4.10 Status of Biological Communities

Impacts of environmental change on fish, higher plants, and the odor or appearance of the water are frequently obvious and of concern to many people. Much less obvious are the impacts on the multitude of other aquatic organisms. Two approaches are being used increasingly to provide indices of community change.

Indicator organisms are used occasionally but it is difficult to find and use a sensitive species. It must be easily identified (many invertebrates cannot be), abundant under its preferred conditions, not migratory, and have a narrow and consistent tolerance of environmental conditions. It may indicate change by disappearance or by a sudden increase in abundance. An example of the latter is prevalence of certain aquatic earthworms in fresh water called tubificids that are commonly associated with gross organic pollution because they tolerate very low concentrations of dissolved oxygen. They are sometimes so abundant that they appear like patchy red carpets.

The usual effect of an environmental change is a decrease in the number of species present, because the impact on an intolerant species is immediate whereas immigration and reproduction of a tolerant new species may take some time. This has been interpreted as degradation of the environment because naturally difficult environments harbor few species, some of which may be present in great abundance. The concept of *diversity indices* holds that a large number of species (species richness) indicates a clean environment and a small number indicate a polluted environment. Numerous scientists have developed indices of species richness that relate to pollution.

But neither indicator organisms nor diversity indices can be used without thorough study or be generally applied. Too little is now known about the tolerance of most organisms to the many environmental factors (Hart and Fuller, 1974).

14.4.11 Prediction and Modeling

The sine qua non of any proposed environmental action is a prediction of the consequences of alternatives including nonaction. Any prediction involves knowledge of a relationship of some kind that can be called a model. It may be a verbal principle, a simple mathematical relationship that can be expressed graphically, a complex system of equations that must be evaluated with a computer, or a physical model. For example, a simple prediction of the allowable level of a toxin might be based on a bioassay of the toxin and the application factor that has been shown to be satisfactory in a typical body of water. Another example might be a semiscale model of an estuary for the purpose of estimating the tidal circulation.

Water quality modeling has been used for many years to predict dissolved material levels and, more recently, to predict algal growth. The usefulness of such modeling has led to ecological models that attempt to predict population change on the basis of interspecific relationships, especially energy flow among populations (see Sections 7.3 and 8.10.4). More recently, efforts have been made to use many kinds of models to answer the public's questions about environmental changes. These include models of waste discharge, physical and chemical changes in the environment, biotic population trends, cost–benefit ratios, and regulatory strategy (Russell, 1975).

Much effort has gone into development and improvement of such models and they have proven to be useful for improving scientific understanding of the environment as well as assisting with management decisions. But every model is only an approximation and it is difficult to model all of the actual variations that may occur in a real system. Nevertheless, models are exceedingly useful and they will be used more and more in environmental management.

14.5 ENVIRONMENTAL IMPROVEMENT

Preventing the abuse of natural waters is only half of the challenge to fishery scientists. The other half is to repair the damage done in the past, to advocate the cause of fisheries during our continuing environmental modifications, and to improve the natural habitats for the fish or the fishing.

This latter half is, of course, a step in the direction of aquacultural controls, in which the aquatic environment is idealized in favor of the fish as much as is practical, but it differs in some important respects. Much of the natural environment is maintained and usually the fish remain a common property resource.

The essence of environmental improvement is to discover and modify appropriately the environmental factors that limit the populations of desirable fish. The investigation of the limiting factors usually requires study of fish behavior and populations relative to the environment. The amelioration of the limiting factors frequently requires engineering and construction that create conditions that are better for the fish.

14.5.1 Salmonid Streams

The unique centuries-old interest in trout and salmon has led to elaborate efforts to improve the streams that they use. These are usually streams in which the salmon or trout predominate and conditions can be improved just for them. Each stage in their lives may be considered, including migratory conditions, spawning areas, and living areas for growing juveniles or adult fish.

Transport and Guidance Devices   Two of the most obvious requirements are freedom for adults to migrate upstream to spawn and safety for young to migrate back downstream. Meeting these needs has led to extensive study of salmonid behavior and construction of elaborate devices to transport and guide them (Bell, 1973).

Upstream migration can be assisted by removing or modifying small natural obstacles such as rock or log barriers that form during floods. If removal is not practical, or if dams are built, fishways can be constructed. These are series of artificial pools separated by small waterfalls that can be easily jumped by the fish.

Guidance of the young fish migrating downstream requires barriers that keep them out of water diversions such as irrigation canals. Such barriers are usually screens that will pass water and divert fish back to the river. Design of such screens has been complex because the screen must withstand and pass trash while having holes smaller than the fish and low water velocities that will not suck the fish against the holes. It is also desirable to keep the fish out of hydroelectric turbines, but satisfactory screening of the huge volumes of water has been difficult. Many efforts have been made to devise electrical, sonic, or other barriers, but with limited success to date. A "traveling" screen that is self-cleaning and minimizes fish impingement has been the most successful design.

A temporary alternative is to transport either adults or young in trucks or barges fitted with tanks and aeration systems that pick up the fish at special trapping devices and move them around the obstacles or hazards.

Spawning Areas   Trout and salmon lay their eggs in gravel through which water must percolate while the eggs hatch and the fry live on the food from the egg yolk. Then the gravel spaces must allow the fry to emerge. A suitable area must not accumulate silt and sand during the gravel life and it must not shift with floods or freeze. Lack of a suitable spawning area is a common factor that limits the production of salmonids in a stream.

Two artificial methods are used to improve conditions. One is to construct a completely new spawning channel in which the kind of gravel, the quality of the water, and the number of spawners can be controlled. The other is to clean gravel in place with hydraulic equipment that flushes out the fine materials, sucks them up, and removes them from the stream.

Living Space Salmonids in streams need places to feed and hide from predators. The feeding places are usually in or below gravel riffles that produce aquatic food organisms and the hiding places are in slower water in pools, behind boulders, or under banks. The size of the hiding place varies with the size of the fish.

Conditions are improved by placing obstructions in the stream that speed up the current to produce a better balance between pool and riffle areas. The devices include small rock and log dams or bankside deflectors that are intended to utilize the flow regime to dig and maintain pools. Suitable design requires substantial experience with and understanding of stream hydraulics if the devices are to work and be permanent.

Undesirable Species Salmonid streams that are infested with predatory fish or other undesirables can be improved by eliminating them. Usually this is done by poisoning, subsequent restocking of desirable species, and preventing return of the undesirables.

Poisons are selected to be as specific as possible. Rotenone, an extract of tropical plants, kills fish in concentrations of about 2 mg/liter, is relatively nontoxic to humans and fish-eating birds, and becomes nontoxic to fish in a few weeks (or can be detoxified immediately by potassium permanganate). Poisons for a few target species of fish have been developed, such as lampricides for lampreys in the Great Lakes of North America and squoxin for squawfish in Pacific Coast streams of Canada and the United States. When these are applied precisely, most other species can survive.

When the stream is free of undesirable fish it should be protected by a dam or weir to prevent their return and restocked with the desired species.

14.5.2 Small Lakes and Reservoirs

Lakes are treasured by many people who enjoy them for their beauty as well as for the waters that sustain many and increasingly varied human uses. Raising fish and fishing are uses that can be enhanced with minimal interference to others.

This discussion is focused on those factors that can be modified for the benefit of fishing other than regulation of the fishing itself. Included will be manipulations of the habitat and the fish populations. These factors can be changed in all reservoirs and in many small natural lakes but they can rarely be changed in very large lakes in which regulation of fishing is the major or only fishery management technique. Not included is consideration of aquacultural ponds in which production of fish rather than fishing is the objective (see Chapter 11).

Improvement of all reservoirs and most lakes is especially challenging because each is different. Each will have its own biology and limnology and each will be undergoing natural changes both seasonally and over the long term. Reservoirs, especially, change rapidly during the first few decades of their life. A few types of lakes, notably the relatively barren lakes of high latitudes or altitudes, can be managed as groups, but usually the study and management must be on an individual basis.

Dissolved Gases and Temperature   Most lakes become thermally stratified during summer and winter. If the lake is relatively fertile, the dissolved oxygen in the bottom water may be used up and fish lost before the lake turns over during the spring or fall. If the lake freezes over, the ice cover prevents any exchange of gases with the atmosphere and, if fertile, even the surface water may lack oxygen, a condition called "winterkill."

If the lake is natural, little can be done except to reduce the fertility (see following section). But in small lakes emergency aeration with mechanical equipment can be undertaken. One method is to vigorously agitate the surface. Another is to pump air to the bottom through a pipe with many small holes. The latter system promotes circulation of water from bottom to top as well as introducing air.

Still another approach has occasionally been practical to prevent death of fish under ice. If the ice is clear, sunlight will penetrate and sustain some photosynthesis underneath, but snow cover will greatly reduce the light penetration. The solution is to remove the snow.

Reservoirs, on the other hand, provide major opportunities for controlling the quality of the deep water by controlling the depth from which the outlet water is taken. Bottom water with little oxygen can be drained through a low outlet and aerated upon release to provide good water downstream. If this is done during the spring runoff, the bottom water can be improved at the start of the summer season.

Temperature of the deep water in reservoirs can also be adjusted to some extent by the depth of the outlet. A surface outlet will draw off the warm surface water in summer leaving the cool deep water below. A deep outlet will draw off the cool water leaving a gradually deepening layer of warm water above. The usefulness of such adjustments must be determined separately for each reservoir because the water quality will depend also on fertility. If the lake is receiving a large amount of organic material, the dissolved oxygen in the deep water may be used so rapidly that adequate dissolved oxygen and low temperature cannot be maintained.

Salmonid living space in reservoirs can sometimes be provided through suitable location of outlets. In the typical reservoir in temperate climates the surface water in summer becomes too warm for salmonids while the deeper water remains at a suitable temperature. If the outlet depth can be varied to draw on the cooler deep water during the spring, to establish a suitable oxygen content, and then draw on the warmer surface water during the summer months, a "two-story" fishery can be established. Trout fishing can be pursued in the deeper water and bass, sunfish, or other warm-water species can be caught near the surface.

All reservoir management will also have an ecological effect on the river below that will depend on the degree to which the flow is controlled (Ward and Stanford, 1979). Even if none of the water is diverted at the dam, the storage will reduce the floods, contain the sediments brought down from upstream, change the channel erosion (which is greatest during floods), and change the life of the organisms that have become adjusted to the floods and the sediments they bring. If the water is all used for hydroelectric power production, the pulses of water released with changing electric demand will change the flow in vastly different ways than the usual seasonal cycle. The most common flow problem below storage reservoirs is maintaining enough water for the fish. If the water is diverted or used in pulses for power production, there will be a conflict over the amount that should be left for fish (Orsborn and Allman, 1976).

Temperature changes in the outlet water during summer may change the nature of the fisheries downstream. A shallow reservoir in a cold-water stream may warm enough at all depths in summer to change a trout stream below into a warm-water fish stream. On the other hand, a deep reservoir in a warm climate may permit the development of a trout fishery below by deep-water releases through the dam during the summer.

Fertility Lakes and reservoirs with more or less natural fish populations vary in their ability to produce fish from about 10 kg/ha in cold, barren, high-latitude or high-altitude locations to about 200 kg/ha in warm, shallow, tropical lakes. The difference is due to relative fertility, length of the growing season, and temperature. The fertility is frequently the major factor and it can be controlled with substantial benefits to the fisheries.

The fundamental fishery problem is to maintain the fertility at a level that produces the maximum quantity of desirable fish without degrading their environment. This may mean adding fertilizer to a barren lake or reducing excess organic material in an overly rich one.

Domestic sewage or agricultural wastes are not problems in moderate quantities—and they may be beneficial. But commonly the quantities are excessive, especially when they include large amounts of phosphorous from detergents or other sources, and then they produce the troublesome symptoms of excessive fertility or eutrophication. Aerobic organisms become so abundant in the deeper waters that they use up the oxygen and die, leaving the zone to the anaerobic organisms, which produce foul odors. Aquatic plants become so abundant at all depths in the photic zone that they become a nuisance, some even toxic to humans and livestock.

The solution for excessive fertility is first to reduce the input of nutrients and, second, if necessary and practical, to mitigate the effects of the accumulated organic materials (Keup, 1979). These may be removed by dredging in extreme cases. Rooted vegetation may be removed by underwater mowing or controlled by introduction of fish that eat it. If the lake is behind a dam, the water flow through the deep water may be increased or the lake drained in order to oxidize the mud. The phosphorous content of the water may be reduced by adding alum, which precipitates it and prevents its use by plants.

Each body of water requires special study of its physical, chemical, and biological characteristics in order to design a successful restoration program.

Living Space   Given water of good quality, much more can be done to physically enhance a lake for desired species. Brush shelters can be submerged in a lake to provide habitat or spawning facilities for some species, or brush and trees can be left in part of a reservoir area to be submerged. Access to a marsh area can enhance spawning for northern pike. Rocky bottoms can be placed for spawning lake trout or walleyed pike. Barrels or similar containers can provide spawning areas for channel catfish. Old auto tires and bodies can provide shelter for many species.

Undesirable Species   Desirable species are the game species and their principal prey. Undesirable species include stunted fish of all species that have become too abundant, bottom-rooting fish, such as carp, that create turbid water, abundant algae, and dense stands of higher plants.

Unwanted fish can be partially controlled by a variety of methods. They can be netted or selectively poisoned when they congregate for spawning or feeding. Stunted fish of desirable species in a reservoir can be reduced in abundance by a major drawdown that kills a large part of the population and leaves a balance to grow rapidly after refilling. Some species that spawn in shallow water can be reduced by a drawdown that is timed to kill the eggs or fry in the nests. But all such techniques produce only temporary control that must be repeated again and again. Total control is desirable but is practical only in small lakes that can be totally drained or poisoned and then restocked with preferred species.

Unwanted plants can be controlled by spot poisoning with copper salts or various herbicides, but all must be used with great care to avoid damage to desirable species, including fish. Better control in reservoirs may be possible through well-timed drawdowns with subsequent drying of the bottom.

14.5.3 Marine Waters

Practical ways of improving the marine environment are more limited than the methods for improving fresh waters, but some are important.

Oyster Bottoms   Many waters that are suitable for growing oysters lack rocks or old shells, to which young oysters must become attached. Providing suitable "cultch" is a regular practice of commercial oyster growers in many areas. Recreational oyster gatherers may be required to shuck their harvest on the beach and leave the shells behind.

Artificial Reefs   Many choice marine fish live near rough bottoms, and the numbers of fish around such bottoms may be much greater than the numbers on underwater sand flats. Underwater structures such as reefs, wrecked ships, and offshore oil rigs are transformed into underwater communities as organisms become attached and numerous species of fish gather around them.

Such knowledge has stimulated many attempts to create artificial reefs where fishing is desired. Surplus ships, old oil drilling platforms, scrap auto tires, auto or streetcar bodies, and concrete scrap can be sunk or dumped to attract rough bottom fish (Stone, 1978). Substantial efforts are being made to discover waste disposal methods that will create the most durable and economical reefs. More elaborate engineered structures are being used in Japan to attract desired species.

Fish Attractant Devices (FADs)   Numerous observations of tuna or other schools near floating debris have led to trials of anchored rafts to see if they would improve tuna fishing. They do. Rafts anchored in several hundred meters of water near Hawaii attract schools of tuna that can be fished with purse seines or live-bait techniques. Similar trials elsewhere in the tropical Pacific have been promising. Studies are under way to improve the construction and anchoring of the rafts and reduce the costs.

"Reforestation"   Many desirable fish and invertebrates congregate around seaweeds or rooted aquatic plants in fresh water. Such plants cannot remain attached in strong currents or to loose shifting bottoms. Currents can be controlled by breakwaters, and sandy or muddy bottoms can be covered with stable rocky structures to which the plants can remain attached and in which animals can find shelter (Mottet, 1981).

14.6 ROLE OF FISHERY SCIENTISTS

The environmental management aspect of fishery management differs in major ways from the regulation of fishing. First, and perhaps most important, the public constituency is much larger than just the fishermen and has more varied interests. It includes people who are concerned about quality and preservation of the environment, and it also includes the diverse, politically powerful, and long-established groups of people who use and abuse water. These people attach quite different social and economic values to the water than do the fishermen.

Second, the public and private organizations that employ fishery scientists include not just the fishery agencies, but water and land management organizations, the engineering firms that design water use systems, and many large water users. All of these kinds of organizations are involved with changing the aquatic environment during their planning or operations and usually have obligations to protect that environment.

Third, the laws with which the fishery scientists must be concerned are far more complex and diverse than the fishery law. They include the old bodies of traditional water law and the new environmental laws with their emphasis on quality.

Fourth, the science and engineering required are much more diverse than those used for fishery regulation. Environmental aspects may require knowledge of river hydraulics, hydrology, limnology, or oceanography. Planning and design may require scientists and engineers with numerous specialties. Monitoring of water quality may require experts in fish pathology or analytical chemistry. Mitigation of damage may require operation of aquacultural facilities.

Fifth, solutions must be sought that accommodate the value judgments of diverse groups of people. The sciences are playing an increasing and vital role, but their use to predict ecological changes is still woefully inadequate.

The diversity of the aquatic environmental problems demands a very broad ecological approach leavened with social and economic sensitivity. Fishery scientists will find that teams of specialists are usually desirable for both design of solutions and implementation of action.

14.7 AN OUTSTANDING SUCCESS

The pink salmon and sockeye salmon of the Fraser River in southern British Columbia were a major resource for Native Americans for centuries, and began to be harvested commercially in the 1870s. The salmon migrate to and from the sea through Canadian and United States waters, and international regulation began to be discussed in 1892. A major decline in the runs occurred after railroad construction in the Fraser Canyon dumped accumulated rock in the river in 1911–1919, which blocked some of the runs, and catches from some declined by 80% or more. Thereafter the run sizes were much smaller, and fishing pressure was increasing. After years of negotiation, and the loss of hundreds of millions of dollars in potential earnings by the fishermen, the two countries agreed on a treaty to form a commission in 1937 to regulate the fishery and recommend environmental improvements.

The problems were technically and politically complex. They involved consideration of sewage effluents, rock slides from railroad construction, pulp mill effluents, fish hatcheries, Native American rights, ocean fishing by fishermen from the two countries, construction of spawning channels and hatcheries, flood control, and diversion of water from Columbia River tributaries to the Fraser. Construction costs totaled about $5 million by the early 1970s when construction was completed. Operating costs rose to about $3 million annually in the early 1980s. Then the commission finished its work and was disbanded. Average annual catches are now two to three times greater than in the early decades of the twentieth century (Roos, 1991).

REFERENCES

Bell, M. C. (1973). "Fisheries Handbook of Engineering Requirements and Biological Criteria." U.S. Army Corps of Engineers, Portland, Ore.

Gore, J. A., and Petts, G. E., eds. (1989). "Alternatives in Regulated River Management." CRC Press, Boca Raton, Florida.

Gucinski, H., Lackey, R. T., and Spencer, B. C. (1990). Global climatic change: Policy implications for fisheries. *Fisheries* 15(6), 33–36.

Haines, T. A. (1981). Acidic precipitation and its consequences for aquatic ecosystems: A review. *Trans. Ames. Fish. Soc.* 110, 669–707.

Hart, C. W., and Fuller, S. L. H., eds. (1974). "Pollution Ecology of Freshwater Invertebrates." Academic Press, New York.

Ischinger, L. S. (1979). Nonpoint source pollution. *Fisheries* 4(2), 50–52.

Keup, L. E. (1979). Fisheries in lake restoration. *Fisheries* 4(1), 7–9, 20.

Mottet, M. G. (1981). Enhancement of the marine environment for fisheries and aquaculture in Japan. *Wash. Fish. Tech. Rep.* 69, 1–96.

Orsborn, J. F., and Allman, C. H., eds. (1976). "Instream Flow Needs," Vols. 1 and 2. Amer. Fish. Soc., Bethesda, Md.

Perkins, E. J. (1974). "The Biology of Estuaries and Coastal Waters." Academic Press, New York.

Roos, J. F. (1991). "Restoring Fraser River Salmon." The Pacific Salmon Commission, Vancouver, Canada.

Ross, J. (1994). An aquatic invader is running amok in U.S. waterways. *Smithsonian* 24(11), 41–51.

Russell, C. S., ed. (1975). "Ecological Modeling in a Resource Management Framework." RFF Working Paper QE1. Johns Hopkins Univ. Press, Baltimore.

Stone, R. B. (1978). Artificial reefs and fishery management. *Fisheries* 3(1), 2–4.

Trelease, F. J. (1970). New water laws for old and new countries. *In* "Contemporary Developments in Water Law" (C. W. Johnson and S. H. Lewis, eds.), pp. 40–54. Center for Research in Water Resources, Univ. of Texas, Austin.

United Nations Environmental Programme (UNEP) (1980a). "Guidelines for Assessing Industrial Environmental Impact and Environmental Criteria for the Siting of Industry," Ind. and Environ. Guidel. Ser., Vol. 1. UNEP, Paris.

United Nations Environmental Programme (UNEP) (1980b). "Environmental Impact Assessment: A Tool for Sound Development," Ind. and Environ. Guidel. Ser., Spec. Iss. No. 1. UNEP, Paris.

U.S. Environmental Protection Agency (EPA) (1976). "Quality Criteria for Water" (The Red Book). Office of Water and Hazardous Materials, U.S. EPA, Washington, D.C.

Ward, J. V., and Stanford, J. A., eds. (1979). "The Ecology of Regulated Streams." Plenum, New York.

15 | Fishery Development and Restoration

Fishery development has been and remains a part of overall socioeconomic goals. It has been pursued in almost all countries of the world with access to fish either as an independent national policy or as part of international assistance. But with the topping out of production from the world's capture fisheries and the destructive fishing practices, the traditional roles of both the capture and culture fisheries are certain to change.

The goals of the capture fisheries have been characteristically complex and broad in concept, touching benefits such as increased food supply, increased earnings of poor people, or increased export earnings. Such goals are inherently good, but their realization has proved to be extraordinarily complex because of the diversity of the fisheries, their linkages to all other uses of the waters, and the leveling off of wild fish production worldwide in the 1990s. The fishermen have probably discovered all of the substantial resources and are attempting to capture more than the sustainable yields in most of them unless they are constrained by government regulation.

The people engaged include the recreational, subsistence, or commercial fishermen who use common property resources and the fish farmers who grow fish as crops. They pursue freshwater and marine resources, domestic and international resources, large-scale industrial and small-scale artisanal operations, use of water in competition with industry and agriculture, and marketing a highly perishable product.

Achievement of development goals must emphasize many kinds of assistance, most of which need to be directed toward small-scale fishermen. International donors have usually supported short- and medium-term programs designed to start a process of modernization within the fishing, processing, or marketing sectors. Most national governments, however, have continuing fishery management programs (Chapters 13 and 14) that already supply some assistance to the small-scale fisheries through allocation of stocks for their exclusive use or fishery regulations that favor them. Furthermore, most national governments in developed countries have assisted parts of their fisheries with continuing subsidies, which may have started as short-term assistance but which have become politically addictive and of indefinite duration.

Fishery development received much more emphasis during the 1980s. First, the worldwide acceptance of the concept of Exclusive Economic Zones in the sea out to 200 miles from the coast (called the Fishery Conservation Zone in the United States) led to the perception by coastal countries that they had wealth in their EEZ ripe for the taking. Second, the higher price of fish resulting from the leveling off of marine production led to increased interest in fish farming and increased production. Third, the rapidly increasing interest in recreational fishing is leading to allocation of resources to, and government development of, recreational fisheries, sometimes at the expense of commercial fisheries.

15.1 ROLE OF GOVERNMENT

Economic development in all countries except those with centrally planned economies is a partnership of government and the private sector. Fishery development is an additional task for governments that already have responsibility for protection and use of the public resources—fish and water. It is a complex task for government because many sectors of society are concerned and the most effective level of intervention in the fish businesses is difficult to find.

All too often government policies have indirectly inhibited development programs. Examples include controlled pricing and marketing of products, controlled imports of essential supplies or raw materials, lack of credit to small-scale enterprises, and lack of decentralization to local administrative entities. Examples in fishery development programs include the widespread expansion of fisheries on stocks already overfished, lack of pollution control, and giveaway fishing agreements with other countries. Especially common in small-scale commercial fisheries programs have been destructive efforts to circumvent the wholesaling and credit services provided by the people who buy fish. It has been easy to criticize buyers for excessive profits (and sometimes profits are), but substitution of government buying has usually been ineffective.

These kinds of mistakes are characteristic of what Owens and Shaw (1974) described as dual societies, which are extensions of the ruler–ruled relationship in which decisions are made at the top and passed down. The ruling class (even in so-called democratic societies) does little to involve the ordinary citizens in the planning, the decisions, the investments, or the profits. Such societies tend to be bureaucratic thickets in which projects are delayed because of trivial purchasing problems, or because no civil servant initiates the action required, or because the local community has no taxing power or authority. A concomitant factor is lack of a sense of loyalty to the central government.

In contrast, Owens and Shaw defined *modernizing governments* as those that work with local institutions and depend on local leaders to work with the people. Only in this way is it possible to deal with the diverse problems of local people, to build the local knowledge and capability, and to develop a public sense of confidence in, and loyalty to, the system.

The most important steps in the government support of sustained fishery production or development are: first, careful determination of the goals and objectives of any program; second, evaluation of alternatives; third, thorough planning to overcome the diverse problems; and fourth, periodic evaluation of progress. This may seem obvious but it requires close adherence to the concept of coordinating all levels of government with the involvement of concerned groups at all stages.

15.2 PERSPECTIVES ON FISHERY DEVELOPMENT

Perhaps the most important characteristic of any development process is its approach to effecting social change. It is a slow process, requiring years, if not decades, for people to reorder their activities around new technical and economic concepts.

A useful approach to planning a fishery development project is to define all of the systems that will be involved and identify those that the development will alter. These will include both ecological and social systems that are already in place and that must adjust if the production is to be stabilized or if development is to succeed. The systems will vary, of course, with the kind of fishery, but prevailing issues are development of artisanal fisheries and destructive overfishing in developed countries.

Such existing fisheries are typically close to a technical and social balance. The resources being used are frequently overfished if the fishery has been long established, because the fishing community has expanded as long as fish are available and governments have not restrained them. The fishing operations use equipment that can be purchased or constructed, operated, and maintained by the people in the community (Fig. 15.1). The catches can be processed in the community before spoiling (Fig. 15.2). The products can then be transported and sold to a group of consumers who depend on them (Fig. 15.3). The entire process has been socially acceptable in the sense that people have known about it and have had no reason to overturn it.

FIGURE 15.1   A fisherman on Madagascar returning to his village of Anororo. (Photo by P. Pittet, courtesy of the FAO.)

15.2.1 Characteristics of Large- and Small-Scale Fisheries

Food fishery enterprises exist at all levels of complexity and investment from one fisher without a boat to a multinational corporation operating fleets of vessels and numerous processing facilities. There is no clear separation between large and small, but a great difference between the extremes.

Large-scale fisheries are similar to large agroindustrial firms in the developed countries. They are capital-intensive, provide higher incomes for more specialized and skilled employees, and use large amounts of energy. They are usually vertically integrated, require sophisticated port facilities, and are lo-

FIGURE 15.2   Kilns for smoke-drying fish on the Ivory Coast. (Photo by A. Defever, courtesy of the FAO.)

cated in urban areas. They seek abundant, well-known species that are acceptable in widespread markets. They usually maintain high standards of quality and produce most of the canned and frozen fish that enter export markets. They also produce almost all of the fish meal and oil, either from fish caught especially for those purposes or from fish-processing wastes. Those that farm fish will be in rural areas but will be similarly capital-intensive and able to maintain high-quality products for distant markets.

Small-scale food fisheries are usually closely allied with rural agriculture. In fact, the most widespread kind of fish farming is usually more efficient as a part of rural agriculture. Both fishing and fish farming are labor-intensive and even the fishermen with motorized boats seldom have mechanical assistance for handling the gear. They usually supply nearby markets with a great variety of whole or cured fish that are handled in simpler ways with substantial losses due to spoilage. Yet they are important suppliers of animal protein to hundreds of millions of people in developing countries.

Another very different kind of small-scale fishery is recreational. These develop amid interesting recreational resources to which people can gain access. They are typical of aquatic recreational areas in developed countries and tourist locations in developing countries. The fishing methods are almost always hook

FIGURE 15.3   Whole fish displayed in a street market in Rio de Janeiro. (Photo by J. Fridthjof, courtesy of the FAO.)

and line from sophisticated boats operating out of comfortable accommodations. They are small scale in the sense that the businesses involved are small, although the economic activity as a whole may be equal to or greater than that of the commercial fishing in some localities.

15.2.2 Fishery Development Constraints

Fishery development may be constrained by many factors that have been most numerous in the small-scale fisheries of developing countries. Those identified for the countries bordering the Bay of Bengal during the 1970s, when rapid growth in fishing occurred in many countries (Indian Ocean Fishery Commission, 1978) were probably typical of assistance that was simply directed at increased fishing effort:

(i) inadequate gear and equipment;

(ii) inadequate technological know-how related to fish production, handling and fish utilization;

(iii) low level mechanization;

(iv) inadequate repair and maintenance facilities;

(v) inadequate landing facilities;

(vi) inadequate post-harvest technology and facilities;

(vii) adverse weather conditions especially for the technology in use;

(viii) low level traditional technology;

(ix) competition between small-scale fisheries on the one hand, and medium or large-scale industrial fisheries on the other;

(x) socio-economic plight of the fishermen;

(xi) too many fishermen in the small-scale fishery sector of some countries;

(xii) institutional weaknesses in (a) administration, (b) training, (c) marketing, (d) credit;

(xiii) poorly structured and inefficient fishermen's organizations;

(xiv) poor communication systems with fishermen, e.g. lack of effective extension services;

(xv) the lack of necessary socio-economic, catch/effort and resources data in usable and useful form;

(xvi) inadequate national financial and technical resources to meet these needs.

The report went on to outline the material and technical assistance that would be required.

15.2.3 Fishery Development Options

Most fishery development projects have had a short-term goal of helping with a single aspect of the development process that has been identified as a constraint. Included are the following activities:

1. Training resource scientists, food scientists, fishery officers, and fishery administrators.
2. Establishing the management process at sustainable production levels.
3. Introducing small-scale fish farming.
4. Providing facilities for vessels, fish marketing, cold storage, ice plants, and other infrastructure.
5. Providing better catching and fish protection equipment.
6. Introducing better processing and preservation methods.
7. Helping to establish cooperatives.
8. Providing low-cost credit, guaranteed loans, and various kinds of subsidies.
9. Encouraging information exchange and collaboration between neighboring communities or countries.

But if the long-term goal is to establish a permanent improvement in the fisheries, it is necessary to strengthen the public institutions involved so that they can convince the fishing and processing businesses of the steps necessary to sustain the productivity of the resources. They may emphasize several of the preceding activities successively, plus others as the constraints change with time.

15.2.4 Large- and Small-Scale Opportunities

Despite the fact that worldwide production from the capture fisheries has leveled off, there are many opportunities for increased production and reallocation of the fish among users. These opportunities can be challenging to both large- and small-scale enterprises.

Reallocation to coastal countries of the fishing in the EEZ may provide both small- and large-scale opportunities. The foreign fleets probably will have consisted of large distant-water vessels and their experience may provide a model for where and how to fish with sophisticated gear. In addition, some of the resources may be harvestable by small-scale gear operating out of coastal villages.

Other new large-scale fishing opportunities may lie in the sea, mostly far from ports, where some large, little-exploited stocks are located. These include the nonconventional species such as krill or small fishes of the middle depths and some more conventional species such as squids that are difficult to harvest. Increasing the harvest of all such resources will probably be risky because of the need for special equipment to catch and process such animals and the need to develop large international markets. Therefore, they will be highly capital-intensive and suitable only for large private organizations that still may need development assistance.

Large-scale fish farming is similar to large-scale agriculture in many respects. It will require large investments in facilities, effective quality control, and vigorous marketing efforts. Opportunities lie in expansion of salmonid and shrimp farming, both of which produce high-value foods, and in other species such as carp, which are easy to farm in both fresh and coastal waters.

Small-scale opportunities will be found around the edges of the sea and in inland waters, either in fishing or fish farming. Most of them will tend to be site-specific using a local resource and producing something for nearby markets. They will produce a large variety of items, each with processing, handling, and prices suited to local conditions. In almost all aspects, the small-scale opportunities will provide a greater challenge for governments. The great variety of people, resources, products, and control systems will require many different programs that include possible options for development.

15.2.5 Fads and Failures

The efforts to expedite development in the lesser developed countries during the 1950s and 1960s began with great hopes that were seldom fulfilled. Owens and Shaw (1974) emphasized that development is a complex historical process that does not have simplistic, quick solutions. They identified four major issues in the development activities of this period.

First was the effort to transfer constitutions and democratic forms of government. This was done in many countries as they became independent, but it did not change the fundamental power of the ruling elite. The lack of local leadership with experience in government, the lack of local understanding of democratic government, and the lack of confidence of illiterate and inexperienced people in any government, all mitigated against the change. It was forgotten that the democratic institutions of Europe and North America had evolved over centuries.

Second was the effort to train people. This was helpful but not a panacea. Much of the training was irrelevant and many university-trained people eventually left their home countries. Some training led to migration of people out of rural agriculture, and it did little to stimulate private entrepreneurship.

Third was the emphasis on industrialization and the idea that benefits would trickle down. It tended to introduce technology prematurely, to produce unemployment, to encourage migration to the cities, and to lead to establishment of inefficient enterprises.

Fourth was transfer of money, in either gifts or credits. This too tended to create more capital-intensive projects, unemployment, and benefits primarily for the well-to-do.

Fishery development has sometimes seemed easy because better equipment can immediately produce such a great increase in catch per person (Table 10.9) and because fishing coasts are close to many poor and hungry people. Yet there have been many disappointments. The following are some examples from this author's experience (with the countries deliberately not identified.) Emmerson (1980) gives many others.

A project to build hundreds of farm fish ponds and a fishery experiment station in an African country did not include the extension service needed by the farmers and after a few trials the ponds lay idle. The experiment station turned to trials of rearing crocodiles for their leather, which never became commercially successful.

A large project in a Latin American country to build fish meal plants failed because of scarce fish resources and left the country responsible for a large multimillion dollar debt. Another country developing elaborate plans for a similar venture was fortunate to obtain a resource survey before construction started. These plans were abandoned.

A fleet of large ocean-going trawlers was provided to an African country. After they proved to be impractical because of their design, fuel requirements, and maintenance difficulties, they were given to an Asian country, where they were still impractical. Nearly 20 years after the "gift" they were awaiting the scrapper's torch without ever having been operated at a profit.

A new, first-class fishery research vessel costing about $3 million was given to a country where it remained almost idle for years because the country had neither the funds nor scientific staff to operate it.

A fleet of hundreds of small coastal trawlers was subsidized in an Asian country. After the vessels began to operate successfully they reduced the stock of fish available to the traditional fishermen and competed with them in the market. The traditional fishermen organized themselves and burned most of the new vessels.

A new modern fish freezing and storage plant was given to a country. It failed completely after four years because of bad management, despite technical assistance from several expatriate experts.

A chain of cold stores that had been operating successfully for several years in a small African country went out of business because of government requirements that fish must be purchased from a minister's brother and that refrigerated transport must be leased from another minister.

A young man employed by a fishery agency in an African country had an innovative idea for oyster rearing and received a grant from Canada. He received the money personally and had his accounts audited by the Canadian consulate because if the money had gone to his agency it would not have reached him.

Such a list could be extended, but without adding to the point that development projects are fragile ventures. They require attention to all government, business, and social matters that may affect them. Initiating them is an art in which the requirements for success are much less well understood than merely encouraging exploitation of a resource.

15.2.6 Prior Disciplines

Part of the difficulty with fishery development projects can be traced to the backgrounds of the principal planners. Emmerson (1980) pointed out that people from four disciplines have been looking at fishery development from very different viewpoints.

Fishery biologists have approached the problem from the standpoint of obtaining the maximum sustainable yield (MSY) or the optimum yield (OY) from the resources without really being able to optimize the human aspects of the operations. They tend to look at it primarily to maintain physical yields, while assuming that the socioeconomic problems will work themselves out.

Resource economists have approached the problem from the standpoint of obtaining the maximum economic yield (MEY) from the fishing sector operations, with scant attention to issues of equity, people's beliefs about rights, and the processing and marketing aspects of the fisheries.

Lawyers have sought ways to use the common property resources more effectively, sometimes from a biological viewpoint, sometimes from an economic viewpoint. They also promoted the 200-mile-wide Exclusive Economic Zone

(EEZ) in which coastal countries have exclusive fishing rights but which may be exercised only after an extraordinary effort (for most countries) in scientific investigation, treaty negotiation, and control of fishing.

The few cultural anthropologists who have concerned themselves with the social aspects of modern fishing villages have described the complexities of the human relationships and some of the consequences of attempts to introduce new gear, new money, and new regulations. But they are seldom involved in project planning and evaluation.

People from other disciplines also tend to emphasize their specialties. The large-scale business person who is familiar with sophisticated processing and foreign markets is likely to underestimate the difficulty of establishing such a system in a small-scale fishery, and fishermen may underestimate the complexities of credit and marketing.

Of course it may be impractical to involve all disciplines in planning and execution of many projects. Rather, it is necessary for the leader from whatever background to be aware of the complex of technical, economic, legal, and social aspects of the projects and to make sure that all are considered.

15.3 NATIONAL ASSISTANCE

Assistance to small-scale fishermen started a few millenia ago when rulers in the eastern Mediterranean forbade distant-water vessels the use of nearshore grounds used by coastal fishermen. It continues as a part of national policy in most countries to this day.

The justification is usually based on the low income and difficult working conditions of the fishermen and the value of their product to the country. The small-scale fisheries are usually composed of individual entrepreneurs or family units as fishermen, processors, and traders, most of whom live in scattered small villages near the fishing grounds. They operate small vessels with limited range, limited capability in rough weather, and less fish-catching ability than large offshore vessels. Many must operate on grounds that are already producing the maximum sustainable yield. The occupation is sometimes considered to be inferior for reasons of low income, irregular employment, or prestige—in India, many fishermen are untouchables.

Many developed countries have substantial assistance programs for their own small-scale fishermen (OECD, 1976). Canada, Norway, France, Germany, Finland, and Sweden all have had assistance programs. Many of the programs included direct subsidies in the form of price supports, low-interest loans on boats, wage supports, assistance in formation of cooperatives, allocation to them of fish stocks or fishing grounds, and provision of port and market facilities.

Japan (OECD, 1976) has had an especially elaborate support system for its inshore fisheries, which totaled more than 200,000 units, about 30% of which were shallow-sea aquaculture. The government subsidized low-rate loans for engines and better fishing equipment, encouraged development of aquaculture, restricted entry into the nearshore fisheries, and assisted with provision of landing, marketing, processing, and cold storage facilities. A major feature of the assistance was encouragement of cooperatives. These held fishing rights and licenses, operated fisheries, and performed the functions of buying, processing, and selling fish, buying supplies and equipment, and providing educational services. The government also deliberately helped the coops to supplant the services of primary fish buyers.

The United States expanded its fishery development program in 1979 to support the industry in its efforts to take advantage of priority fishing rights in the EEZ (U.S. Dept. Commerce, 1980–1981). The major impediments to expansion were identified as the fragmented and dispersed industry, insufficient levels and imbalances of capital investment, inefficient market and distribution systems, and a narrow consumer market base. The country was a major importer of fish and recognized that many foreign operations were heavily subsidized. A major step was to foster seven regional nonprofit Fisheries Development Foundations. Their projects have been developed around an appraisal of each region's priority needs and directed mostly at problems of processing both at sea and ashore, fishing resources and gear, domestic and export marketing, and safety or quality aspects.

This program has been in addition to a National Sea Grant Advisory Service that provides information from more than 30 offices, assistance in the formation of more than 100 fishery cooperatives, extensive resource and market information services, and low-cost loans or loan guarantees for vessels and shore facilities.

15.4 INTERNATIONAL ASSISTANCE

Most of the international assistance in fisheries has three goals: increase in a major protein food supply, economic growth, and reduction of poverty. The assistance is usually financial and technical and is given for a specific term after careful planning (see also Section 3.2.6).

Three kinds of organizations provide fishery assistance: private foundations, individual nations through bilateral agreements, and United Nations or regional organizations through multilateral agreements. The foundations, of which the Ford and Rockefeller foundations of the United States and the Ecumenical Development Cooperative Society of The Netherlands are examples, have relatively limited resources and usually focus on development constraints that have been neglected by bilateral or multilateral agreements. An example is formation and limited support by the Rockefeller Foundation of the International Center for Living Aquatic Resources Management (ICLARM) in Manila.

Bilateral agreements have been supported by many developed nations, the oil-producing nations, and some lesser developed nations. Such agreements are usually characterized by benefits to both donor and recipient. The benefits to the donor may be political, military, or trade. Such benefits are not realized or are minimal if the development is supported through the multilateral organizations.

Multilateral agreements have been supported by the Regional Development Banks (Asian, African, and Inter-American), regional organizations such as the European Development Fund of the European Economic Community, and United Nations agencies. One of these is the World Bank with its three branches (World Bank, 1982): the International Bank for Reconstruction and Development (IBRD), the International Development Association (IDA), and the International Finance Corporation (IFC). The other major agency is the United Nations Development Programme (UNDP). The Fishery Department of the Food and Agriculture Organization of the United Nations provides some assistance with its own funds but most of the development projects that it operates are financed by UNDP, the development banks, or a few national donors. Much of its assistance is technical advice, training for various specialists, information, and special studies—all precursors to financial assistance (see Section 3.3.6).

15.4.1 Financial Assistance

Financial assistance has been provided in several ways by the various donors, but the largest amounts and the most comprehensive assistance have come from the Regional Development Banks and the three branches of the World Bank. These branches perform quite different financial functions. The IFC works with the private sector in developing countries and it has supported a few large-scale fishery projects. The IBRD works primarily with governments of developing countries and supplies funds at 0.5% above its cost of borrowing. The IDA works in parallel with IBRD to supply funds to developing countries without interest and with 50-year repayment periods. It levies an annual administrative charge of less than 1.0% of the loan but permits the start of repayment to be delayed for 10 years.

The Regional Banks and the World Bank have worked closely with each other, with other donors, and with other financial institutions. The World Bank has maintained resident representatives who regularly participate in coordination of assistance projects in most recipient countries. Part of the work of these banks has been to assist local lending institutions with funds for refinancing and with advice on the operations and credit needs of small-scale entrepreneurs—including fishermen and fish farmers.

The fishery projects of these banks, although a small percentage of their total lending, have been significant. During fiscal years 1964 to 1981, the World Bank (1982) loaned US$259 million, which was a little more than half of the total project costs. Of this total, about 32% was for vessels, 27% for fishing ports, 15% for aquacultural ponds, 13% for technical assistance, and the balance for working capital and other facilities. A substantial increase in lending occurred for the 1982–1986 period, when US$540 million was programmed, mostly for small-scale fishing and aquacultural development.

In the 1980s and 1990s the Inter-American Development Bank has promoted and financed dozens of technical assistance projects and development projects in fisheries. It has financed about half of the total project costs of several hundred million dollars (U.S.).

15.4.2 Technical Assistance

Technical assistance is supplied by most donors in many ways and on all aspects of the projects. Fishery projects may be especially complex and involve many specialists. The major international agency funding technical assistance is the UNDP. It has worked in more than 150 countries and maintained offices in 114 of them. Its annual budget has been on the order of US$700 million plus matching funds. Part of its funds have been used to support fishery projects in many countries, most of which have been executed by the Fishery Department of FAO.

Marine science development is a part of technical assistance that has received rapidly increasing support. It consists principally of help to individual scientists and laboratories in the form of equipment, libraries, fellowships, and support for scientific conferences. UNESCO is the principal multilateral agency funding this area.

15.5 PROJECT PLANNING AND EXECUTION

Each donor has a system of project management and a good example is the Project Cycle system that has been used by the World Bank. It consists of six phases:

1. Identification: Project alternatives that support national development strategies in the sector under discussion and that agree with the Bank's priorities are identified and a selection is made.

2. Preparation: The borrowing country or agency examines the technical, institutional, social, economic, and financial aspects of the proposed project. Advisory and financial assistance for the preparation may be provided by the Bank, UNDP, the FAO, or other assistance agencies. This stage may take as long as two years to complete.

3. Appraisal: The Bank staff makes a comprehensive and systematic review of all aspects of the project. This may require three to five weeks in the field. The appraisal report is reviewed at Bank headquarters and serves eventually as a basis for negotiations with the borrower.

4. Negotiations: Representatives of the borrowing country and the Bank meet to discuss the measures that must be undertaken to ensure effectiveness of the project and develop covenants for the loan documents. Following agreement and approvals, the loan becomes effective.

5. Implementation and Supervision: The borrower is responsible for implementation of the project but the Bank expects progress reports and undertakes periodic field visits. Procurement of goods and services for the project must follow Bank guidelines. Supervision is continued throughout the project period.

6. Evaluation: After the project is completed, a report is prepared by operational staff of the Bank and is reviewed by the Operations Evaluation Department. This department may also conduct an independent audit.

These procedures, which for the IDA part of the World Bank were especially designed to help poorer people, help to ensure project performance as well as financial performance by the developing country. The continuing guidance is intended to improve institutional performance and expedite modification of the project as necessary.

15.6 INSTITUTIONAL DEVELOPMENT

A critical aspect of most development projects is the strengthening or formation of institutions that sustain the projects and continue the development process after the projects are completed. Usually the government institutions that play key roles in planning and management of the development process are already in existence and need strengthening. For the fishery institutions, this is accomplished by special training in the sciences and government administration. But new institutions are frequently considered to be essential, because of the complexity of the social problems of development.

15.6.1 Fishery Cooperatives

Most developing countries encourage formation of cooperatives. They may have ministries or departments of cooperatives to provide advice on organization, training for leaders, or financial encouragement. Fishery cooperatives may perform a variety of functions, including purchase of gear and supplies, provision of credit, processing and marketing of fish, ownership of exclusive fishing rights, ownership of vessels or facilities, training and advice, and social activities. They have been thought to be ideal for development of small-scale fishing or fish farming.

But fishery cooperatives have proved to be fragile organizations. More have failed than have succeeded. Transfer of the European concept of cooperatives to other societies has been very difficult.

Some aspects of the lives of small-scale fishermen make formation of fishery cooperatives more difficult than that of agricultural cooperatives. Fishing tends to be a seasonal activity that may alternate with farming or other work. A large proportion of fishermen may be young people who fish only a few years until they find other employment. Fishermen have a low social status in some countries, and most fishermen are notably independent.

But some fishery cooperatives have been successful and long-lived, as in Italy, Japan, the Philippines, Malaysia, United States, and Belize. The fishery cooperatives in Belize have been identified as models and FAO chose the country as a site for training in the organization of cooperatives.

Successful fishery cooperatives have some common characteristics that seem to be essential for success. They include people with strong group loyalties, most of whom perceive the cooperative as beneficial to everyone in a group in which no individuals have a special advantage. The cooperative services, in marketing for example, provide more benefits to the members than they can get elsewhere. The business functions in administration, buying, selling, and accounting are performed openly and effectively. And the cooperative leadership is oriented toward problem solving rather than dogmatism.

Successful cooperatives have usually been organized by the members themselves, frequently with some government assistance but not by government fiat. They have not been constrained by an inflexible government bureaucracy, nor protected by government paternalism. Their formation may be the most essential and difficult part of a development program.

15.6.2 Fishery Extension Systems

Neither fishermen nor fish farmers will voluntarily change their practices unless they are convinced that they will benefit through greater earnings, less work, or less risk. Both small-scale fishing and fish farming in developing countries are usually practiced in rural areas to which transportation may be difficult. The fishery people usually follow a traditional life based on family and community practices. They are frequently illiterate or little educated and suspicious of strangers. In many developing countries, they may speak a language other than the national language.

Innovations in their practices must be accomplished by communicating ideas that they recognize as beneficial and convincing them that they can make changes. Such communication must involve people familiar with their language, customs, and problems, and this is an area in which cultural anthropologists can play a significant role. This must be done at the person-to-person level in the village, market, or farm. The most convincing communications are demonstrations of actual benefits obtained from new methods.

The problems of the development planner are to discover the methods and equipment that can be recognized as beneficial, to communicate with the target community, and to persuade the people to change. These problems have been approached in agriculture in many countries by research stations that work on the farmer's immediate and long-range problems, and by bidirectional extension services, which communicate the research findings to the farmers and the farmers' problems to the research stations. An alternative and supplement to the research station output is the experience of the best farmers that can be used by the poorer farmers.

Foreign advisors, even though skilled and sympathetic, have difficulties with the language and customs, the living and transport conditions, and the continuity of long-term projects. Furthermore, they are expensive; the cost of maintaining a foreign fishery advisor in a remote village with necessary transport, office, housing, and allowances was budgeted at US$75,000 to $100,000 annually in the early 1980s.

Organization of extension services in developing countries has been difficult in agriculture, where there has been much more experience than in fisheries. A method that has worked well is described by Benor and Harrison (1977). In this example, a unified service under a single agency is established to employ local full-time extension agents. The agents, called village extension workers in this case, are recruited from respected local residents with practical experience. They are given systematic training interspersed with regular visits to their clients. Six to eight of them are supervised by extension officers, who are in turn supervised by district officers and supported by subject matter specialists. Close liaison is maintained with a research station, and special efforts are made to demonstrate new methods on the farms of innovative farmers.

A research station–extension system has been formed in the Philippines to assist fish farmers. The country has been divided into four areas with different farming conditions and practices. Each area has a research station and extension service focusing on the special problems of the area.

Another example is the National Sea Grant Advisory Service, which assists fishermen in the United States. It is a communication link between the fishery communities and 31 Sea Grant programs at universities in almost all of the coastal states and Puerto Rico. In the 1980s, it employed about 350 people who worked through personal contact, conferences, and publications. They stay in close contact with the many government agencies concerned with the fisheries and provide information on regulations and government services. They assist with diverse problems such as improved port facilities, new vessel equipment, pollution, quality control, tax issues, and weather forecasting services.

15.6.3 Joint Ventures

Whenever a country has fishery resources that may supply export products or needs people experienced in management of fish businesses, it may be desirable to bring together in a single business venture the local knowledge of the people, practices, and resources with foreign knowledge of markets, processing, product standards, and financing (Crutchfield *et al.*, 1975). Many variants of such cooperation are possible between the extremes of a licensed concession for foreign operations and a domestic enterprise using foreign technical or managerial expertise on a short-term contract basis. The middle ground is a continuing partnership divided about equally between foreign and domestic interests. The latter may be either governmental or private.

The successful joint ventures have been very carefully planned and sensitively managed. Of special importance is solution of the many local financial, legal, and political problems, as well as the solution of scientific and technical problems.

15.6.4 Ownership of the Right to Fish

One of the most successful fishery management programs in the United States has been the recent Alaskan salmon management plan. The commercial fishery began in the late nineteenth century, and production rose to an average of about 200,000 metric tons annually in the 1960s.

Alaska had been a territory under U.S. federal management until statehood in 1959. Then in the early 1970s the state government formed a Commercial Fishery Entry Commission with a professional staff. It started a program in which the maximum number of fishing units of each type in each area needed to harvest the maximum runs was determined. Then ownership permits were issued to fishermen who owned a right to fish that could be sold or willed to their children. The system was contested but approved by a statewide ballot. The result has been nearly a tripling of salmon catch levels to about the high average level of 20 years earlier (Royce, 1988).

An older system of fishing property rights has been in effect in Japan, where more than 5000 coastal villages have fishermen's cooperatives, the rights of which are kept in balance with the nearshore rights of some of the largest fishing companies in the world. The result has been a good record of conserving the inshore stocks.

Other systems without property rights in the United States have been destructively overfished. The New England example is more or less typical. Here the ocean banks (under federal management) have been fished more and more intensively until the population levels and production of the major species (cod, haddock, and flounders) declined to only about 20 to 25% of the earlier levels in the late 1980s, where it remains in the middle 1990s. Better management requires a reduction of catch levels for a number of years to allow increased growth and reproduction, but this has not been politically acceptable to the Congress, the regional fishery councils, many local businesses, the Native Americans, or the fishermen.

15.7 ROLE OF FISHERY SCIENTISTS

Fishery development, if it is to be successful, must usually include professional consideration of a greater range of scientific, technical, social, economic, political, financial, and managerial issues than any other fishery science activity (Section 15.2.2). But fishery scientists will usually be working for government and it is essential to recognize the limitations of governments. The comments of Tussing (1971) with respect to fishery development priorities in Indian Ocean countries are especially cogent.

> A universally conspicuous shortcoming in the fisheries development efforts of the Indian Ocean Countries is in the fields of economic analysis, planning, and business management. There, shortcomings are general, from the national planning level to the level of enterprise operation. In many cases the responsible officials and managers have a disturbing lack of awareness of even the notions of productive efficiency, cost effectiveness, or opportunity cost, particularly with

respect to capital. Almost all institutions and (government) enterprises in fisheries development are under the direction either of scientists or of civil service generalists. I was generally impressed by the ability in their specialties and by the dedication of those I met, but any training, background, or inclination for business management is very rare among them.

After noting the massive obstacles to development caused by currency controls, Tussing continued:

> Another commonplace problem is the multifunctional government Fisheries Corporation, directed by managers with the job security of civil servants, without explicit standards of performance, and with accounts from its various functions hopelessly intermingled so that the profit or loss of various activities cannot be distinguished.

Tussing also commented on the technical and scientific orientation of the fishery administrations and noted that technical and scientific training and assistance projects were easily identified and commonplace, whereas projects requiring common financial and business orientation were seldom sought.

Scientific and technical problems were indeed given priority by representatives of 25 countries in a 1978 survey (Craib and Ketler, 1978). Their needs were expressed in terms of general problem areas, with stock assessment, fresh and brackish water aquaculture, reduction of postharvest losses, and improvements in processing usually given higher priority in a list of research assistance needs. Socioeconomic studies were usually given low priority, if mentioned at all, and research on business needs was not mentioned.

Another increasingly important scientific and technical activity is associated with environmental management by governments of lesser developed countries. Water is almost always scarce and in demand for industry and irrigation, so its use for fish must compete with other uses. Furthermore, pollution from both agricultural and industrial sources is a major problem in many countries.

Scientific and technical assistance projects will always be part of development programs and will usually be planned on a case-by-case basis. But the fishery scientist who is involved in planning or execution of a project should try to make sure that it is part of a balanced approach to the overall development goals. If the project is not so planned it may well become one more item on the list of failures.

The problems in Europe during the early 1990s are probably typical of many of the world's more populated areas (Dill, 1993):

> In summing it up, one can say with respect to the inland fisheries of Europe that the commercial fisheries are declining as recreational and aquacultural fisheries grow, that new technologies are replacing the old, that the role of private fishing is declining, that environmental considerations are more important than ever before, and that there is a growing realization that many of the factors influencing fisheries are quite external to the fishing itself. These phenomena, and the other changes that have been mentioned, are not, however, unique to Europe. This is the way of the world today.

REFERENCES

Benor, D., and Harrison, J. Q. (1977). "Agricultural Extension: The Training and Visit System." World Bank, Washington, D.C.

Craib, K. B., and Ketler, W. R., eds. (1978). "Fisheries and Aquaculture. Collaborative Research in Developing Countries: A Priority Planning Approach," Contract report prepared for U.S. Agency for International Development, U.S. Dept. of State, Washington, D.C.

Crutchfield, J. A., Hamlish, R., Moore, G., and Walker, C. (1975). "Joint Ventures in Fisheries." Indian Ocean Fishery Commission, FAO, Rome.

Dill, W. A. (1993). "Inland Fisheries of Europe," EIFAC Tech. Paper. FAO, Rome.

Emmerson, D. K. (1980). "Rethinking Artisanal Fisheries Development: Western Concepts, Asian Experiences," World Bank Staff Working Paper No. 423. World Bank, Washington, D.C.

Indian Ocean Fishery Commission (1978). "Project for the Development of Small Scale Fisheries in the Bay of Bengal—Preparatory Phase," Vol. 1, Report; Vol. 2, Working Papers. Indian Ocean Fishery Commission, FAO, Rome.

Organization for Economic Cooperation and Development (OECD) (1976). "Economic State and Problems of Small Scale Fisheries." OECD, Paris.

Owens, E., and Shaw, R. (1974). "Development Reconsidered: Bridging the Gap between Government and People." Lexington Books. Lexington, Mass.

Royce, W. F. (1988). Alaskan salmon management: An outstanding success. *Pac. Fish. Rev.* **15**, 17–33.

Tussing, A. R. (1971). "Economic Planning for Fishery Development," IOFC/DEV/71/19. Indian Ocean Fishery Commission, FAO, Rome.

U.S. Department of Commerce (1980–1981). "Fisheries Development Report." National Marine Fisheries Service, Washington, D.C.

World Bank (1982). "IDA in Retrospect: The First Two Decades of the International Development Association." World Bank, Washington, D.C.

Appendix A
A Code of Professional Ethics

Most organizations whose members use science directly to serve clients have adopted or recommended an ethical code. Included are most of the medical societies, engineering societies, The Society of American Foresters, The Wildlife Society, the National Association of Environmental Professionals, and the American Fisheries Society. The following code was adopted by the American Fisheries Society in 1966.

CODE OF PRACTICES

The following code of practices is promulgated to be considered as a general guide and not as a denial of the existence of other duties, equally imperative, but not specifically included.

Section I

Relations with the Public

1. The members of the American Fisheries Society will professionally advise on, design, or supervise only such projects as their training, ability, and experience render them professionally competent.

2. The member will not associate himself with, or allow the use of his name by an enterprise of doubtful character.

3. When serving as an expert witness before a court, commission, or other tribunal, the member will express an opinion only when it is founded on adequate knowledge and honest conviction.

4. The member will refrain from expressing publicly an opinion of fisheries subjects unless he or she is informed as to the facts relating thereto.

5. The member will not knowingly permit the publication of his or her reports, or parts of them in a manner calculated to mislead.

Section II

The Dignity and Well-Being of the Profession

1. The members will cooperate in extending the effectiveness of the fisheries profession by interchanging information and experience with other professionals and students, and by contributing to the work of the professional societies, schools, and the scientific press.

2. A member will not advertise in an unprofessional manner by making misleading statements regarding his or her qualifications or experience.

3. The member will endeavor to protect the fishery profession collectively and individually from misrepresentation and misunderstandings.

4. The member will give personal approval only to those plans, reports, and other documents for which he or she is responsible or in agreement with, and which have been prepared under his direction or authority.

Section III

Relation with Clients and Employers

1. The member will act in professional matters for each client or employer as a faithful agent or trustee.

2. He or she will not disclose any information concerning the business affairs or the technical processes of clients or employers without their consent.

3. He or she will engage, or advise his client or employer to engage, and will cooperate with, other experts and specialists whenever the client's or the employer's interests are best served by such service.

4. He or she will present clearly the consequences to be expected from deviation proposed if professional judgment is overruled by other authority in cases where he or she is responsible for the technical adequacy of fisheries work.

5. He or she will not accept compensation, financial or otherwise, from more than one interested party for the same service or from services pertaining to the same work without the consent of all interested parties.

Section IV

Relationships with Other Professionals

1. The member will take care that credit for professional work is given to those to whom credit is properly due.

2. The member will uphold the principle of appropriate and adequate compensation for those engaged in professional work including those in subordinate capacities as being in the public interest and maintaining the standards of the profession.

3. The member will endeavor to provide opportunity for professional development in advancement of fisheries professionals in his employ.

4. The member will not attempt to injure falsely or maliciously, the professional reputation, prospects, or business of another fisheries professional.

5. The member will refrain from criticizing another professional's work in public, recognizing that the Fisheries Society and the Fisheries Press provide the proper forum for technical discussions and criticisms.

6. The member will not use the advantages of a salaried position to compete unfairly with another professional.

The foregoing duties are hereby imposed on all members of the American Fisheries Society.

Appendix B
Excerpts Pertaining to Fisheries from the Law of the Sea

After centuries of piecemeal development of the Law of the Sea, a process was begun in 1969 under the direction of the United Nations to prepare a comprehensive legal code. It was agreed to cover 25 main subjects, most involving several issues. The subjects pertaining especially to fisheries included No. 5, the continental shelf; No. 6, exclusive economic zones beyond the territorial sea; No. 7, coastal state preferential rights or other nonexclusive jurisdiction over resources beyond the territorial sea; No. 8, the high seas; No. 9, landlocked countries; No. 10, rights and interests of shelf-locked states and states with narrow shelves or short coastlines; No. 12, preservation of the marine environment; and No. 13, scientific research.

After several arduous meetings, a convention was adopted on April 30, 1982 [K. R. Simmonds (1983). "United Nations Convention on the Law of the Sea, 1982." Oceana Publications, Dobbs Ferry, N.Y.]. It is a complex document containing 320 articles plus annexes that deal with virtually all human activities outside the territorial sea. When it was presented to member countries for a vote, a large majority approved, several abstained, and four, including the United States, declined to ratify. (The United States disagreed with the provisions regarding marine mining.) Many matters remain difficult or abstruse and debate will continue, but the provisions with respect to fisheries have been widely agreed upon. Some major articles (much abbreviated) are as follows:

56. *Rights, jurisdiction and duties of the coastal State in the exclusive economic zone*. In the exclusive economic zone, the coastal State has: (a) sovereign

rights for the purpose of exploring and exploiting, conserving and managing the natural resources, whether living or non-living. . . .

57. *Breadth of the exclusive economic zone.* The exclusive economic zone shall not extend beyond 200 nautical miles from the baselines from which the breadth of the territorial sea is measured.

61. *Conservation of the living resources.*

1. The coastal State shall determine the allowable catch of the living resources in its exclusive economic zone.

2. The coastal State, taking into account the best scientific evidence available to it, shall ensure through proper conservation and management measures that the maintenance of the living resources in the exclusive coastal State and competent international organizations, whether subregional, regional or global, shall cooperate to this end.

3. Such measures shall also be designed to maintain or restore populations of harvested species at levels which can produce the maximum sustainable yield, as qualified by relevant environmental and economic factors, including the economic needs of coastal fishing communities and the special requirements of the developing states, and taking into account fishing patterns, the interdependence of stocks and any generally recommended international minimum standards whether subregional, regional or global.

4. In taking such measures the coastal State shall take into consideration the effects on species associated with or dependent upon harvested species with a view to maintaining or restoring populations of such associated or dependent species above levels at which their reproduction may become seriously threatened.

5. Available scientific information, catch and fishing effort statistics, and other data relevant to the conservation of fish stocks shall be contributed and exchanged on a regular basis through competent international organizations, whether subregional, regional, or global, where appropriate, and with participation by all states concerned, including States whose nationals are allowed to fish in the exclusive economic zone.

62. *Utilization of the living resources.*

1. The coastal State shall promote the objective of optimum utilization of the living resources in the exclusive economic zone. . . .

2. The coastal State shall determine its capacity to harvest the living resources of the exclusive economic zone. Where the coastal State does not have the capacity to harvest the entire allowable catch, it shall give other States access to the surplus of the allowable catch.

3. In giving access to other states . . . the coastal State shall take into account all relevant factors including the economy of the coastal State concerned and its other national interests . . . the requirements of developing States . . . and the need to minimize economic dislocation in states whose nationals have habitually fished in the zone or which have made substantial efforts in research and identification of the stocks.

4. Nationals of other States fishing in the exclusive economic zone shall

comply with the conservation measures and regulations of the coastal State. These may relate to the following:

(a) Licensing of fishermen, fishing vessels, and equipment including payment of fees. . . ;

(b) . . Species which may be caught . . . and quotas of catch . . . during any period;

(c) Regulating the seasons and areas of fishing; the types, sizes and amount of gear; and the numbers, sizes and types of fishing vessels that may be used;

(d) Fixing the age and size of fish . . . that may be caught;

(e) Specifying information required of fishing vessels, including catch and effort statistics and vessel position reports.

(f) Requiring . . . fisheries research programmes . . . and reporting of associated scientific data;

(g) The placing of observers or trainees on board such vessels. . . ;

(h) The landing of all or any part of the catch . . . in the coastal State;

(i) Terms and conditions relating to joint ventures or other cooperative arrangements;

(j) . . . Training of personnel and the transfer of fisheries technology, including fisheries research;

(k) Enforcement procedures.

63. *Stocks occurring within the exclusive economic zones of two or more coastal States or both within the exclusive economic zone and in an area beyond and adjacent to it.*

   1. Where the same stock or stocks of associated species occur within the exclusive economic zones of two or more coastal States, these States shall seek to agree upon the measures necessary to ensure the conservation and development of such stocks. . . .

64. *Highly migratory species.*

   1. . . . States whose nationals fish . . . for the highly migratory species . . . shall cooperate with a view to ensuring conservation . . . and optimum utilization of such species . . . both within and beyond the exclusive economic zone.

65. *Marine mammals.*

Nothing . . . restricts the right of a coastal State . . . to regulate the exploitation of marine mammals more strictly than provided for in this part. States shall cooperate with a view to the conservation of marine mammals.

66. *Anadromous stocks.*

   1. States in whose rivers anadromous fish originate shall have the primary interest in and responsibility for such stocks.

77. *Rights of the coastal State over the continental shelf.*

   4. The natural resources referred to in this Part consist of the mineral and other non-living resources of the sea-bed together with living organisms belonging to sedentary species, that is to say, organisms which, at the harvestable stage, either are immobile on or under the sea-bed, or are

unable to move except in constant physical contact with the sea-bed or the subsoil.

207. *Pollution from land-based sources.*

    1. States shall . . .prevent, reduce, and control pollution from . . . rivers, estuaries, pipelines and outfall structures. . . .

210. *Pollution by dumping.*

    1. States shall . . . prevent, reduce and control . . . dumping.

Bibliography

The preceding survey of fishery science encompasses dozens of sciences, each of which has grown to a size and complexity that defy the ability of any one person to completely digest it. The scientific literature useful to fisheries has become immense, whereas the citations at the end of each chapter refer, by and large, only to some details that have been used to make special points. The more general references, which have also been consulted, may not have been cited because broad principles have usually evolved, over time, from the work of many people and cannot fairly be attributed to any one source. Therefore, there is a need to identify additional major scientific sources and policy papers. Such a listing, which is provided here, may also assist the more advanced student in further exploration of the trends in a particular field of fishery science, even though only a small fraction of the total references can be included.

A part of the exponential growth of scientific publication is the proliferation of journals in many countries. These are searched most easily through use of the abstracting services. The most important service to fisheries is *Aquatic Science and Fisheries Abstracts,* which is prepared through collaboration of FAO, UNESCO, and nine countries including Canada and the United States. Others include *Biosis Previews* (formerly *Biological Abstracts*), *Aquaculture, Oceanic Abstracts, Pollution Abstracts, Environmental Bibliography,* and *Food Science and Technology Abstracts.* Most of these are available in print, as well as to computer searches.

Especially useful listing of journal contents are the *Marine Science Contents*

Tables, the *Freshwater and Aquaculture Contents Tables* published monthly by FAO, and the American Fisheries Society indices in the *Journal of Fishery Management.* The 1994 volumes listed the contents of about 200 scientific journals, most of which are important to fishery scientists. A few of those journals have been listed in this bibliography with the 1994 volume number.

Chapter 1 Expansion of Fishery Problems

Angermeier, P. L., and Williams, J. E. (1994). Conservation of imperiled species and reauthorization of the Endangered Species Act of 1973. *Fisheries* **19**(1), 26–29.

Botkin, D. B. (1990). "Discordant Harmonies: A New Ecology for the 21st Century." Oxford Univ. Press, New York.

Buchholz, R. A., Marcus, A. A., and Post, J. E. (1991). "Managing Environmental Issues: A Case Book." Prentice–Hall, New York.

Clark, W. C. (1989). Managing planet earth. *Sci. Amer.* **261**(3), 47–54.

Environmental Law Institute (1977). "The Evolution of National Wildlife Law." Council on Environmental Quality, U.S. Govt. Printing Office, Washington, D.C.

Food and Agriculture Organization of the United Nations (FAO) (Annually). "The State of Food and Agriculture." FAO, Rome.

Hardin, G. (1993). "Living within Limits. Ecology, Economic, and Population Taboos." Oxford Univ. Press, New York.

Harville, J. P. (1992). Three decades of expanding fishery management horizons. *In* "Fishery Management and Watershed Development" (R. H. Stroud, ed.), Amer. Fish. Soc. Sympos. No. 13, pp. 27–38. Amer. Fish. Soc., Bethesda, Md.

Huth, H. (1957). "Nature and the American: Three Centuries of Changing Attitudes." Univ. of California Press, Berkeley.

Jones, S. (1994). Endangered species act battles. *Fisheries* **19**(1), 22–25.

Larkin, P. A. (1988). The future of fisheries management: Managing the fisherman. *Fisheries* **13**(1), 3–9.

Marsh, G. P. (1864). "Man and Nature." Republished in 1965 by Belknap Press of Harvard University Press, Cambridge, Mass. (See Chap. 4, "The Waters," pp. 281–381.)

Meybeck, M., ed. (1990). "Global Freshwater Quality: A First Assessment." Basil Blackwell, Cambridge, Mass.

Royce, W. F. (1989). A history of marine fishery management. *CRC Crit. Rev. Aquat. Sci.* **1**(1), 27–44.

Russell, E. S. (1942). "The Overfishing Problem." Cambridge Univ. Press, London/New York.

Shapiro, S., ed. (1971). "Our Changing Fisheries." U.S. Govt. Printing Office, Washington, D.C.

U.S. Government (1981). "The Global 2000 Report to the President," 3 vols. U.S. Govt. Printing Office, Washington, D.C. [Reprint of "Fisheries Projections" from Vol. 2 in *Fisheries (Bethesda)* **6**(5) (1981).]

Chapter 2 Work of Fishery Scientists

American Meterological Society and the University Corporation for Atmospheric Research (1992). "Curricula in the Atmospheric, Oceanic, Hydrologic, and Related Sciences for Colleges and Universities in the United States, Puerto Rico, and Canada."

Becker, C. D. (1991). "Fisheries Laboratories of North America." Amer. Fish. Soc., Bethesda, Md.

Carigma, M. A. A., and Morales, R. G. (1989). "Directory of Educational and Training Opportunities in Fisheries and Aquaculture." FAO, Rome.

Darnay, B. T., and Young, M. L. (1988). "Life Sciences Organizations and Agencies Directory," 1st ed. Gale Research Co., Detroit.

Drucker, P. F. (1974). "Management." Harper & Row, New York.

Food and Agriculture Organization and Intergovernmental Oceanographic Commission of Unesco (1983). "International Directory of Marine Scientists." UNESCO, Paris. (Identifies about 2500 institutions and 18,000 specialists in the marine sciences.)

Hull, E. W. S. (1979). "National Sea Grant College Program: The First Ten Years." Office of Sea Grant, U.S. Dept. of Commerce, Washington, D.C.

Hunter, J. R., ed. (1990). "Writing for Fishery Journals." Amer. Fish. Soc., Bethesda, Md.

Idyll, C. P. (1979). FAO: A force in world fisheries. *Fisheries (Bethesda)* 4(3), 2–4, 26.

McCain, K. W. (1994). Islands in the stream: Mapping the fisheries and aquatic sciences literatures. *Fisheries* 19(10), 20–27.

National Wildlife Federation (Annually). "Conservation Directory: A List of Organizations, Agencies, and Officials concerned with Natural Resource Use and Management." National Wildlife Federation, Washington, D.C.

Stroud, R. H., ed. (1992). "Fisheries Management and Watershed Development," Amer. Fish. Soc. Sympos. No. 13. Amer. Fish. Soc., Bethesda, Md. (See especially W. G. Gordon, Fishery scientists and managers for the 21st century, pp. 19–25; and J. P. Harville, Three decades of expanding fishery management horizons, pp. 27–38.)

Fisheries. American Fisheries Society. Vol. 19.

Chapter 3 Professional Careers

Adelman, I. R. (1990). Technical specialization and a broad foundation in an undergraduate fisheries degree. *Fisheries* 15(5), 26–27.

Adelman, I. R, Griswold, B. L., Herring, J. L., Menzel, B. W., Nielson, L. A., Noble, R. L., Schram, H. L., Jr., and Winter, J. D. (1990). Criteria for evaluating university fisheries programs. A report of the University Program Standards Committee. *Fisheries* 15(2), 13–16.

Angermeier, P. L., Neves, R. J., and Nielsen, L. A. (1991). Assessing stream values: Perspectives of aquatic resource professionals. *N. Amer. J. Fish. Mgmt.* 11(1), 1–10.

Bronstein, D. A. (1993). "Law for the Expert Witness." CRC Press, Boca Raton, Fla.

Brouha, P. (1993). A professional initiative for natural resource professionals. *Fisheries* 18, 32.

DeAngelis, L. P., Basler, S. C., and Yeager, L. E. (1989). "The Complete Guide to Environmental Careers." Island Press, Washington, D.C.

De Witt, G., ed. (1968). Session VII. Education and training. *In* "The Future of the Fishing Industry of the United States." Univ. of Washington Press, Seattle.

Donaldson, J. R. (1979). Fisheries education from the state perspective. *Fisheries (Bethesda)* 4(2), 18–21.

Hayman, M. A., and Brouha, P. (1991). A comparison of state fisheries biologists salaries, 1983, 1987, 1990. *Fisheries* 16(4), 5–11.

Hunter, R. G. (1984). Managerial professionalism in state fish and wildlife agencies: A survey of duties, attitudes, and needs. *Fisheries,* 9(5), 2–7.

Kennedy, J. J., and Thomas, J. W. (1992). Exit, voice, and loyalty of wildlife biologists in public natural resource agencies. "American Fish and Wildlife Policy: The Human Dimension" (W. R. Mangun, ed.), pp. 221–238. Southern Illinois Univ. Press, Carbondale.

Lackey, R. T. (1979). Fisheries education in the 1980's: The issues. *Fisheries* 4(2), 16–17.

Norgaard, R. B. (1992). Environmental science as a social process. *Environ. Monitoring and Assessment* **20**, 95–100.

Oglesby, R. T., and Krueger, C. C. (1989). Undergraduate fisheries education: Technical specialization or broad foundation? *Fisheries* **14**(5), 17–21.

Royce, W. F. (1984). A professional education for fishery scientists. *Fisheries* 9(3), 12–14, 16–17.

United Nations Educational, Scientific and Cultural Organization (UNESCO) (1981). "Fishery Science Teaching at the University Level," Report of a workshop on university curricula in fishery science, UNESCO, Paris, 5–8 May, 1980, UNESCO Reports in Marine Sciences No. 15. UNESCO, Paris.

Warner, D. J. (1992). "Environmental Careers. A Practical Guide to Opportunities in the 90's." CRC Press, Boca Raton, Fla.

Wilbur, R. L. (1990). Gray literature: A professional dilemma. *Fisheries* **15**(5), 2–6.

Wooster, W. S. (1988). Immiscible investigators: Oceanographer, meteorologists and fishery scientists. *Fisheries* **13**(4), 18–21.

See especially:

"Becoming an Environmental Professional, 1990." The CEIP Fund, Inc., 68 Harrison Ave., Boston.

The Environmental Professional. Vol. 16.

The Journal of Environmental Education. Vol. 25.

Marine Fisheries Review. Vol. 50(4) (1988). Anniversary issue. More than 30 historical review articles on fishery development and management.

Panel on Scientific Responsibility and the Conduct of Research, National Academy of Sciences, National Academy of Engineering, and Institute of Medicine (1992). "Responsible Science. The Integrity of Scientific Conduct."

Transactions of the American Fisheries Society. Vol. 124(1) (1995). Guide for authors, pp. 151–157.

Chapter 4 The Aquatic Environment

Boon, P. J., Calow, P., and Petts, G. E. eds. (1992). "River Conservation and Management." Wiley, New York.

Brewer, P. G., ed. (1983). "Oceanography: The Present and Future." Springer-Verlag, Berlin/New York.

Calow, P., and Petts, G. E. (1992, 1994). "The Rivers Handbook," Vols. 1 and 2. Blackwell Scientific, Cambridge, Mass.

Coker, A., and Richards, C. (1992). "Valuing the Environment. Economic Approaches to Economic Evaluation." CRC Press, Boca Raton, Fla.

Crompton, T. R. (1992). "Comprehensive Water Analysis," Vol. 1, "Natural Waters;" Vol. 2, "Treated Waters."

Dennison, M. S., and Berry, J. F. (1993). "Wetlands. Guide to Science, Law, and Technology." Noyes Publ., Park Ridge, N.J.

Dietrich, G., Kalle, K., Kraus, W., and Siedler, G. (1980). "General Oceanography: An Introduction," 2nd ed. Wiley, New York.

Duxbury, A. C., and Duxbury, A. B. (1994). "An Introduction to the World's Oceans," 4th ed. Wm. C. Brown, Dubuque, Iowa.

Freedman, B. (1994). "Environmental Ecology. The Ecological Effects of Pollution, Disturbance, and Other Stresses." Academic Press, San Diego, Calif.

Gobas, A. P. C., and McCorquod, J. A. (1992). "Chemical Dynamics in Freshwater Ecosystems." CRC Press, Boca Raton, Fla.

Gordon, N. D., McMahon, T. A., and Finlayson, B. L. (1992). "Stream Hydrology: An Introduction for Ecologists." Wiley, New York.

Kennish, M. J., ed. (1994). "Practical Handbook of Marine Science," 2nd ed. CRC Press, Boca Raton, Fla.

Libes, S. M. (1992). "An Introduction to Marine Biochemistry." Wiley, New York.

Loeb, S. L., and Spacie, A. (1994). "Biological Monitoring of Aquatic Systems." Lewis Publ., Boca Raton, Fla.

Meehan, W. R. (1991). "Influences of Forest and Rangeland Management on Salmonid Fishes and Their Habitats," Spec. Publ. No. 19. Amer. Fish. Soc. Bethesda, Md.

Meybeck, M., ed. (1990). "Global Freshwater Quality: A First Assessment." Basil Blackwell, Cambridge, Mass.

Millero, F. J., and Sohn, M. L. (1991). "Chemical Oceanography." CRC Press, Boca Raton, Fla.

Morel, F. M. M., and Hering, J. G. (1993). "Principles and Application of Aquatic Chemistry." Wiley, New York.

Pankow, J. F. (1991). "Aquatic Chemistry Concepts." CRC Press, Boca Raton, Fla.

Salo, E. O., and Cundy, T. W. (1987). "Streamside Management: Forest and Fishery Interactions." Univ. of Washington Press/Inst. of Forest Resources, Seattle.

Sears, M., and Merriman, D., eds. (1980). "Oceanography: The Past," *Proc. Int. Congr. on the History of Oceanography,* 3rd ed. Springer-Verlag, Berlin/New York.

Sullivan, T. E. P., ed. (1992). "Directory of Environmental Information Sources," 4th ed. Government Institutes Inc., Washington, D.C. (Section 1, Federal Government Resources Agencies; Section 2, State Government Resources Agencies; Section 3, Professional Scientific and Trade Organizations; Section 4, Reference Books and Periodicals; Section 5, Computerized Data Bases.)

Sverdrup, H. U., Johnson, M. W., and Fleming, R. H. (1942). "The Oceans: Their Physics, Chemistry, and General Biology." Prentice-Hall, Englewood Cliffs, New Jersey.

Thayer, G. W., ed. (1992). "Restoring the Nation's Marine Environment." Maryland Sea Grant College.

Tu, A. T. (1988). "Handbook of Natural Toxins," Vol. 3, "Marine Toxins and Venoms." Marcel Dekker, New York/Basel.

U.S. Environmental Protection Agency (EPA) (1992). "Created and Natural Wetlands for Controlling Nonpoint Source Pollution." CRC Press, Boca Raton, Fla.

van der Leeden, F., Troise, F. L., and Todd, D. K. (1990). "The Water Encyclopedia," 2nd ed. CRC Press. Boca Raton, Fla.

Wetzel, R. G. (1983). "Limnology," 2nd ed. Saunders, Philadelphia.

Acta Oceanographica Sinica. Chinese Society of Oceanography (in English). Vol. 13.
Australian Journal of Marine and Freshwater Research. Vol. 45.

Journal of Applied Ecology. British Ecological Association. Vol. 31.

Bulletin of the Japanese Society of Fisheries Oceanography (in English). Vol. 57.

Bulletin of Marine Science. University of Miami. Vol. 51.

Canadian Water Resources Journal. Vol. 19.

Chinese Journal of Oceanology and Limnology (in English). Vol. 12.

Current Contents. Agriculture, Biology, and Environmental Sciences. Institute for Scientific Information, Philadelphia and Uxbridge, U.K. Vol. 25.

Environmental Impact Assessment and Review. Elsevier. Vol. 14.

Environmental Science and Technology. American Chemical Society. Vol. 28.

The Environmentalist. Professional Science and Technology Letters. Middlesex, U.K. Vol. 14.

Estuaries. Estuarine Research Foundation. Vol. 17.

Journal of Applied Ecology. British Ecological Society. Vol. 31.

Journal of Aquatic Ecosystem Health. Klewer Academic Publishers. Vol. 3.

Journal of Freshwater Ecology. Vol. 9.

Journal of Marine Science. International Council for the Exploration of the Sea. Vol. 51.

Limnology and Oceanography. American Society of Limnology and Oceanography. Vol. 39.

Marine Pollution Bulletin. Pergamon Press. Vol. 23.

New Zealand Journal of Marine and Freshwater Research. Vol. 28.

Oceanology. Russian Academy of Sciences (in English). Vol. 33.

Reviews in Aquatic Sciences. CRC Press. Vol. 5.

Chapter 5 Food Chain and Resource Organisms

Dawes, C. J. (1981). "Marine Botany." Wiley, New York.

George, D., and George, J. (1979). "Marine Life: An Illustrated Encyclopedia of Invertebrates in the Sea." Wiley, New York.

Hart, J. L. (1973). "Pacific Fishes of Canada," Bulletin No. 180. Fisheries Research Board of Canada.

Jordan, D. S., and Everman, B. W. (1896–1900). The fishes of North and Middle America. *Bull. U.S. Nat. Mus.* **47.**

Lee, D. S., *et al.* (1980). "Atlas of North American Freshwater Fishes." North Carolina State Museum of Natural History, Raleigh.

Leim, A. H., and Scott, W. R. (1966). Fishes of the Atlantic Coast of Canada. *Bull. Fish. Res. Board Can.* **155,** 1–485.

Lindberg, G. U. (1974). "Fishes of the World: A Key to Families and a Checklist." Halstead Press, New York. (Translated from Russian by Hilary Hardin for Israeli Program for Scientific Translations.)

Lowe-McConnell, R. H. (1968). Identification. *In* "Methods for Assessment of Fish Production in Fresh Waters" (W. E. Ricker, ed.), Int. Biol. Progr. Handb. No. 3, pp. 46–77. Blackwell, Oxford, U.K.

Merritt, R. W., and Cummins, K. W., eds. (1978). "An Introduction to the Aquatic Insects of North America." Kendall–Hunt, Dubuque, Iowa.

Moyle, P. B., and Cech, J. J., Jr. (1988). "Fishes, an Introduction to Ichthyology." Prentice–Hall, New York.

Rosowski, J. R., and Parker, B. C., eds. (1982). "Selected Papers in Phycology II." Phycolog. Soc. Amer., New York.

Schram, F. R. (1986). "Crustacea." Oxford Univ. Press, New York.

Scott, W. B., and Crosssman, E. J. (1973). "Freshwater Fishes of Canada," Bulletin No. 184. Fisheries Research Board of Canada.

Scott, W. B., and Scott, M. G. (1988). "Atlantic Fishes of Canada," Can. Bull. Fish. Aquat. Sci. No. 219.

Thorp, J. H., and Covich, A. P. eds. (1991). Ecology and Classification of North American Freshwater Invertebrates." Academic Press, San Diego, Calif.

Turgeon, D. D., *et al.* (1988). "Common and Scientific Names of Aquatic Invertebrates from the United States and Canada. Mollusks." Amer. Fish. Soc., Bethesda, Md.

Williams, A. B., *et al.* (1989). "Common and Scientific Names of Aquatic Invertebrates from the United States and Canada. Decapod Crustaceans." Amer. Fish. Soc., Bethesda, Md.

Botanica Marina. Walter de Gruyer, Berlin. Vol. 37.

Canadian Journal of Fisheries and Aquatic Sciences. National Research Council of Canada. Vol. 51.

Chapter 6 Biology of Aquatic Resource Organisms

Atema, J., Fay, R. R., Popper, A. N., and Tavolga, W. N., eds. (1988). "Sensory Biology of Aquatic Animals." Springer-Verlag, New York/Berlin.

Bliss, D. E., editor-in-chief (1982–1985). "The Biology of Crustacea" Vols. 1–10. Academic Press, San Diego, Calif.

Carlander, K. D. (1969, 1977). "Handbook of Freshwater Fishery Biology," Vol. 1, "Life History Data on Freshwater Fishes of the United States and Canada, Exclusive of the Perciformes"; Vol. 2, "Life History Data on Centrarchid Fishes of the United States and Canada." Iowa State Univ. Press, Ames.

Cobb, J. S., and Phillips, B. F., eds. (1980). "The Biology and Management of Lobsters," Vol. 1, "Physiology and Behavior." Academic Press, New York.

Evans, D. H., ed. (1993). "The Physiology of Fishes." CRC Press, Boca Raton, Fla.

Hara, T. J., ed. (1982). "Chemoreception in Fishes." Elsevier, Amsterdam.

Hoar, W. S. and Randall, D. J., eds. (1969–1992). "Fish Physiology," Vols. I–XII. Academic Press, San Diego, Calif.

Schreck, C. B., and Moyle, P. B., eds. (1990). "Methods for Fish Biology."

Schuijf, A., and Hawkins, A. D., eds. (1976). "Sound Reception in Fish." Elsevier, Amsterdam.

Thorpe, J. E., ed. (1978). "Rhythmic Activity of Fishes." Academic Press, New York.

Wilbur, K. M., ed. (1983). "The Mollusca," Vols. 1–12. Academic Press, New York.

Aquatic Botany. Elsevier. Vol. 45.

Bulletin of the Japanese Society of Scientific Fisheries (in Japanese and English). Vol. 60.

Canadian Journal of Fisheries and Aquatic Sciences. National Research Council of Canada. Vol. 51.

Crustaceana. International Union of Crustacean Research. Vol. 16.

Crustacean Biology. Crustacean Society. Vol. 14.

Journal of Experimental Marine Biology and Ecology. Vol. 168.

Transactions of the American Fisheries Society. Vol. 123.

Chapter 7 Ecological Concepts

Bartell, S. M., Gardner, R. H., and O'Neill, R. V. (1992). "Ecological Risk Estimation." CRC Press, Boca Raton, Fla.

Botkin, D. B. (1990). "Discordant Harmonies: A New Ecology for the Twenty-First Century." Oxford Univ. Press, New York.

Edmondson, W. T., and Winberg, G. G., eds. (1971). "Methods for Assessment of Secondary Productivity in Fresh Waters," Int. Biol. Progr. Handb. No. 9. Blackwell, Oxford, U.K.

Gerking, S. D. (1994). "Feeding Ecology of Fish." Academic Press. San Diego, Calif.

Giller, P. S., Hildrew, A., and Raffaelli, D. G. (1994). "Aquatic Ecology: Scale, Pattern, and Process." Blackwell Scientific, Cambridge, Mass.

Hutchinson, G. E. (1978). "An Introduction to Population Ecology." Yale Univ. Press, New Haven, Conn.

Kitchell, J. F., ed. (1992). "Food Web Management: A Case Study of Lake Mendota." Springer-Verlag, New York.

Le Cren, E. D., and Lowe-McConnell, R. H., eds. (1980). "The Functioning of Freshwater Ecosystems." Cambridge Univ. Press, Cambridge, U.K.

Meffe, G. K. (1986). Conservation genetics and the management of endangered species. *Fisheries* 11(1), 14–23.

Nikolsky, G. V., (1963). "The Ecology of Fishes." Academic Press, New York. (Translated from Russian by L. Birkett.)

Real, L. A., and Brown, J. H. eds. (1991). "Foundations of Ecology. Classic Papers with Commentaries." Univ. of Chicago Press, Chicago/London.

Steele, J. H., ed. (1970). "Marine Food Chains." Univ. of California Press, Berkeley.

U.S. National Academy of Sciences (NAS) (1975). "Productivity of World Ecosystems," Proceedings of a Symposium Sponsored by U.S. Nat. Committee for the Int. Biol. Prog. NAS, Washington, D.C.

Weatherley, A. H. (1972). "Growth and Ecology of Fish Populations." Academic Press, San Diego, Calif.

Ecology. Ecological Society of America. Vol. 75.
Environmental Biology of Fishes. Klewer. Vol. 39.
Estuaries. Estuarine Research Foundation. Vol. 17.

Chapter 8 Analysis of Exploited Populations

Beverton, R. J. H., and Holt, S. J. (1957). On the dynamics of exploited fish populations. *Fish. Invest. (London)* 19, 1–533.

Graham, M., ed. (1956). "Sea Fisheries, Their Investigation in the United Kingdom." E. Arnold, London.

Johnson, J. E., and Walters, C. J. (1992). "Protected Fishes of the United States and Canada." Amer. Fish. Soc., Bethesda, MD.

Kohler, C. C., and Hubert, W. A., eds. (1993). "Inland Fisheries Management in North America." Amer. Fish. Soc., Bethesda, Md.

Nielsen, L. A. (1992). "Methods of Marking Fish and Shellfish." Amer. Fish. Soc., Bethesda, Md.

Nielsen, L. A., and Johnson, D. L., eds. (1983). "Fisheries Techniques." Amer. Fish. Soc., Bethesda, Md.

Nikolski, G. V. (1969). "Theory of Fish Population Dynamics as the Biological Background for Rational Exploitation and Management of Fishery Resources." Oliver & Boyd, Edinburgh. (Translated from Russian by J. E. S. Bradley.)

Parker, N. C., *et al.*, eds. (1990). "Fish Marking Techniques." Amer. Fish. Soc., Bethesda, Md.

U.S. Department of Commerce (1945–*et seq.*), Fisheries of the United States. *Current Fisheries Statistics Series*. U.S. Dept. of Commerce, Washington, D.C.

Bulletin of the Japanese Society of Scientific Fisheries (in Japanese with English summaries). Vol. 60.

Canadian Journal of Fisheries and Aquatic Sciences. National Research Council of Canada. Vol. 51.

Fisheries Research. Elsevier. Vol. 18.

Chapter 9 Aquacultural Sciences

Ackefors, H., Huner, J. V., and Konikoff, M. (1994). "Introduction to the General Principles of Aquaculture." Food Products Press, Binghamton, N.Y.

Bardach, J. E., Ryther, J. H., and McLarney, W. O. (1972). "The Farming and Husbandry of Freshwater and Marine Organisms." Wiley–Interscience, New York.

Barg, U. C. (1992). "Guidelines for the Promotion of Environmental Management of Coastal Aquaculture Development," Fish. Tech. Paper No. 328. FAO, Rome.

Barnabe, G., ed. (1990). "Aquaculture," Vols. 1 and 2. Ellis Horwood, Ltd., West Sussex, U.K. (Translated from the French edition, 1989.)

Boyd, C. E. (1979). "Water Quality in Warmwater Fish Ponds." Agr. Exp. Sta., Auburn Univ., Auburn, Ala.

Creswell, R. LeR. (1993). "Aquaculture Desk Reference." Van Nostrand–Reinhold, New York.

Dore, I. (1990). "Salmon: The Illustrated Handbook for Commercial Users." Van Nostrand–Reinhold, New York.

Fallu, R. (1991). "Abalone Farming." Fishing News Books, London.

Food and Agriculture Organization (FAO) (1992). "Aquaculture Production, 1984–1990," Fish. Circ. No. 815, Rev. 4. FAO, Rome.

Halver, J. E., ed. (1989). "Fish Nutrition." Academic Press, San Diego, Calif. (Also Oxford and IBH Publishing Co., 66 Janpath, New Delhi, (India. 1992.)

Hardy, D. (1991). "Scallop Farming." Fishing News Books, London.

Huner, J. V., ed. (1994). "Freshwater Crayfish Aquaculture in North America, Europe, and Australia." Food Products Press, Binghamton, N.Y.

Huner, J. V., and Brown, E. E., eds. (1985). "Crustacean and Mollusc Aquaculture in the United States." AVI Publ. Co., Westport, Conn.

Jolly, C., and Clonts, H. (1993). "Economics of Aquaculture." Food Products Press, Binghamton, N.Y.

Kirpichnikov, V. S. (1981). "Genetic Bases of Fish Selection." Springer-Verlag, Berlin/New York. (Translation and revision of Russian edition. 1979)

Landau, M. (1992). "Introduction to Aquaculture." Wiley, New York.

Lawson, T. B. (1995). "Fundamentals of Aquacultural Engineering." Chapman and Hall, New York.

Manzi, J. J., and Castagna, M. eds. (1989). "Clam Mariculture in North America." Elsevier, Amsterdam.

Menzel, W., ed. (1991). "Estuarine and Marine Bivalve Mollusk Culture." CRC Press, Boca Raton, Fla.

Mills, D., ed. (1993). "Salmon in the Sea, and New Enhancement Strategies." Fishing News Books, London.

Muir, J. F., and Roberts, R. J. "Recent Advances in Aquaculture," Vol. 1 (1982), Vol. 2 (1985), Vol. 3 (1988), and Vol. 4 (1993). Blackwell Scientific, London.

Owens, E., and Shaw, R. (1974). "Development Reconsidered: Bridging the Gap between Government and People." Lexington Books, Lexington, Mass.

Perkins, F. O., and Cheng, T. C. eds. (1990). "Pathology in Marine Science." Academic Press, San Diego, Calif.

Pillay, T. V. R. (1990). "Aquaculture: Principles and Practices." Fishing News Books, London.

Reinertsen, H., et al. (Research Council of Norway) (1993). "Fish Farming Technology." A. A. Balkema/Rotterdam.

Roberts, R. J., ed. (1978). "Fish Pathology." Bailliere Tindall, Philadelphia.

Ryman, N., and Utter, F. eds. (1987). "Population Genetics and Fishery Management." Univ. of Washington Press, Seattle/London.

Schaperclaus, W. (1979). "Fischkrankheiten," 4th ed. Akademie-Verlag, Berlin.

Sindermann, C. J. (1990). "Principal Diseases of Marine Fish and Shellfish," 2nd ed., Vol. 1, "Fish" and Vol. 2, "Shellfish." Academic Press, San Diego, Calif.

Smith, L. J., and Peterson, S. (1982). "Aquaculture Development in Less Developed Countries: Social, Economic, and Political Problems." Westview Press, Boulder, Colo.

Snieszko, S. F., ed. (1970). "A Symposium on Diseases of Fishes and Shellfishes," Spec. Publ. No. 5, pp. 1–526. Amer. Fish. Soc., Bethesda, Md.

Sparks, A. K. (1973). "Invertebrate Pathology." Academic Press, San Diego, Calif.

Spotte, S. H. (1970). "Fish and Invertebrate Culture: Water Management in Closed Systems." Wiley, New York.

Stoskopf, M. K. (1993). "Fish Medicine." Saunders, Philadelphia.

Swift, D. R. (1993). " Aquaculture Training Manual," 2nd ed. Fishing News Books, London.

Tave, D., and Anderson, R. O. (1993). "Strategies and Tactics for Management of Fertilized Hatchery Ponds." Haworth Press, Binghamton, N.Y.

Tave, D., and Tucker, C. S., eds. (1993). "Recent Developments in Catfish Aquaculture." Food Products Press, Binghamton, N.Y.

U.S. National Academy of Sciences (NAS) (1973). "Nutritional Requirements of Domestic Animals," No. 11, "Nutritional Requirements of Trout, Salmon, and Catfish." Subcommittee on Fish Nutrition, National Academy Press, Washington, D.C.

Aquaculture. Elsevier. Vol. 112.

Aquaculture and Fisheries Management. Blackwell. Vol. 25.

Freshwater and Aquaculture Contents Tables. FAO. Vol. 13.

Journal of Applied Aquaculture. Food Products Press. Vol. 3.

Journal of Shellfish Research. National Shellfisheries Association. Vol. 13.

Progressive Fish Culturist. Amer. Fish. Soc. in cooperation with U.S. Dept. of Interior. Vol. 56.

Oxford and IBH Publishing Co., New Delhi/Bombay/Calcutta (1992). "Aquacultural Research Needs for 2,000 A.D." (An extraordinary compedium with 33 chapters by different authors from presentations at a symposium held in New Delhi in November, 1988.)

Chapter 10 The Capture Fisheries

Ben-Tuvia, A., and Dickson, W., eds. (1968). "Proceedings of the Conference on Fish Behavior in Relation to Fishing Techniques and Tactics," Fish. Rep. No. 62, Vols. 1–3. FAO, Rome.

Canadian Fisheries (Annually). "Annual Statistical Review." Canadian Govt. Publishing Center, Supply and Services, Hull, Quebec, Canada.

Copes, P., and Knetsch J. L. (1981). Recreational fisheries analysis: Management modes and benefit implications. *Can. J. Fish Aquat. Sci.* **38**(5), 559–570.

Dement'eva, T. F., and Zemshaya, K. A., eds. (1969). "Methods of Assessing Fish Resources and Forecasting Catches." U.S. Dept. of Commerce, Springfield, Va. (Translated from Russian.)

Dill, W. A. (1978). Patterns of change in recreational fisheries: Their determinants. *In* "Recreational Freshwater Fisheries: Their Conservation and Development" (J. S. Alabaster, ed.), pp. 1–22. Water Research Center, Stevenage Laboratory, Hertford, U.K.

Dill, W. A. (1993). "Inland Fisheries of Europe," European Inland Fisheries Advisory Committee (EIFAC), EIFAC Tech. Paper, 52 Suppl. FAO, Rome.

Flick, W. A., and Webster, D. A. (1992). Standing crops of brook trout in Adirondack waters before and after removal of non-trout species. *N. Amer. J. Fish. Mgmt.* **12**, 783–796.

Food and Agriculture Organization (FAO) (Annually). "Yearbook of Fishery Statistics," in two series, "Catches and Landings" and "Fishery Commodities." FAO, Rome.

Food and Agriculture Organization (FAO) (1978). "FAO Catalogue of Fishing Gear Designs." FAO, Rome.

Freeman, B. L., and Walford, L. A. (1974–1976). "Anglers Guides to the United States Atlantic Coast." U.S. Govt. Printing Office, Washington, D.C.

Nishimura, M., and Vestnes, G. (1980). "Echo Sounding and Sonar for Fishing." Fishing News Books, New York.

Radcliffe, W. (1926). "Fishing from the Earliest Times," 2nd ed. Dutton, New York.

Robinson, M. A. (1982). "Prospects for World Fisheries to 2000," *Fish. Circ.* No. 722, Rev. 1. FAO, Rome.

Roos, J. F. (1991). Restoring Fraser River salmon. *In* "A History of the International Pacific Salmon Fisheries Commission, 1937–1985." Pacific Salmon Fisheries Commission. Vancouver, Canada.

Royce, W. F. (1989). A history of marine fishery management. *CRC Crit. Rev. Aquat. Sci.* **1**(1).

Squire, J. L., and Smith, S. E. (1977). Anglers Guides to the United States Pacific Coast." U.S. Govt. Printing Office, Washington, D.C.

U.S. Department of Commerce (Annually). "Fisheries of the United States." Current Fisheries Statistics Series. U.S. Govt. Printing Office, Washington, D.C.

U.S. Fish and Wildlife Service (1993). "1991 National Survey of Fishing, Hunting, and Wildlife-Associated Recreation." U.S. Govt. Printing Office, Washington, D.C.

U.S. Fish and Wildlife Service. (1994). "1980–1990 Fishing, Hunting, and Wildlife-Associated Recreation Trends. State and Regional Trends," Addendum to 1991 National Survey of Fishing, Hunting, and Wildlife-Associated Recreation. U.S. Govt. Printing Office, Washington, D.C.

Fishing Vessels of the United States. 1994 ed. Nautilus Publ. Co., Seattle.

North American Journal of Fisheries Management. American Fisheries Society. Vol. 14.

Ocean Development and International Law. The Journal of Marine Affairs Vol. 25 (1994).

Chapter 11 The Culture Fisheries

Axelrod, H. R., Burgess, W. E., Pronek, N., and Walls, J. G. (1985). "Atlas of Freshwater Aquarium Fishes." TFH Publ., Neptune City, N.J.

Bardach, J. E., Ryther, J. H., and McLarney, W. O. (1972). "The Farming and Husbandry of Freshwater and Marine Organisms." Wiley–Interscience, New York.

Barnabe, G., ed. (1990). "Aquaculture," Vols. 1 and 2. Ellis Technology and Horwood, Ltd., West Sussex, U.K. (Translated from the French edition of 1989.)

Brown, E. E. (1983). "World Fish Farming: Cultivation and Economics," 2nd ed. AVI Publ., Westport, Conn.

Committee on Assessment of Technology and Opportunities for Marine Aquaculture in the United States (1992). "Marine Aquaculture. Opportunities for Growth." National Academy Press, Washington, D.C.

Coshe, A. G. (1984). "Aquaculture in Fresh Waters. A List of Selected Reference Books and Monographs, 1951–1984." FAO, Rome.

Creswell, R. LeRoy (1993). "Aquaculture Desk Reference." Van Nostrand–Reinhold, New York.

Fallu, R. (1991). "Abalone Farming." Fishing News Books, London.

Halver, J. E., ed. (1989). "Fish Nutrition," 2nd ed. Academic Press, San Diego, Calif.

Hanson, J. A., and Goodwin, H. L. (1977). "Shrimp and Prawn Farming in the Western Hemisphere." Dowden, Hutchinson, & Ross, Stroudsburg, Penn.

Hardy, D. (1991). "Scallop Farming." Fishing News Books, London.

Heen, K., Monahan, R. L., and Utter, F. (1993). "Salmon Aquaculture." Fishing News Books, London.

Huet, M. (1970). "Traite de Pisciculture," 14th ed. Wyngaert.

Huner, J. V., and Brown, E. E., eds. (1985). "Crustacean and Mollusk Aquaculture in the United States." AVI Publ., Westport, Conn.

Imai, T., ed. (1977). "Aquaculture in Shallow Seas: Progress in Shallow Sea Culture." Koseisha Ioseiku, Tokyo. (English translation from 1971 Japanese version. Available from U.S. Dept. of Commerce, Natl. Tech. Info. Serv., Springfield, Va.)

Korringa, P. (1976a). "Farming Marine Organisms Low in the Food Chain; A Multidisciplinary Approach to Edible Seaweed, Mussel, and Clam Production." Elsevier, Amsterdam.

Korringa, P. (1976b). "Farming Marine Fishes and Shrimps; A Multidisciplinary Treatise." Elsevier, Amsterdam.

Korringa, P. (1976c). "Farming the Cupped Oysters of the Genus *Crassostrea;* A Multidisciplinary Treatise." Elsevier, Amsterdam.

Korringa, P. (1976d). "Farming the Flat Oysters of the Genus *Ostrea;* A Multidisciplinary Treatise." Elsevier, Amsterdam.

Landau, M. (1992). "Introduction to Aquaculture." Wiley, New York.

McVey, J. P. (1983). "Handbook of Mariculture," Vol. I, "Crustacean Aquaculture." CRC Press, Boca Raton, Fla.

McVey, J. P. (1991). "Handbook of Mariculture," Vol. II, "Finfish Aquaculture." CRC Press, Boca Raton, Fla.

Menzel, W., ed. (1991). "Estuarine and Marine Bivalve Mollusc Culture." CRC Press, Boca Raton, Fla.

Nosho, T., and Freeman, K., eds. (1993). "Marine Fish Culture and Enhancement," Conference Proceedings, Seattle, Wash. Washington Sea Grant Program, Univ. of Washington Press, Seattle.

Pacific Science Congress (13th) (1976). Aquaculture symposium. *J. Fish. Res. Board Can.* **33**(4), Part 2, 875–1119.

Pillay, T. V. R. (1990). "Aquaculture, Principles and Practices." Fishing News Books, London.

Pullin, R. S. V., Rosenthal, H., and McLean, J. L. (1993). "Environment and Aquaculture in Developing Countries." ICLARM, Manila, The Philippines.

Sarma, B. C. (1992). Social issues in aquaculture, No. 33, *In* "Aquacultural Research Needs for 2,000 A.D., pp. 399–405. Oxford and IBH Publ. Co., New Delhi/Bombay/Calcutta.

Spotte, S. H. (1970). "Fish and Invertebrate Culture: Water Management in Closed Systems." Wiley, New York.

Sterba, G., ed. (1983). "The Aquarium Encyclopedia." The MIT Press, Cambridge, Mass.

Stickney, R. R. (1979). "Principles of Warmwater Aquaculture." Wiley–Interscience, New York.

Stickney, R. R., ed. (1991). "Culture of Salmonid Fishes." CRC Press, Boca Raton, Fla.

Stickney, R. R., ed. (1993). "Culture of Non-salmonid Freshwater Fishes." CRC Press, Boca Raton, Fla.

Stickney, R. R. (1994). "Principles of Aquaculture." New York.

Swift, D. R. (1993). "Aquaculture Training Manual," 2nd ed.

US–India International Symposium (1992). "Aquaculture Research Needs for 2,000 A.D." Oxford and IBH Publ. Co., New Delhi.

U.S. National Research Council (1992). "Marine Aquaculture: Opportunities for Growth." U.S. Govt. Printing Office, Washington, D.C.

Chapter 12 Food Fishery Products

Ahmed, F. E., ed. (1991). "Seafood Safety." National Academy Press, Washington, D.C.

Burgess, G. H. O. (1971). The alternative uses of fish. *FAO Fish. Rep.* **117**, 1–28.

Codex Alimentarius Commission (1970–1978). "Recommended International Standard for Quick Frozen Gutted Pacific Salmon," CAC/RS 36-1970. See also the International Standards for "Fillets of Cod and Haddock," CAC/RS 50-1971; "Fillets of Flat Fish," CAC/RS 91-1976; "Shrimps or Prawns," CAC/RCP 17-1976; "Fillets of Hake," CAC/RS 93-1978; and "Lobsters," CAC/RS 95-1978.

Codex Alimentarius Commission (1976–1978). "Recommended Code of Practice for Fresh Fish," CAC/RCP 9-1976. See also the Codes of Practice for "Canned Fish," CAC/RCP 10-1976; "Frozen Fish," CAC/RCP 16-1978; "Shrimps or Prawns," CAC/RCP 17-1978; and "Molluscan Shellfish," CAC/RCP 18-1978.

Connell, J. J. (1975). "Control of Fish Quality." Fishing News Books, New York.

Food and Agriculture Organization (FAO) (1975). Report of the FAO/NORAD round table discussion on expanding the utilization of marine fishery resources for human consumption. *FAO Fish. Rep.* **175**, 1–47.

Gilbert, D., ed. (1968). "The Future of the Fishing Industry of the United States." Univ. of Washington Press, Seattle.

Hamlisch, R., and Kerr, A. A., eds. (1965). Report of the meeting on business decisions in fishery industries. *FAO Fish. Rep.* **22**, Vols. 1–3.

Jay, J. M. (1978). "Modern Food Microbiology," 2nd ed. Van Nostrand, New York.

Kaul, P. N., and Sindermann, C. J., eds. (1978). "Drugs and Food from the Sea: Myth or Reality?" Univ. of Oklahoma Press, Norman.

Keup, L. E. (1979). Fisheries in lake restoration. *Fisheries* 4(1), 7–9, 20.

Kreuzer, R., ed. (1969). "Freezing and Irradiation of Fish." Fishing News Books, New York.

Kreuzer, R., ed. (1971). "Fish Inspection and Quality Control." Fishing News Books, New York.

Kreuzer, R. (1974). "Fishery Products." Fishing News Books, New York.

Larkin, E. P., and Hunt, D. A. (1982). Bivalve mollusks: Control of microbiological contaminants. *Bioscience* 32, 193–197.

Mottet, M. G. (1981). Enhancement of the marine environment for fisheries and aquaculture in Japan. *Wash. Fish. Tech. Pap.* 69, 1–26.

Russell, F. E. (1969). Poisons and venoms. *In* "Fish Physiology" (W. S. Hoar and D. J. Randall, eds.), Vol. III, pp. 401–449. Academic Press, New York.

Singh, R. P., and Heldman, D. R. (1993). "Introduction to Food Engineering." Academic Press, San Diego, Calif.

Stansby, M. E., and Dassow, J. A., eds. (1963). "Industrial Fishery Technology." Reinhold, New York.

U.S. Department of Commerce (Annually). "Fisheries of the United States." Natl. Tech. Info. Serv., Springfield, Va.

U.S. Department of Commerce (Triweekly, weekly, monthly). *Fishery Market News Reports.* National Marine Fisheries Service, Boston/New York/New Orleans/Terminal Island/Seattle.

U.S. Department of Commerce (1975). "Baseline Economic Forecast of the U.S. Fishing Industry to 1985." Natl. Tech. Info. Serv., Springfield, Va.

U.S. Food and Drug Administration (FDA), Center for Food Safety and Applied Nutrition, Shellfish Sanitation Branch, National Shellfish Sanitation Program. "Manual of Operations." U.S. Govt. Printing Office, Washington, D.C.

Journal of Aquatic Food Product Technology. Food Products Press. Vol. 3.

Chapter 13 Regulation of Fishing

Barber, W. E., and Taylor, J. N. (1990). The importance of goals, objectives, and values in the fisheries management process and organization: A review. *N. Amer. J. Fish. Mgmt.* 10(4)365–373.

Bronstein, D. A. (1993). "Law for the Expert Witness." CRC Press, Boca Raton, Fla.

Groth, P. G. (1981). Effective use of sociocultural data in fisheries management: A case study. *Fisheries* 6(2), 11–16.

Gulland, J. A. (1978). Fishery management: New strategies for new conditions. *Trans. Amer. Fish. Soc.* 107, 1–11.

Lackey, R. T., and Nielsen, L. A. (1980). "Fisheries Management." Blackwell, Oxford, U.K.

Maiolo, J. R., and Orback, M. K. (1982). "Modernization and Marine Fisheries Policy." Ann Arbor Science, Ann Arbor, Mich.

Miles, E. L., ed. (1989). "Management of World Fisheries: Implications of Extended Coastal State Jurisdiction." Univ. of Washington Press, Seattle/London.

Pearce, P. H. (1988). "Rising to the Challenge. A New Policy for Canada's Freshwater Fisheries." Canadian Wildlife Federation.

Ricker, W. E., and Weeks, E. P., eds. (1974). Technical conference on fishery management and development. *J. Fish. Res. Board Can.* 30(12), Part 2, 1921–2537.

Wise, J. P. (1991). "Federal Conservation and Management of the Marine Fisheries of the United States." Center for Marine Conservation, Washington, D.C.

Wright, S. (1992). Guidelines for selecting regulations to manage open-access fisheries for natural populations of anadromous and resident trout in stream habitats. *N. Amer. J. Fish. Mgmt.* **12**(3),517–527.

Chapter 14 Aquatic Environmental Management

Becker, C. D., and Neitzel, O. A., eds. (1992). "Water Quality in North American River Systems." Battelle Press, Columbus, Ohio.

Benner, C. S., and Middleton, R. W., eds. (1991). "Fisheries and Oil Development on the Continental Shelf." Amer. Fish. Soc., Bethesda, Md.

Bennett, G. W. (1962). "Management of Artificial Lakes and Ponds." Reinhold, New York.

Boon, P. J., Calow, P., and Petts, G. E., eds. (1992). "River Conservation and Management." Wiley, New York.

Calabrese, E. J., and Baldwin, L. A. (1993). "Performing Ecological Risk Assessments." CRC Press, Boca Raton, Fla.

Colt, J., and White, R. J., eds. (1991). "Fisheries Bioengineering Symposium." Amer. Fish. Soc., Bethesda, Md.

Cothern, C. R. (1993). "Comparative Environmental Risk Assessment." CRC Press, Boca Raton, Fla.

Decker, D. J., *et al.* (1992). Toward a comprehensive paradigm of wildlife management: Integrating the human and biological dimensions. *In* "American Fish and Wildlife Policy: The Human Dimension." (W. R. Mangun, ed.), Chap. 2. Southern Illinois Univ. Press, Carbondale.

Erickson, P. A. (1994). "A Practical Guide to Environmental Impact Assessment." Academic Press, San Diego, Calif.

Esch, G. W., and McFarlance, R. W., eds. (1976). "Thermal Ecology II." U.S. Energy Research and Development Administration, Washington, D.C.

Gibbons, D. R., and Salo, E. O. (1973). "An Annotated Bibliography of the Effects of Logging on Fish of the Western United States and Canada," Gen. Tech. Rep. PNW 10. U.S. Dept. of Agriculture, Forest Service, Portland, Ore.

Glantz, M. H. (1990). Does history have a future? Forecasting climate change effects on fisheries by analogy. *Fisheries* **15**(6), 39–44. (See also several other articles about climate change in the same issue.)

Gorelick, *et al.* (1993). "Groundwater Contamination. Optimal Capture and Containment." CRC Press, Boca Raton, Fla.

Graney, R. L., Kennedy, J. H., and Rodgers, J. H., eds. (1994). "Aquatic Mesocosm Studies in Ecological Risk Assessment." Lewis Publ., Boca Raton, Fla.

Haefele, E. T. (1973). "Representative Government and Environmental Management." Johns Hopkins Univ. Press, Baltimore, Maryland.

Hairston, N. G., Jr. (1989). "Ecological Experiments: Purpose, Design, and Execution." Cambridge Univ. Press, New York.

Hildebrand, S. G., and Cannon, J. B., eds. (1993). "Environmental Analysis. The NEPA Experience." CRC Press, Boca Raton, Fla.

Howard, P. H., *et al.* (1991). "Handbook of Environmental Degradation Rates." CRC Press, Boca Raton, Fla.

Kennish, M. J. (1994). "Practical Handbook of Marine Science." CRC Press, Boca Raton, Fla.

King, R. B., Long, G. M., and Sheldon, J. K. (1992). "Practical Environmental Remediation." CRC Press, Boca Raton, Fla.

Laws, E. A. (1993). "Aquatic Pollution, an Introductory Text," 2nd ed. Wiley, New York.

Meehan, W. R., ed. (1991). "Influences of Forest and Range Land Management on Salmonid Fishes and Their Habitats," Spec. Publ. No. 19. Amer. Fish. Soc., Bethesda, Md. (An outstanding compendium on the environmental factors and their effects.)

Merriman, D., and Thorpe, L. M. (1976). "The Connecticut River Ecological Study: The Impact of a Nuclear Power Plant," Monogr. No. 1. Amer. Fish. Soc., Bethesda, Maryland.

Naiman, R. J., ed. (1992). "Watershed Management. Balancing Sustainability and Environmental Change." Springer-Verlag, New York.

Pimentel, D., and Edwards, C. A. (1982). Pesticides and ecosystems. *Bioscience* **32**, 595–600.

Pollution Committee and Socioeconomics Section (1992). "Investigation and Valuation of Fish Kills." Amer. Fish. Soc., Bethesda, Md.

Rosenberg, D. M., *et al.* (1981). Recent trends in environmental impact assessment. *Can. J. Fish. Aquat. Sci.* **38**, 591–624.

Ryding, S. O. (1992). "Environmental Management Handbook." CRC Press, Boca Raton, Fla.

Schlesinger, W. H. (1991). "Biogeochemistry. An Analysis of Global Change." Academic Press, San Diego, Calif.

Shinehelder, C. L. (1992). "Handbook of Environmental Contaminants." CRC Press, Boca Raton, Fla.

Southwick Associates (1993). "Sourcebook for Investigation and Valuation of Fish Kills." Amer. Fish. Soc., Bethesda, Md.

Stickney, R. R. (1984). "Estuarine Ecology of the Southeastern United States and the Gulf of Mexico." Texas A & M Univ. Press, College Station.

Stroud, R. H., ed. (1992). "Fisheries Management and Watershed Development." Amer. Fish. Soc., Bethesda, Md.

Thayer, G. W., ed. (1992). "Restoring the Nation's Marine Environment." Maryland Sea Grant College.

Thurston, R. V., *et al.* (1979). "A Review of the EPA Red Book: Quality Criteria for Water." Amer. Fish. Soc., Bethesda, Md.

See also ten feature articles in *Fisheries,* Vol. 11, No. 2, 1986, on the complexities of regulating, assessing, and managing the introduction of aquatic organisms.

Marine Pollution Bulletin. Pergamon. Vol. 27.

Ocean Engineering. Pergamon. Vol. 21.

Pesticide Science. Elsevier. Vol. 40.

Chapter 15 Fishery Development and Restoration

Allsopp, W. H. L. (1985). "Fishery Development Experiences." Fishing News Books. Farnam, Surrey, U.K.

Barg, U. C. (1992). "Guidelines for the Promotion of Environmental Management of Coastal Aquaculture Development." Fish. Tech. Paper No. 328. FAO, Rome.

Indo-Pacific Fishery Commission (1980). "Symposium on the Development and Management of Small Scale Fisheries, Kyoto, Japan, 21–23 May, 1980." FAO Regional Office, Bangkok, Thailand.

Jackson, R. I., and Royce, W. F. (1986). "Ocean Forum: An Interpretative History of the International North Pacific Fisheries Commission." Fishing News Books, Farnham, Surrey, U.K.

Lele, U. (1975). "The Design of Rural Development: Lessons from Africa." World Bank, Washington, D.C.

Royce, W. F. (1987). "Fishery Development." Academic Press, Orlando, Fla.

Smith, L. J., and Peterson, S. (1982). "Aquaculture Development in Less Developed Countries: Social, Economic, and Political Problems." Westview Press, Boulder, Colo.

Index

Abundance 193
 cycles, 203–205
Acid rain, *see* Environmental protection
 strategy, acid rain
Adaptation, 211–213
Age determination, 176–184
 method
  annual ring, 177–183
  length frequency, 177
  marked animal, 176
Age group, *see* Cohort
American Fish Culturist's Association,
 28
American Fisheries Society (AFS)
 code of practices, 418–420
 formation, 28
Anchovy, 117
Application factor (AF), toxic materials,
 383
Aquacultural environment
 beds and racks, 271
 control, 268–269
 enclosures, 271
 open waters, 271–272

 ponds, 272
 water quality, 269–270
Aquacultural systems, 272–273
 carnivorous animals, 321–322
 filter-feeding mollusks, 322–324
 omnivorous animals in ponds, 318–
  320
 ornamental plants and animals, 325–
  326
 plants, 324–325
Aquaculture
 growth, 8
 nonfood, 9
 policy issues, 332
 recent production trends, 316–318
 scope, 315–316
Availability, 239

Bathythermograph, 55
Bedload, 45
Benguela current, 88
Bioassay, 382
Biochemical oxygen demand (BOD),
 382–384

Bio-geo-chemical cycle, *see* Upwelling
Biomass, 210

Canada's International Development Research Centre (IDRC), 19
Canadian International Development Agency (CIDA), 19
Carp, 119, 129
Catch disposition, 294
Catch per unit of effort, 252
Catch statistics (~1992)
  major fishing grounds, 302
  major fishing nations, 300
  major species groups
    Canada, 298
    United States, 297
    world, 297
Catch trend
  in a new fishery, 218–220
  world and regional, 304–306
Catfish, 120–121
Char, 118
Cichlid, 127–128
Classification
  international statistical, 99–100
  scientific, 96–97
Cod, 122–123
Codex Alimentarius Commission, 340
Cohort, 195
Common property resources, 21
Community succession, 205–207
Compensation depth, 73
Competition, 202–203
Continental shelf, 50–51
Continental slope, 50
Coriolus force, 76
Crab, 106–107
Croaker, 124
Current
  cause, 75–77
  measurement, 78–79

Density dependent mortality, *see* Population, dynamics
Disease, 274
  control, 324
  identification of organisms causing, 281

kinds of, 282–285
prevention, 330–332
Dynamic height, 80

Ecosystem, 198
  controlling factors, 199–200
  fishery type, 200–201
  production
    primary, 210
    secondary, 205
Eggs, incubation, 155
Electronarcosis, 141
Electrotaxis, 141
Endangered Species Act of 1973, 10
Endangered Species Conservation Act of 1969, 9
Energy flow, 2
Environmental change
  acids and alkalis, 377
  bottoms and basins, 374
  communities, 379
  entrainment, 376
  excess dissolved gas, 378
  flow, 375
  mixing zone, 377
  multiple changes, 379
  nutrient materials, 376
  oil, 377
  oxygen depletion, 378
  suspended and settleable solids, 375
  temperature, 375
  toxicants, 376
Environmental Impact Assessment, 373–374
Environmental Impact Statement (EIS), 10
Environmental improvement, 389–395
  artificial reefs, 395
  dissolved gases, 392
  fertility, 393
  fish attractant devices (FADS), 395
  living space, 394
  oyster bottoms, 394
  reforestation, 395
  spawning areas, 390
  temperature, 392
  transport and guidance devices, 390
  undesirable species, elimination, 391

Environmental protection strategy, 380–389
  acid rain, 385–386
  allocation of water use, 381
  biological tests, 382–384
  criteria and standards, 382
  incremental effects, 387
  organic waste treatment, 384
  policy and organizational development, 373, 380
  prediction and modeling, 389
  reduction of risk of accidents, 386
  status of biological communities, 388–389
  waste management, 384
Equatorial undercurrent, 86
Estuary, 48–49
  bar-built, 49
  circulation, 89
  deltiac, 49
  drowned river mouth, 49
  fjord, 49
  tectonic, 49
Eutrophic lake, 48
Exclusive economic zone, 11

Fecundity, 156
Fish, competition for water, 7
Fish attractant devices (FADS), *see also* Environmental improvement, fish attractant devices
  external anatomy, 95
  in human diets, 4
  in international trade, 299–301
  marketing trends, world, 303–306
  senses, *see* Sensory reception
  temperature tolerance, 52
  transport and guidance devices, *see* Environmental improvement, transport and guidance devices
  usage of word, 25
  world production, 5, 297
Fish flesh
  bioactive compounds, 337, 338
  composition, 336
  contaminants, 338
  deteriorative processes, 340–341

  disease organisms, 340
  nutritive value, 337
Fish marketing strategy
  consumer preferences, 352
  long-term trends, 350–351
  market intelligence, 351–352
Fish preservation for market
  chilled or frozen, 342–343
  cured, 342
  heated, 344
  irradiated, 344
  live, 34
Fisheries, recreational, 6
Fisheries Academy of U.S. Fish and Wildlife Service, 20
Fishery Conservation and Management Act of 1976 (FCMA), 11
Fishery development, 398–417
  constraints, 403–404
  fads and failures, 405–407
  opportunities, 405
  options, 404
  perspectives, 400–408
  prior disciplines, 407–408
  project planning, 411–412
  support
    financial, 410
    international, 408–410
    national, 398–399
    technical, 401
Fishery institutional development
  cooperatives, 412
  extension systems, 413
  joint ventures, 414
Fishery product, 291–293, 311–346
  quality
    assessment, 346
    codes of practice, 348
    control, 345
    inspection systems, 348–349,
    standards, 346–347,
Fishery research, scope, 27
Fishery science
  definition, 25
  institutional activities, 14–20
Fishery scientist
  career planning, 36
  designation, 26

Fishery scientist (*continued*)
  education, 37
  employers, 14–20
  employment, 30
  function, 13–15
  peer recognition, 26
  role
    aquaculture, 333–344
    aquatic environmental management, 395–396
    fish processing and marketing, 352–353
    fishery development, 415–416
    fishery regulation, 368–370
    fishing business, 311–313
Fishery statistics, collection, 234–239
  biological data, 237–239
  catch data, 234–235
Fishing gear, kinds, 290–295
  longline, 274
  otter trawl, 292
  purse seine, 293
Fishing systems
  commercial, 309–310
  production levels, 307–308
  recreational, 310–311
  subsistence, 308–309
Flatfish, 130–132
Food and Agriculture Organization of the United Nations (FAO)
  formation, 29
  function, 18, 400
Food habits
  bottom filterer, 151
  bottom forager, 152
  determination, 150
  giant filterer, 153
Food web, 208
Freshwater bass, 127

Gear selectivity, 234–236
Geomagneticelectrokinetrograph (GEK), 80
Grouper, 124
Growth
  allometric, 171–172
  endogenous factors, 165
  estimation, 251–252
  exogenous factors, 166–167
    heterogonic, 171
    isogonic, 171
    patterns, 165–168
    rate, 172
Gulf Stream, 87

Hakes, 122
Hearing, *see* Sensory reception, sound
Herring,
  menhaden, 115–116
  sardine, 114–115
  sea herring, 114
  shad, 114
  sprat, 114
Humboldt current, 88
Hydrograph, 43
Hydrographic cast, 79
Hydrographic cycle, *see* Water cycle
Hydrology, 41

Insect, aquatic, 102
Intergovernmental fishery commissions and councils, 16
International Commission on Micro-biological Specifications for Foods, 347
International development agencies, 17, 409

Jack, 125

Krill, 103
Kuroshio current, 87

Lagoon, 48
Lake basin, 47
  index of shape, 48
  index of shoreline development, 48
Lake circulation, 83
Larval development, 159, 160
Law of the Sea, 5, 10, 368, 421–421
Length–weight relationship, 172–173
Lobster, 107–108

Mackerel, 128
Marine Mammal Protection Act of 1972, 10
Marketing trends, world, 299–300
Marks and Tags, 223–227

Material cycle, 208–209
Maturity, 154–155
Migration, 221–222
Minnow, 119–120
Mollusk
  clam, 108
  mussel, 109
  octopus, 111–112
  oyster, 110–111
  scallop, 111
  squid, 111–112
Mortality, estimation, 245–247

National Environmental Policy Act of
  1969 (NEPA), 9, 23, 365
National Food Processors Association,
  345
National Oceanic and Atmospheric, Administration (NOAA), 26
National Shellfish Sanitation Program,
  United States, 347
Norwegian Agency for International Development, 19
Nutrition of fish, 277–279
  larval feeding, 279–281
  requirements, 277–281

Ocean bottom, 49
Ocean circulation, 74–90
  gyres, 86
  surface, 74, 86–90
    convergence, 88
    divergence, 88
Ocean surface temperature, 51
Oligotrophic lake, 48
Osmoregulation, 145
Oxygen consumption, 145

Peace Corps, 19
Perch, 126
Photosynthesis, 72–74
  macroplankton, 100
  nannoplankton, 100
  net plankton, 100
Pompanos, 125
Population
  age structure, 195–196
  distribution patterns, 189–192
  dynamics, 192–193

density, 192–193
  mortality, 193–195
growth, 197–198
limiting factors, 186–189
single species, 186
Porgies, 125
Predation, 201–202
Professional canons of behavior, 32
Professional ethics, 415
Professional status, 25
Pyramid of mass, 211

Rating curve, 44
Recreational fishing, 299
Recruitment, 241–247
Regulation of fisheries
  decision factors, 361–364
    large oceanic fishery, 363
    multiple domestic fisheries, 363
  objectives
    allocation, 359
    conservation, 358–359
    orderly fishing, 360
    prevention of waste, 360
    protection of public health, 361
  origins of public policy, 355, 356
  principles of public action, 356–357
Regulatory regime, new
  implementation, 366–367
  legislation, 368
  policy development, 364–365
Reproduction, 154–166
  Atlantic mackerel, 159–160
  oyster, 160
  shrimp, 161–162
  sockeye salmon, 164–166
Rockfish, 129–130

Salmon, 104, 117–119
Saurie, 121–122
Schooling, 147
Scorpion fish, 129–130
Sea bass, 123
Sea Grant College Program, 20
Seaweed, 104
Secchi disk, 74
Sediment
  classification, 47
  transport, 44–47

Seiche
  internal, 77
  surface, 77
Selective breeding, 274–275
  control of spawning, 275–277
  hybridization, 275
Sensory reception, 136–145
  chemical, 138–140
  electrical, 143–145
  light, 136–138
  mechanical, 141
  pressure, 142
  sound, 142–143
  temperature, 141
Sex reversal, 157
Sexual characteristics, 156
Shark, 112–114
Shrimp
  northern, 106
  tropical, 105
Smell, see Sensory reception, chemical
Smithsonian Institution, 37
Snapper, 124
Spawning
  behavior, 158
  time, 157
Species
  definition, 93
  description, 94
Squid, 112
Standing crop, 211
Stock
  correlated populations, 249
  definition, 220
  direct counts, 248–249
  identification, 222–223
  marked members, 250–251
  size, 247–252
Stream
  catchment area order, 32
  drainage area, 43
  network, 43
  order, 44
Sunfish, 127
Survival, estimation, 252–254
Suspended load, 45
Swedish International Development Authority (SIDA), 19

Taste, see Sensory reception, chemical
Thermocline, 55
Tides, 77
Trophic efficiency, 205–210
Trophic relationship, 207–211
Trout, 117–119
Tuna, 128–129

United Nations Educational Scientific and Cultural Organization, (UNESCO), 18
United Nations Environmental Programme UNEP, 19
United States Agency for International Development (AID), 19
United States Commission on Fish and Fisheries, 5
Upwelling, 71

Vision, see Sensory reception, light

Wash load, 44
Waste management, see Environmental protection strategy
Water
  density, 58
  dissolved gases, 65
  dissolved salts, 69
  electric conductance, 63
  light penetration, 64
  light transmission, 63
  nutrient elements, 70
  pH, 68
  sound transmission, 62
  temperature measurement, 55
Water cycle, 41
Water use policy
  allocation, 372
  environmental quality, 372–373
  trends, 372
Wave size and speed, 78
Whale, 132–133

Yield models, 259–265
  recruitment, 261–263
  surplus production, 261
  trophic, 263

Zooplankton, 102–103